SECRET WARRIORS

SECRET WARRIORS

The Spies, Scientists, and
Code Breakers of World War I

Taylor Downing

PEGASUS BOOKS
NEW YORK LONDON

SECRET WARRIORS

Pegasus Books LLC
80 Broad Street, 5th Floor
New York, NY 10004

ISBN: 978-1-60598-694-4

10 9 8 7 8 6 5 4 3 2 1

Printed in the United States of America
Distributed by W. W. Norton & Company, Inc.

For my grandfather
William Forward John Downing
Who operated a Vickers machine gun in the First World War
and survived

Contents

Prologue   1

1 New Century, New World   13

Part One – Aviators

2 The Pioneers   41
3 The New Science   62
4 Observing the War   75

Part Two – Code Breakers

5 Room 40   101
6 The Great Game   124

Part Three – Engineers and Chemists

7 The Gunners' War   149
8 The Yellow-Green Cloud   173
9 Breaking the Stalemate   191

Part Four – Doctors and Surgeons

10 The Body   213

11 The Mind   241

Part Five – Propagandists

12 The War of Words   269

13 The War in Pictures   290

14 Masters of Information   315

Epilogue – The First Boffins   335

Who's Who of Secret Warriors   359

Acknowledgements   383

Endnotes   385

Bibliography   405

Index   413

About the Author   439

Prologue

Soon after midnight in the early hours of Tuesday 5 August 1914, the captain of the CS *Alert*, a cable-laying ship moored at Dover harbour and owned by the General Post Office, received the special coded telegram he had been expecting. He immediately ordered the *Alert* to slip out of harbour and head north-east. In the early dawn, the captain drew up the *Alert* in the grey waters of the North Sea a few miles off the German port of Emden, near that country's border with Holland. It was only about six hours since Britain had formally declared war on Germany at 11 p.m. London time, midnight in Berlin. Having taken up his position, the captain of the *Alert* ordered the ship's grappling equipment to be dropped into the murky waters. The crew of the 1000-ton cable ship were highly skilled in laying and maintaining the undersea cables which, since the mid-nineteenth century, had crossed the bed of the oceans to link continents via telegraphic and, more recently, telephonic communications. They knew exactly how to find the cables to repair them. But this time the vessel's mission was destructive. Although the *Alert* was a civilian ship, the captain was about to engage in an act of war.

After a short period of dredging along the bottom of the sea, the grappling hooks were hauled to the surface bringing with them giant strands of thick, insulated cable that flailed like giant, underwater snakes. Dripping with water and covered in mud and seaweed they were dragged on to the deck of the *Alert*. The crew sawed and hacked the cables, breaking through them, and then tossed them back overboard into the sea. They then repeated the whole process on four further cables, dragging on board and cutting each one before throwing the ends into the sea. The operation, which took about four hours, was fraught with danger for the unarmed British vessel. The captain and lookouts scanned the horizon for any sign of German ships coming out to see what was going on. As the dawn became brighter the sea grew rougher and a heavy rain squall passed over the ship.

The captain of the *Alert* was carrying out Britain's first offensive act of the First World War. The five German cables that ran across the North Sea and down the English Channel linked Germany with France and Spain and then went out into the Atlantic to Africa and the Americas. They were now severed. Germany could no longer send telegrams or cables to its colonies or to the United States. Cut off from the US and much of the rest of the world, the country's telegraphic links were now limited to its immediate neighbours across land borders. Berlin had lost its top secret communication link with the new world. From now on, any communication would have to be via radio. And there was one major problem with sending signals by radio. Anyone with a receiver could also tune in and listen to them. This would have significant consequences as the war progressed.[1]

Just over a week later, at dawn on 13 August, another group left Dover. On this day it was the turn of the aviators of the recently formed Royal Flying Corps. For the first time, Britain was sending aircraft to war to accompany its ground troops. It was a chance for the youngest addition to the military to prove itself. Among the small group of fliers gathered was Captain Philip Bennett Joubert de la Ferté. Just twenty-seven years of age, he was typical of the first wave of military fliers. He had been in the artillery when he heard about the formation of the Royal Flying Corps and was one of the first officers to join. He had to pay to learn to fly himself (with his father's support) as the army did not then have funds available to train pilots. Joubert quickly took to flying, although the aircraft were so light and fragile that they needed a lot of care. Most flying took place in the early dawn before the wind had got up and Joubert, like most pilots, had experienced the embarrassment of actually being blown backwards when trying to fly into a strong wind. On one flight he had ended up seven miles behind his point of take-off. When war was declared, Joubert was in command of 'C' Flight in 3 Squadron, flying a French aircraft built by the Blériot company.

For the journey, Joubert, like most of the other pilots, was accompanied by his mechanic. It was the quickest way of transporting to France the men who were essential to keep the aeroplanes flying. Joubert was briefed at about 5.30 a.m. and given maps of France and Belgium and sealed orders. When he opened them, the orders contained details of his destination. Along with the others, Joubert was given a revolver, a set of field glasses and a spare pair of goggles. The mechanics were issued with a tool kit. Emergency rations of biscuits, a bar of chocolate and a pack of soup concentrate were handed out in a haversack. Advance parties at Dover had acquired a large number of cast-off inner tubes. Each man carried one of these, to be inflated if the aircraft came down into the sea and used as a makeshift lifebelt. But the pilots' instructions were to ascend to 3000 feet before starting their Channel crossing so if an engine failed they should have enough height to glide across the Channel. There was no planned sea rescue.

The people of Dover cheered as Joubert and his fellow pilots climbed into their aircraft on the hills above the cliffs dominating the town. The contraptions they climbed into consisted of wooden frames held together with wire and covered in linen canvas, powered by large combustion engines that sat imposingly near the centre of the structure. Today they look as ancient as the dinosaurs, but to the crowds gathered on that August morning these craft were the very cutting edge of modernity. Only five years before, Louis Blériot had made the first Channel crossing by air. Now Joubert and the other pilots in their flying machines were planning to carry out a similar journey and to take up their position alongside the British Expeditionary Force.

Soon after 6.25 a.m. on what proved to be a beautiful, clear August day, the first aeroplanes taxied across the grass and soon got up to speed. One after another the pioneers in their Blériots, BE2s, BE8s and Henri Farmans took off and rose high into the sky to reach the planned altitude. Then, in a line, each aircraft, powered by an engine that could only muster a few horsepower, headed off across the Channel at roughly two-minute intervals. Their course was to hit the French coast at Boulogne, fly down the coast to the Somme estuary and then head inland to Amiens. Not everything went to plan. A few aircraft were damaged when they came down in a ploughed field. One pilot got lost and had to land and ask an astonished passer-by where he was. On landing in France, another pilot was arrested by officials who could not understand what language he was speaking and thought he must be a spy. It took three days to get the pilot released from prison. Yet another aircraft was delayed as its pilot flew around the Cap Gris-Nez lighthouse and tried to drop his inner tube, like a quoit, on to the spiky top, as though in a fairground.

For Joubert, the flight from Dover to Amiens took just two hours. The aerodrome at Amiens was a simple affair, just a cut grass field with a few large sheds known as hangars at one end. At this point, the RFC had almost nothing in the way of ground transport, were desperately short of spares and had barely any reserves. Having made the journey, Joubert and the pilots of his squadron came to rest along the side of a field as there was not enough hangar space for all the British machines. As the morning passed an enthusiastic crowd gathered, waving flags and shouting 'Vive l'Angleterre'. The French had been doubtful as to whether the British would join them in their war against the Germans. But here they were, and the Gallic reception included throwing flowers and even fruit in a tremendous welcome. That afternoon, Joubert and his fellow aviators received another visitor, General Sir John French, commander-in-chief of the British Expeditionary Force. A cavalry man who traditionally relied on scouts riding on horses for reconnaissance, he had little idea how effective this new fighting force would be in a similar role, but he wanted to come and see the men and their machines. French was reassured by the sight of forty-nine aircraft from three squadrons lined up along the side of the aerodrome. There was a sense of excitement and jollity about the whole event.

That night Joubert was put up at one of the best hotels in Amiens, the Hotel Belfort. Not expecting billets with comfortable beds and fitted sheets, he had brought no pyjamas with him. Along with several other pilots, he had to borrow a nightdress from the hotel owner. It was the last time for many months that these men would need pyjamas. And as they cavorted along the hotel corridors in borrowed nightdresses down to their ankles they did not look much like a group of men who represented the very latest in the science of war.[2]

The late autumn sun shone brightly through the tall sash windows in the splendid first-floor meeting room of the Royal Society in the East Wing of Burlington House. Wood-panelled and lined with shelves of books, the meeting room overlooked the courtyard just off London's busy Piccadilly with its continuous bustle of motor traffic. But the room was surprisingly quiet as the clock struck eleven on the morning of 12 November 1914. Precisely on the last stroke, a clerk opened the large, heavy door and the President of the Royal Society led into the room a procession of ten distinguished gentlemen, the youngest of whom was in his forties, the eldest in his eighties. First behind the President was John William Strutt, Baron Rayleigh, a Fellow of Trinity College, Chancellor of Cambridge University, a previous President of the Society and one of the most distinguished scientists in Britain. Famous for his work on optics and acoustics, as a young scientist he had come up with an explanation for why the sky is blue, while as a physicist at Cambridge he had helped to determine the absolute values of the ohm, the ampere and the volt. He had been awarded a Nobel Prize for the discovery of argon, an inert gas. He was close to government and served as chair of the explosives committee of the War Office and as president of a key committee on aeronautics. He was in his seventies but still lively and energetic, and he saw the war as an opportunity for scientists to demonstrate how their work could assist in bringing victory.

Rayleigh and his colleagues looked immaculate in their suits, waistcoats, wing collars and ties. One of them was in the full dress uniform of an admiral. The group included several more of the most eminent men of science in Britain. Two were leading physicists and four were prominent chemists. One of them had discovered thallium, another helium. One had pioneered wireless telegraphy at sea. Among them were two engineers, a mathematician, and the Director of the National Physical Laboratory at Teddington. Seven were knights and Rayleigh was a peer of the realm.

The Royal Society was the leading organisation of scientists in Britain. It had been founded in 1660 by Christopher Wren, Robert Boyle, John Evelyn and other prominent 'natural philosophers' as a forum to witness and discuss scientific experiments. Two years later the Society was awarded a Royal Charter by Charles II. It had gone into a decline during the late eighteenth and early nineteenth centuries when it became little more than a gentlemen's debating club, only about one-fifth of whose members were active practitioners of science. But in the second half of the nineteenth century the Royal Society had transformed itself into an influential professional academy of four to five hundred Fellows, all of whom were distinguished scientists.

When war had been declared over the early August bank holiday, the widespread feeling was that it would be a quick war, fought by professional armies on distant fields, possibly with a naval engagement at sea and that without doubt it would all be 'over by Christmas'. But by the time of the November meeting, it was clear that this was no longer the case. News reports from the front were strictly censored, but Rayleigh and his fellow scientists could see that the European armies were lining up for what would be a much longer war than anyone had expected. Accordingly, these men of science had agreed with Rayleigh that they must make some gesture of support, some indication that the scientific establishment was ready to rally behind the war effort.

Sitting at the central meeting table, Sir William Crookes, the President of the Royal Society, took the chair. As the others fell silent, Crookes began to speak. After the President, Rayleigh spoke and a few others joined the discussion. In less than twenty minutes they had reached unanimous agreement that they should form a committee that would be known as the War Committee of the Royal Society. Its purpose would be to organise assistance to the government and the armed services with any scientific questions that arose. With their prominent connections across the universities of Britain and within the manufacturing and technical industries of the nation, the senior members of the Royal Society would be supremely well placed to know whom to approach, what to ask and how to help.

After a short further discussion, it was agreed that the Secretary of the new War Committee should write to the War Office, the government department from where the British Army was governed, to the Admiralty, the department of state that ruled the Royal Navy with the largest fleet of warships in the world, and to the Board of Trade, the government department that was closest to industry. The letters should express 'the readiness of the Committee to organise assistance to the Government in conducting or suggesting scientific investigations in relation to the war'. It was further decided to form two sub-committees so that the men of science could immediately start to investigate possible scientific applications relevant to the war. The first would look into the new technologies of telephony and wireless telegraphy, and into the broad field of 'General Physics'. The second would investigate the field of chemistry, and would send letters to the directors of the chemical laboratories at every university and college in the country, inviting their assistance in undertaking the manufacture of chemicals, most especially drugs or other medicaments, 'the supply of which is inadequate in consequence of war conditions'.

There being no further business, the date of the next meeting was set for the same time exactly one week later. The chairman declared the meeting closed and the committee members stood and followed him out of the meeting room. Some of them, like Lord Rayleigh, had urgent business elsewhere. Others settled in the reading room for an hour or so before heading off for lunch. The meeting had been a low-key affair, but it marked an historic step. For the first time, the nation's leading scientists had come forward to offer the support and assistance of the entire scientific community to the government and the armed forces. It was already clear to Rayleigh and his colleagues that in the twentieth century a war would have to be fought drawing on all the advances that physics, chemistry, medicine and mathematics could offer. This war would be conducted in laboratories and scientific workshops, as well on the battlefields and across the oceans of the world.[3]

These three scenes illustrate in very different ways how the conflict that would become known as the 'Great War' would involve the world of science and the scientists of the day. Thanks to the cutting of undersea cables, the German government had to use wireless telegraphy, radio, for much of its long-distance communications. The Germans' use of code would challenge the Allies, first to find a way to intercept the signals and second to decipher the intelligence they contained. This demanded the application of new technologies and the development of new code-breaking techniques. After the first aircraft went to war in 1914, the new science of aviation would advance in leaps and bounds over the next few years as each side tried to outdo the technology of its enemy. In four years' time, aircraft engines would be unrecognisable in their power and output and aircraft designs would have advanced beyond anything imaginable in 1914. Finally, the fact that the most eminent men of science were offering to assist the army, the navy and the flying corps was a recognition of the vital role science would play in a modern, industrial and technological war.

Much writing about the Great War concentrates on the troglodyte world of the trenches that made up the 450 miles of battle lines known as the Western Front. This extended from the English Channel, through the western tip of Belgium, across the industrial north-east of France and down through the chalklands of Picardy, then ran east through the Champagne district and circuited the great fortress defences of Verdun, finally turning south into the Vosges mountains and the Alps on the Swiss border. The First World War is often seen exclusively as a war fought by armies of millions living in the subterranean world of the trenches, slogging it out in human wave assaults and being slaughtered in dreadful numbers. The Western Front in which the French, British and Commonwealth armies faced the German army dominates the popular image of the war, although much fighting took place on other fronts and against other enemies.[4]

The role usually given to science in the First World War is that all it did was to introduce ghastly new inventions to the arsenal of war, including powerful new high explosives and hideous clouds of poison gas. It has been written that it was a 'chemists' war of poison gases and explosives'.[5] This brutal industrial-scientific war, conducted by means of the long-range artillery shell, the machine gun and newly-formulated chemical weapons, it is argued, led to killing on a vast, appalling, unprecedented scale. In many histories of the war, the contribution of science goes no further than this. It was seen as murderous, destructive and entirely negative.

Beyond this common view of the war, however, it is possible to see how engineers, chemists, physicists, doctors, psychologists, mathematicians, intelligence gatherers and propagandists were taking part in an unknown struggle that made a more positive contribution to what happened at home, at the front, at sea and in the air. They helped to fight a war that was won by scientific advantage,

achievement and breakthrough in many fields that helped to transform life after the conflict ended. Many of the foundations of
scientific progress in the 1920s and 1930s in fields such as radio technology and medicine, aviation and psychology were laid in the four
years of war. In Britain, the new skills developed in aeronautics and
intelligence gathering would live on for much of the rest of the century. The expertise that went on to produce the jet engine, a powerful
aviation industry and the supersonic airliner was developed during
the First World War. In addition, Britain is still today, for good or ill,
a nation renowned for its intelligence-gathering capabilities and is
often referred to as 'the surveillance state'.[6]

However, at the opening of the twentieth century the world of science was itself deeply divided. Several men of science (the word
'scientist' was still relatively little used) thought that their discipline
was best pursued in academic isolation through pure research.
Fighting for professional status and independence in the universities, they believed the only valuable science was pure science. They
looked down on the world of applied science, just as gentlemen as
a class looked down on those who dirtied their hands with business
or worked in industry. But a growing group of scientists began to
feel that this was a false distinction, that the principal role of science
was to improve the lot of men and women. For instance, Professor
John Haldane of New College, Oxford, a leading physiologist, used
his understanding of poisons and of respiration to work tirelessly for
the improvement of industrial health in coal mines. Professor John
Ambrose Fleming, a top electrical engineer at University College,
London, in addition to his academic duties also worked as consultant and scientific adviser to commercial companies and invented the
first electric valve. In the world of medicine, of course, there never
was a distinction between pure and applied science. All medical
research was for practical purposes and an immense amount of it
took place in the early twentieth century. But although medicine had
been the exception rather than the rule, there was unquestionably a
shift towards finding the practical application of science and of using

scientific principles to understand and improve the electrical and mechanical world that was developing fast in the early years of the century.

As part of this transition, new organisations came into being in Britain in the years before the war to advise on the science behind many of the great technological changes that were transforming the age. The Advisory Committee on Aeronautics was established in 1909 and the Medical Research Committee in 1913. They and other such bodies bred a new type of scientist who helped to link the universities with the practical world, the academy with government and industry. These scientists helped develop new forms of intelligence gathering, helped to save lives by developing new medicines, advanced the understanding of how the human mind worked, and brought dramatic breakthroughs in code breaking, naval warfare and the war in the air. In 1915 Professor John Ambrose Fleming summed up the impact of these changes in a public lecture when he said: 'It is beyond any doubt that this war is a war of engineers and chemists quite as much as of soldiers.'[7]

I have written extensively about the boffins, backroom scientists and mavericks of the Second World War. In a sense this book is the prequel to these histories, looking at the boffins of the Great War (not that the word 'boffin' was ever used during that war). But it is not just those obviously defined as scientists who provide the subject matter for this book. This is the story of many individuals with specific skills, often from the universities or from industry or the arts, who contributed to the war effort. In the top secret world of the code breakers, the Admiralty recruited men and women with specific and vital linguistic skills and brought in classical scholars who were experienced in piecing together the full meaning of a manuscript from fragments of text. In the censorship of the press and in the new medium of the cinema, the War Office recruited a broad range of writers including one of the greatest novelists of the day, while the army engaged film-makers, cinema distributors, photographers and artists to help depict the war for the public at home and abroad. A general awareness of the existence of a 'Home Front' came into being and slowly it was realised that in a modern war a vital relationship would exist between what people thought at home and the general level of support needed to sustain the fighting abroad. So a wide net was cast to recruit the skills needed to manage this relationship and to win the war.

Secret Warriors takes in aviation, intelligence and code breaking, engineering and gunnery, chemistry and medicine, as well as censorship and propaganda. It follows the work of some extraordinary individuals who became part of the first 'total war' in which all the resources of the state were involved. This was not just a war fought by sailors, soldiers and airmen, but one in which public opinion would play a central role in supporting the fighting front. As the war progressed it drew in an ever-widening group of experts, scholars, scientists and literary figures. They were the secret warriors of the Great War. A huge group of brilliant men (and they were mostly men, although there were a few women) were drawn into the titanic struggle. For the first time in this nation's history some of the finest brains from the university laboratories, colleges, factories and hospitals of Britain willingly came forward not to do their bit on the battlefield but to contribute their expertise at a time of national crisis.

As the Great War was transformed from a conflict fought by professional armies on continental fields into a national struggle that affected most aspects of the life of the nation, so the scientific and intellectual establishments were drawn in. As the army and the air force, and to a degree the navy, became more professional and more prepared for a long, bitter, attritional war, so the best and the brightest were called upon to contribute. A few of those who helped to solve the problems of aeronautics, to carry out extraordinary new forms of surgery or to write the first narratives of the war, became household names. Most, however, remained unknown and returned to relative obscurity in their laboratories, libraries or university departments. Many remarkable individuals appear in these pages. This is their story. It is an unusual one.

1

New Century, New World

The year 1914 is often seen as the end of an era. It has been described as the last year of the 'long' nineteenth century, and as marking the final break between the old world order and the modern era. As a consequence, Edwardian Britain is often depicted as a sort of Indian summer, the last decade of the old world before everything was engulfed in the Great War. It is sometimes portrayed as an elegant, golden age shimmering in the distant light of country house parties, with public schoolboys playing cricket on long hot summer afternoons, the navy in great battleships proudly ruling the waves, the pomp and glory of imperial durbars ... and so on. It is presented as a period of stability and continuity before the tsunami of war washed everything away.[1]

However, this is to see life in the first decade of the twentieth century through the looking glass of hindsight. Most evidence shows that the Edwardians believed they were living through years of immense promise and potential, years of dramatic change in the present that offered exciting new possibilities for the future. They saw the Victorian era ending symbolically with the death of the old queen and the coming of the new century. New ideas, new technologies, dramatic changes in workers' rights, the provision of state pensions and the big debate about 'Votes for Women' were the characteristics of their age. The word 'new' itself became one of the most fashionable words of the age: people spoke of the 'new art', the 'new morality' and the 'new woman'. All of this generated great debate. Edwardians argued intensely as to whether so much 'newness' was a good or a bad thing. But they were not wrong about the scale of change they were living through. One of the young technological pioneers of the age, John Moore-Brabazon, summed up the spirit he and his friends felt when he wrote, 'I think we were all a little mad, we were all suffering from dreams of such a wonderful future.'[2]

Britain was still a deeply divided land. In London there was a glittering West End and an impoverished East End. In the countryside there were superbly wealthy country homes and bleak village hovels. Workers had formed trade unions to battle for their rights and for a better livelihood; and rich owners were determined to give them neither. Industrial unrest was widespread; in 1912, for example, forty million working days were lost in strikes. Suffragettes demanded rights for women; the establishment wanted to preserve the status quo. Change was not happening evenly or necessarily fairly. But it was happening.

In the rarefied world of pure science nothing short of a revolution was taking place in the twenty years before the Great War. At Berlin University in 1900, Max Planck discovered quantum theory and a new basis for theoretical physics. At Zurich in 1905, Albert Einstein proposed his 'special theory of relativity'. These ideas were to transform the intellectual landscape of the twentieth century, utterly changing views on space, time and matter. At Manchester University in 1911 Ernest Rutherford discovered the nucleus of the atom and nuclear physics was born. At Cambridge from 1910 to 1913, Bertrand Russell and A.N. Whitehead revolutionised the foundations of modern mathematics with their *Principia Mathematica*. Meanwhile in Vienna, Sigmund Freud laid down the basis for psychoanalysis as a formula for the treatment of psychological problems through dialogue between patient and psychoanalyst. A new science of genetics was established. Incredible advances were made in understanding the activities of microbes in the new science of bacteriology. These fundamental changes in a brief period of time, as Eric Hobsbawm has observed, utterly transformed 'man's entire way of structuring the universe'.[3] But of course, only a tiny number of people picked up these revolutionary ideas. In 1910, there were barely 700 members of the British and the German Physical Societies combined. The total number of pure scientists in the world in 1914 could be counted in only the thousands. And mostly they researched and worked in Western Europe with only very small numbers in either Russia, the United States or elsewhere. In 1901 the Swedish Academy of Sciences first awarded the Nobel Prize to scientists who had made major advances in their field. By 1914, of the first seventy-five highly prestigious awards, all but ten were made to scientists in northern Europe, mostly Britain, Germany, France and the Netherlands.

Most Edwardians knew nothing of these seismic changes and few would have understood them had they known of them. But if merely a few hundred advanced thinkers felt the earthquakes in the world of pure science, pretty well every Edwardian would have been aware of the massive technological changes that were impacting on almost every aspect of their lives. Developments in electricity, the spread of the internal combustion engine, the advance of the chemical industries, huge improvements in medical science, dramatic developments in communications technologies were all bringing about what many saw as a new age. Some people even went as far as to hope that these new scientific technologies could soon eradicate altogether the traditional problems of poverty, disease and war. Others were more pessimistic and feared that traditional values and long-standing social relationships would disintegrate as a result of all this turmoil.

Many of the scientific changes that were taking place had their foundations in the Victorian era, although the consequences were only being felt in the early part of the new century. The first, electricity, had already begun revolutionising life at the end of the

nineteenth century. Just as coal and steam power had been at the
heart of the Industrial Revolution, so electricity was at the foundation of the new era. It had been known since ancient times that
electricity existed, but it was not until the seventeenth and eighteenth centuries that men of science analysed and began to
understand concepts like electrical currents and electric fields. In
1879, the American Thomas Edison invented the electric light bulb –
or at least designed the first incandescent bulb that could be mass
produced, in which a metal filament glowed white within a glass
bulb. Two years later he built the first modern electric power station
in New York to supply the electricity for his light bulbs. Within two
decades, by 1900, a recognisably modern form of the electrical
power industry was beginning to emerge. Electricity was produced
in large generating stations sited near the main centres of demand.
In Britain, the Ferranti company built one of the first of these giant
generating stations in 1889 at Deptford, only a few miles east of
London. Electricity travelled at a high voltage along cables from
power stations to the local user where it was stepped down to a low
voltage through a transformer.

The principal use for electricity at the end of the nineteenth century was to replace gas lighting in public streets in order to provide
a cleaner and safer form of illumination. This central fact of demand
determined the shape of the early electricity supply industry. In
Britain by 1900 there were about 250 separate companies supplying
electricity in a range of different voltages from 100 to 480 volts. There
was no uniformly accepted standard of supply. At least half of the
companies were owned by local municipalities and their task was
merely to light the streets of their town or city. Even by 1914 relatively few households in Britain – only about one in ten – had access
to electricity. And the 10 per cent of houses connected to the electricity supply were clearly the wealthiest homes in the bigger towns
and cities. During the first decades of the century the numbers grew
dramatically, partly fed by the huge growth in the electrical industries as new manufacturers like the General Electric Company

(GEC), whose slogan was 'Everything Electrical', developed into industrial giants. Manufacturers produced a vast array of electrically powered domestic gadgets, from the telephone to the electric fire, from gramophones to vacuum cleaners. By the time of the First World War, these household items were only just beginning to revolutionise the home, but they pointed towards the future.

In the wake of electricity came a huge growth in new electronic industries like those of the telephone and radio, both founded on the development of the electric telegraph. This was another nineteenth-century industry. In 1844, in America, Samuel Morse had demonstrated a code that became a universal system for translating letters into dots and dashes, which could then be sent as electrical pulses along telegraph wires. Tens of thousands of miles of telegraph cables were soon in place, crossing countries and continents. The first underwater cable linking Britain and France was laid in 1851 and the first transatlantic cable began operating in 1866. British scientists soon established themselves as leaders in the technology of insulating copper wires in a rare tree sap and wrapping them in protective steel wire. The Eastern Telegraph Company dominated the process of linking all parts of the British Empire and by the 1870s telegraphic cables extended to Hong Kong and Australia. Initially used by diplomats and news agencies, the telegraphic cables made the world a genuinely smaller place. News, information or dispatches that would have taken weeks to transport around the globe by ship, now arrived in minutes. Officials headquartered in their capitals could send orders to generals and admirals in the field. The first ever world wide web of telegraphic cables was created in a single generation in the second half of the nineteenth century.[4]

In 1876 this went a step further when an American, Alexander Graham Bell, patented his invention of the telephone, basically a telegraph but able to carry the electromagnetic signal of a human voice. It took a while for telephones or 'speaking telegraphs' to catch on as they depended upon a complex and costly infrastructure of local exchanges and telephone operators. Moreover, surprising as it may seem today, few people could see the point of the telephone; official and business users already had the telegraph and could send telegrams worldwide. The telephone seemed nothing but a frivolous extension to this service. It remained largely an urban device and by 1914 there were still only about 1500 exchanges in Britain, of which the vast majority had fewer than 300 subscribers.

By the 1880s, the German physicist Heinrich Hertz had established the existence of electric waves travelling at the speed of light. Hertz's work, though, was purely theoretical and academic. It took others to make some practical application of the discovery. The principal figure in the development of the use of radio waves to send Morse signals was Guglielmo Marconi, an Italian who settled in London and during the 1890s made a series of inventions that created the new technology of wireless telegraphy. Marconi's principal interest was in improving and developing long-distance wireless communication with ships at sea. In March 1899 he sent a radio signal from Britain to France and in 1901 succeeded in sending a signal from Britain to America. In 1909, aged only thirty-five, he was awarded a Nobel Prize for physics for his work on electric telegraphy. At about the same time came the invention of the thermionic valve. Two electrodes were placed inside a glass vacuum tube, enabling an electric current to pass in one direction but not in the other. Advances on this principle followed rapidly, creating one of the first truly electronic components. The use of valves made it easier both to transmit a more powerful radio signal and to amplify a signal once it had been received, improving the transmission of the human voice as well as of Morse code.

In 1900 the Marconi Wireless Telegraph Company was formed in order to establish land-based radio stations that could communicate with radio operators on ships at sea. Marconi tried to enforce a monopoly by not allowing any of his radio operators to communicate with operators from rival companies, a prohibition that soon became impossible to sustain as other companies like Telefunken in Germany developed their own systems. For a while there was a sort of anarchy of the air waves, and before long an international agreement was needed to standardise the bandwidths that could be used. In 1906 a conference in Berlin created some sort of international order in the spread of long-range radio communications. Land-based stations transmitting out to sea had to accept certain uniform agreements about the use of wavelengths and in addition agreement was reached on the sending of distress signals. The Morse signal for SOS (three dots, three dashes and three dots) was approved as an international sign of distress, the spoken equivalent being the word 'Mayday' – based on the French *m'aidez*, help me.

Over the next few years, several incidents headlined the value of the wireless telegraph at sea. In 1910 the murderer Dr Crippen escaped from Britain by sea but was arrested on arriving in Canada when the captain of the ship he was sailing on became suspicious and telegraphed Scotland Yard with his suspicions. And in April 1912, when the *Titanic* hit an iceberg and went down with the loss of 1500 lives, the need for every ship to carry a radio was dramatically highlighted. Only one ship in its vicinity, the RMS *Carpathia*, heard the *Titanic*'s distress SOS, yet that vessel was able to rescue more than 700 passengers who otherwise would have perished. Other ships within range were not equipped with radio and so did nothing to help, and hundreds of passengers drowned as a result.

By the early twentieth century, key developments in the nineteenth had led to the development of another huge industry – the chemical industry. Formerly the province of small-scale local manufacturers, the production of sulphuric acid and bleaching powder had become an industrial process early in the nineteenth century. Then, in 1856, the English chemist William Perkin produced the first synthetic dye, mauve. Two years later, a German chemist synthesised the dye magenta. These and other new colours proved hugely popular in the production of textiles for the fashion industry. In the 1860s another English chemist, Alexander Parkes, invented cellulose, one of the first synthetic plastics. In order to produce these synthetic colours and materials a new chemical industry grew up using organic chemicals, that is compounds that contain the highly versatile element carbon. Although many of the original discoveries had been made in Britain, by the early years of the twentieth century Germany dominated the industry. And coal tar, produced in Britain in vast quantities as a by-product of the conversion of coal into coke or coal gas, was nearly all exported to Germany. Here, giant chemical companies like BASF (Badische Anilin- und Soda-Fabrik), Bayer and Hoechst acquired almost a world monopoly in the manufacture of chemical products derived from coal tar. The chemical industry was making an ever broader range of products, including those needed for the refining of sugar and petroleum; for the manufacture of glass, paint and cement; for photographic materials, cleaning compounds and agricultural fertilisers; and for medicinal and pharmaceutical products.[5] For instance, on the cusp of the new century, in 1897, Bayer invented aspirin. It would soon be described as the new wonder drug.

Furthermore, in the years before the war, the German chemist Fritz Haber invented a process for producing ammonia, a compound containing nitrogen and hydrogen, by synthesising the two elements from the atmosphere using iron as a catalyst. Another chemist in Germany, Carl Bosch, went a step further by developing a brand new high temperature, high pressure process for the bulk industrial production of ammonia. This became known as the Haber-Bosch process. Meanwhile a third German chemist, Friedrich Ostwald, developed a process for turning ammonia into nitric acid. All three chemists were to win Nobel Prizes for their work. The production of these chemicals was intended for the use of agricultural fertilisers, but both ammonia and nitric acid had a further application. They could be used to make explosives.

Dramatic changes also took place at the beginning of the twentieth century in the field of medicine. The use of anaesthetics had begun in the early nineteenth century and ushered in a revolution in surgery, enabling the surgeon to carry out more radical operations than had been possible before. This had coincided with a growing understanding of the role of bacteria as the cause of infection. The thorough sterilisation of equipment to be used in surgery and the ability to maintain strict standards of cleanliness turned the operating theatre into the modern, clinical space familiar today. Antiseptics were developed in the decades before the war to treat bacterial infections, although there was still a general feeling that many bacteria were too powerful to be treated by drugs. In 1909, Paul Ehrlich, a specialist in the new science of bacteriology, developed a drug called salvarsan that provided a treatment for syphilis. German chemists were pioneers in many developments in antiseptics and by 1914 most of these drugs were produced by the booming pharmaceutical industry in Germany.

Another item familiar to modern medicine that came into wide use just before the First World War was the X-ray, invented accidentally by another German, Wilhelm Röntgen, in Würzburg in the 1890s. On the brink of the new century the Curies discovered radium but failed to appreciate that exposure to it could be fatal, and Marie Curie herself died of leukaemia caught from over-exposure. Together, radiology and X-rays began a new era for diagnosing fractures and malformations. The development of film and plates for X-rays came out of the photographic industry and the manufacture of X-ray equipment became another branch of the electrical industries. Out of these industries also emerged the development of the electrocardiograph at the University of Leyden in 1903. Invented as a device to record the electrical activity of the heart muscle, it greatly helped the diagnosis of heart disease.

In addition to the more accurate diagnosis and treatment of disease came substantial improvements in preventative medicine; advances in the supply of clean water and the disposal of waste and sewerage were coupled with the introduction of mass immunisation projects. The incidence of diseases like cholera and typhoid that had been the scourge of large, crowded nineteenth-century cities went into a marked decline in the early twentieth century. The whole concept of public health and of the need for the state, or local authorities, to make provision for improved sanitation and the chemical purification of water supplies became recognised in Europe and North America in the decades before the war. With this went a decline in death rates, a marked drop in infant mortality and a further growth in population.

Another sign of the broadening interest in and support for public health measures in Britain was the establishment in 1913 of the Medical Research Committee (the forerunner of today's Medical Research Council). The government set up this committee with funds raised by National Insurance contributions, payable since Lloyd George's radical budget two years earlier. Every worker in the country paid a penny towards a fund to build sanatoria to treat tuberculosis, one of the great killers of the day, especially among the poor in overcrowded cities. The contributions were known as the 'TB Penny'. Part of this money – a sum of £57,000 per year – was allocated for research. The role of the Medical Research Committee was largely to coordinate such work, although it was itself allowed to carry out research into all aspects of medicine. A group of nine leading scientists, chaired by Lord Moulton and supported by an Advisory Council made up of representatives from the universities, were to formulate plans for medical research that would be funded by the 'TB Penny'. The creation of the Medical Research Committee marks another important step in the state's growing interest in the health of the nation, although its work had barely begun when the war refocused its attention in a different direction.

One of the major new medical concerns in the years before the war was what were loosely called 'nervous conditions'. Indeed, some doctors spoke of nervous breakdowns as *the* disease of the era, brought on by the speed, noise and pressures of modern urban life. Although the medical profession gave a great deal of attention to nervous conditions, the forms of treatment were limited and were largely determined by the class to which the patient belonged.

Doctors with private practices were paid large sums to advise wealthy clients suffering from various psychoses. The general name for this discipline was neurology.

Neurologists had high status and were paid well. At the National Hospital for Nervous Diseases in Queen Square, London, probably the most famous group of specialist doctors in the country studied and consulted on the physical changes that affected the brain in conditions like epilepsy, paralysis and tumour. However, the rest of the population who suffered from nervous breakdowns were classed as lunatics or simply 'the insane'. They were assigned to lunatic asylums where lowly paid medical officers and psychiatrists tried to look after them. For many people, being registered as insane was nothing less than a life sentence. There was little attempt to treat them or solve their problems inside institutions that were often squalid and overcrowded.

New academic work being done in Britain at Cambridge University led the way in the study of the new science of psychology. The university inaugurated a department of psychology at the end of the nineteenth century and in 1912 Charles Samuel Myers established there the country's first experimental laboratory in psychology. But Britain still trailed behind the continent when it came to studying and treating neurotic diseases. In France the study of the mind was far in advance of Britain, and Paris had acquired an international reputation for the treatment of mental conditions. Joseph Babinski in his clinic at La Pitié hospital believed that cases could be cured by a stern approach, using isolation and counter-suggestion to reverse the psychological process that had caused the problem. Across the city at the Saltpetrière hospital, Jules Dejerine, on the other hand, launched the new science of psychotherapy to help a patient understand his or her own illness. Meanwhile, in Vienna, Sigmund Freud had already concluded that many psychological problems were down to the repression of memories and emotions in the unconscious mind. Only by bringing these emotions to the surface could a patient be cured. While the French doctors had a great deal of influence, however, Freud was not widely read or understood in Britain before the war.

Another of the many developments taking place in the early twentieth century, and one that every Edwardian would have been very aware of, was in the field of mass communications. Newspapers that were once read only by the toffs, by the upper wealthy section of society, were becoming available to all. Furthermore, the railways had dramatically improved the speed of distribution and some of the major city newspapers like *The Times* of London and the *Manchester Guardian* became national dailies. Developments in printing with the advent of rotary presses, keyboard-operated type composition machines and roll newsprint paper enabled newspapers to be produced in much greater numbers than before. Following the educational reforms of 1870, literacy spread widely, especially among the lower middle classes, and by 1900 a new generation of clerks, shop assistants, artisans and white-collar office workers formed a reading public eager to buy something more interesting, shorter and more accessible than the traditional newspapers. As literacy levels shot up, prices came down. The numbers of penny journals, illustrated magazines and daily papers soared during the final decades of the nineteenth century. And then came a new breed of newspaper that spread like wildfire, consuming everything before it.

Alfred Charles William Harmsworth was the eldest of eleven children born to a middle-class family. His alcoholic father struggled to keep up appearances for the family growing up in north London and, aged sixteen, Harmsworth left school with two passions, journalism and cycling. In 1886 he combined the two and became editor of *Bicycling News* for a Coventry-based newspaper group. Harmsworth had a brilliant intuitive sense of what the vast reading public wanted and he soon started to publish his own cheap and hugely popular papers and magazines, including *Answers, Comic Cuts* and *Forget-Me-Not*, a popular journal for women. By the end of 1894 his papers and journals were selling more than two million copies weekly and he began to amass a fortune from his growing publishing operations. Two years later he founded the *Daily Mail* with the tag line 'A Penny Newspaper for One Halfpenny'. He changed the layout of newspapers by using bolder typefaces and large eye-grabbing headlines. All the items were short, including political and business reports. There was plenty of sport and daily racing tips. And there were several columns dedicated to women, reporting on fashion, food, cookery and general matters relating to the home. Today this would be seen as gender stereotyping, but in the 1890s Harmsworth judged his readership perfectly and in no time the *Daily Mail* was selling half a million copies a day, rising to over a million during the Boer War. In 1903, Harmsworth went further and founded the *Daily Mirror*, which specialised in the reproduction of photographs using a new half-tone printing process.

Harmsworth helped to project a new form of popular tabloid journalism into Britain just as the numbers of readers with time, money and the interest to buy newspapers exploded dramatically. With the wealth he built up, he went on to acquire established titles like *The Observer*, and in a coup he anonymously bought *The Times*, by now almost bankrupt, thereby gaining control of a central organ of the British establishment. His newspapers took clearly established positions on the big issues of the day. For instance, he repeatedly railed against the commercial, naval and political rivalry of Germany and he regarded the British establishment as being asleep to the threat posed by this growing, expansive nation. The Kaiser himself even complained of the hostility to Germany expressed in the pages of *The Times* and the *Daily Mail* as 'doing the most harm' to Anglo-German relations.[6] Although entirely a self-made man, Harmsworth mixed with leading politicians and writers and, partly thanks to his support of the Unionist cause, in 1905 he was created Baron Northcliffe. As the owner of one of the biggest publishing empires in the world he was able to exert immense influence not only over British politics but also over the popular culture of the nation.

Alongside this, an entirely new type of popular entertainment created at the very end of the nineteenth century would become one of
the great cultural forms of the new century. Like so many inventions,
that of the moving picture cinematograph involved bringing
together many existing or recently discovered technologies. The concept of the 'persistence of vision', that the eye retains an image
flashed only briefly in front of it, until a following image is presented, thus maintaining a sort of visual continuity, had been
understood for some time. It was at the core of many Victorian children's parlour games based around the Zoetrope. The lantern show,
meanwhile, was a more common form of entertainment for people
of all ages in which slides were projected on to a screen. When the
precision technology of the moving camera was combined with
George Eastman's creation of flexible roll film – made from the new
synthetic plastic 'celluloid' – that could be pulled through the
camera on sprockets, all the elements needed to make the cinematograph were in place.

Thomas Edison, the inveterate US inventor, and the Lumière
Brothers in France both laid claim to having been the first to screen
moving pictures to a paying audience in 1895. During the first
decade and a half of the twentieth century, the cinema spread from
being a sideshow attraction at fairgrounds to becoming a form of
popular entertainment in its own right. People could now go and
pay to see a series of short, moving picture entertainments in new,
custom-built 'electric cinema palaces'. By 1914, there were already
4500 cinemas in Britain.

The bicycle was yet another invention of the mid-nineteenth century,
this time bringing together a mechanical means of propulsion with
an effective suspension system. When the mechanical framework
was combined with the invention of the pneumatic tyre, the modern
safety bicycle was born. In the early twentieth century bicycles began
to offer young people opportunities for travel and a means to break
up the insular, parish basis of rural life. Boys could find girlfriends

outside the village or community in which they lived. Girls could cycle to find work away from the streets in which they had grown up. The bicycle offered both men and women an invitation to adventure and an opportunity for freedom.

The bicycle also created a new generation of mechanical processes: first design and manufacture, then constant maintenance and repair. Mechanically minded young men began to open cycle repair shops in every town and city. In Britain, Coventry became the centre of bicycle manufacturing as other mechanically based industries there sought to diversify. The Coventry Sewing Machine Company was one of many operations that made a gradual shift to the manufacture of bicycles, and the growth of the industry would have a major effect upon the face of the West Midlands from the 1860s onwards.

The development of the internal combustion engine was another revolutionary technology, a great force for change in the twentieth century that had seen its origins in the last decades of the nineteenth. The German engineer Nikolaus Otto produced a gas engine in 1876. When Gottlieb Daimler adapted this to run on petrol and a suitable chassis produced by Karl Benz was added in 1889, a period of huge innovation and change followed. Although the motor car was 'invented' in Germany, it was developed far more quickly in France where, in the early 1890s, the likes of Armand Peugeot and Panhard Lavassor soon began to make money by selling large numbers of cars. Britain was characteristically late to the party as the Light Locomotives on Highways Act kept the speed limit on Britain's roads to a pitiful 4 mph, and had initially insisted that a man carrying a red flag should walk in front of every vehicle. It was only after the Act's repeal in November 1896 that motoring in Britain began to spread.

For many years, most motor vehicles in Britain were imports from France or Germany, although a young British engineer began to change that. Frederick Lanchester was another tireless inventor, constantly frustrated by the slow pace at which technical ideas were taken up. In his youth he built a laboratory in his family's home in

Balham, south London. Aged only twenty, having not yet completed his formal education, he so impressed the owner of an engine manufacturing company in Birmingham that he was offered the job of assistant works manager. Here, Lanchester started to register the first of the 426 patents he would take out during his life. In the mid-1890s, he built the first British motor boat and then the first all-British four-wheel petrol-powered motor car. It had a single-cylinder engine and was the first car designed to run on pneumatic tyres. In 1899 he founded the Lanchester Engine Company, helping to advance engine design and develop a native motor industry in the West Midlands.

In the first decade of the twentieth century, however, motoring was only for the wealthy and the determined. Cars were expensive and difficult to run: even starting the engine required a whole sequence of activities which included opening the bonnet, filling the radiator with coolant, adjusting oil pressures, priming the carburettor, swinging the starter handle and advancing the ignition. And when driving, the motorist was likely to suffer from a variety of mechanical failures and punctures. It was no wonder that most owners needed a trained chauffeur-mechanic to assist. In addition, road surfaces were extremely poor, often rutted, throwing up clouds of dust when it was dry or running with water when wet. Pneumatic tyres and basic suspension provided some relief from the bumps but punctures were frequent and were slow to repair. Finding supplies of petrol could also be difficult. Moreover, intrepid early motorists needed an entire new wardrobe of clothing: goggles, leather helmets, gabardine smocks, tweed driving jackets, and for ladies gauze hoods and veils to protect them from the endless dust.

But, for many, driving was about the thrill of speed, and almost from the beginning, enthusiasts started to race motor cars. The Gordon Bennett races began in France in 1900 and transferred to Ireland three years later. Across Europe further long-distance races like the Paris–Bordeaux–Paris and the Circuit des Ardennes began to draw big crowds. The emphasis on speed and long-distance endurance also had the effect of helping to improve engines and increase performance.

However, the manufacture of motor cars relied upon many processes in addition to the creation of an efficient and reliable engine, and car factories were in the early days (as they remain today) largely assembly plants that relied on many separately produced components. The building of a suitable chassis is an essential element, and so many coach manufacturers were drawn into early automobile production. Gearboxes, steering mechanisms, braking controls and powerful suspension systems all had to be produced. In Britain, development of the motor car was the logical next step from bicycle production and so many of the engineer-mechanics who followed Lanchester into the British automobile industry were concentrated in the West Midlands. Alongside came the need for a new level of precision engineering. For instance, pistons had to be made to an accuracy of one hundredth of a millimetre, an exactness that had never been needed in industry before.

Slowly, motoring spread down from the extremely wealthy to the moderately well-off. In some towns or villages the local doctor was sometimes seen driving to visit his patients in a motor car. Garages began to replace blacksmiths and arrays of new machine tools were produced to repair and maintain automobiles and all the components that kept them moving. In 1913, an electric starter system was first used and electric lights began to replace the original acetylene gas powered lamps mounted by the radiator. In 1914, the introduction of hand pumps, rather than two-gallon cans, to fill fuel tanks ushered in the development of an entirely new energy supply industry. Along with the motor car came the development of the articulated lorry and the motor bus. The motorised lorry began to revolutionise the transportation of goods and materials, while most people's first encounter with motoring was to travel in one of the motor omnibuses which rapidly transformed transport within towns and cities. By 1914, London alone had more than two thousand motor buses, a good proportion of which were double deckers. With motor cars spreading everywhere and buses clogging up the streets, the face of the city was changing rapidly.

A new generation of enthusiasts were inspired by the advent of the internal combustion engine. Thousands of young men, fascinated by the lure of speed, took on the mechanical challenge of designing, building and maintaining engines. Several small companies, founded by engineer mechanics who had raised some cash and turned entrepreneurs, were established in Britain in the 1900s. Sometimes they consisted of little more than a workshop or an assembly shed. This created an industry with too many small, uneconomic producers manufacturing too many different designs, and before long some of the more successful companies began to take over the smaller operations. Names like Napier, Morris, Triumph, Rolls, Rover, Hillman, Humber, Wolseley, Swift, Singer and others began to dominate the new motor industry. By 1910, Herbert Austin employed one thousand workers at his Longbridge plant to mass produce a 7 hp car. Austin, alongside Lanchester, was the most visionary of the automobile pioneers and the pair saw that by standardising components they could speed up production and keep down costs. Both their companies prospered. In the United States Henry Ford led the way forward when he finessed the concept of mass production at his giant Highland Park factory in Michigan. Ford broke down the manufacturing process into the smallest possible units, so that a worker would carry out only one of these operations as the production line moved along at waist height. The Ford plant was dedicated to manufacturing a single standardised vehicle, the Model T, the 'Tin Lizzie'. In 1909, just over 12,000 Model Ts came off the factory line; in 1915 more than one million were produced. By introducing the factory production line, Ford unleashed a new era in capitalism and economic growth in which the workers ultimately became the consumers of the new products.

Lighter-than-air balloons had been built and flown in France since the 1780s and great developments in the science of ballooning had followed over the next hundred years. In a separate development, during the nineteenth century many of the basic principles of aerodynamics had been studied and understood. Sir George Cayley, a Yorkshire engineer, produced a treatise on what he called 'Aerial Navigation' and worked out the shape needed for a wing to provide the lift necessary for an aircraft. With a curved top surface and a flat lower surface, air would travel faster over the top of the wing providing the 'lift' to sustain flight. The first man-carrying glider flew across Brompton Dale in Yorkshire in 1853 supposedly piloted by Cayley's footman. Cayley also calculated the power necessary to achieve lift with a given weight of airframe. But no engine at that point had a high enough power output in relation to its weight to make flight possible. It was only with the coming of the petrol-fuelled internal combustion engine that engines both powerful and small enough became available. From that moment on it was inevitable that powered flight would be both possible and practical. But powered flight itself was to be a twentieth-century phenomenon.

There were several key challenges to be solved before an aircraft would be able to fly. Having mastered the correct design of the wing to provide lift, a suitable airframe had to be constructed. This was the relatively easy part, and although many gliders were of a monoplane variety it was soon discovered that extra lift could be guaranteed by constructing biplanes. They had, effectively, double the wingspan, providing double support and lift. Another problem was the provision of controls to manage the craft when it was in the air. This took much experimentation and practice, and was achieved through trial and error rather than by the application of scientific or mathematical principles. The next and far more challenging task was to build an engine with the correct power-to-weight ratio, small and light enough to generate the power needed to get a craft into the air and then enable it to remain airborne. Thanks to the enormous advances taking place with the development of the internal combustion engine for motor vehicles in the early years of the twentieth century, the power-to-weight problem was finally solved with an aluminium engine that produced about 12 hp of energy.

Wilbur and Orville Wright were the first to achieve powered flight at Kitty Hawk in 1903. They were bicycle manufacturers from Dayton, Ohio, illustrating how the mechanical technologies had advanced from bicycles, through automobiles and on to aeroplanes. The brothers moved their experimental work to the North Carolina coast where steady and regular winds provided the extra boost needed to create lift. From 1900, for three years, they built a series of gliders to master the techniques of flight. Then, employing their own basic biplane design constructed with a spruce wood frame and muslin wing coverings, along with controls they had devised themselves and a small engine, they finally launched the era of powered flight on 17 December 1903. Barely anyone noticed this historic achievement and it was a couple of days before even the local press reported it.

However, the Wright brothers quickly improved both their aircraft and their understanding of the basics of flight. When Wilbur arrived in France in the summer of 1908 to give a set of public demonstrations of their aircraft, he was celebrated as the first great pioneer of modern aviation. Leading politicians and captains of industry flocked to see him flying. Lord Northcliffe, the newspaper baron, took Arthur Balfour, the leader of the Opposition, to watch the demonstrations. King Edward VII even went to see Wright fly.

It is often claimed that Britain lagged behind the rest of the world in aviation, although this was definitely *not* the case with military aviation as we shall see in the following chapters. Nevertheless, Britain at the beginning of the twentieth century is usually seen as a technophobe nation and there is certainly some truth to this. Much of it came down to the education system. The public schools that produced the elite of British society were totally focused on providing a classical education, and gave far more importance to Homer and Virgil than to mechanics and physics. A good education was thought to provide not only the ability to read Greek and Latin, but also to compose verse in these two dead languages. Pupils learned to look down on science, while engineering was thought to be beneath the dignity of a gentleman. The Edwardian public schools were most certainly anti-technology.

It has often been said that men with little or no higher education had created the Industrial Revolution that shaped modern Britain, *despite* the educational system. However, in the early twentieth century the university sector was slowly changing. It is difficult today to imagine a university system as tiny and exclusive as that of Edwardian Britain. The oldest and most prestigious universities were filled by pupils from the public schools and so were still dominated by the classics, or by a combination of classical studies and mathematics. But even Oxford and Cambridge, and definitely the newer redbrick civic universities often sited in the great industrial cities, offered a growing mix of applied scientific studies to the increasing number of middle-class students. Liverpool, Leeds and Manchester Universities were forging ahead in this regard. Osborne Reynolds, Professor of Engineering at Manchester, did important work in fluid mechanics and his successor Joseph Ernest Petavel worked on aeronautics. At Leeds there was a professorship dedicated to the coal and gas industries. And there were signs of a new interest in science even at Edwardian Cambridge. Lord Rayleigh at Trinity College, one of the most famous and respected physicists in the country, and Bertram Hopkinson, Professor of Applied Mechanics, did important work on explosions, on the internal combustion engine and on metal fatigue. The engineering school at Cambridge produced the largest number of graduate engineers in Britain, many of them trained in the sort of complex mathematics that was essential for research in new sciences like aeronautics. However, the emphasis in British universities was predominantly on pure sciences and few graduates were encouraged to dirty their hands by going into industry. The war would transform this.

Britain's industrial supremacy was looking decidedly shaky in the first decade of the twentieth century. Its key industrial rivals possessed a great advantage in the provision of vocational training in science and engineering. In Germany, alongside the old universities which, as in Britain, looked down on industry, there was a new generation of technical schools, the excellent *Technische Hochschulen*. Here the professors maintained close links with industry so the schools could carry out vital research needed for industrial progress. As a consequence, the major engineering and manufacturing companies had a ready supply of trained graduates. In the United States of America, where many universities were also close to industry, the establishment during the nineteenth century of such important centres as the Massachusetts Institute of Technology and the Stevens Institute of Technology brought together the highest academic standards with the needs of rapidly developing technology.

There were technical schools in Britain, although they did not fare well in the educational reforms of the Edwardian era and suffered from chronic under-funding. Nevertheless, these colleges were to play an important role in the development of the new sciences. For instance, the technical colleges at Finsbury and at Crystal Palace in London provided a suitable education for young men who were enthusiastic about the new mechanical technologies. They also offered courses in electrical engineering and motor sciences. Many of the pioneers of aviation went to these schools. Sylvanus Thompson at Finsbury was tutor to both Frederick Handley Page and Richard Fairey. Geoffrey de Havilland went to Crystal Palace. Some colleges also offered evening classes for students who were in employment but keen to improve the academic basis of their work. Their graduates were to drive forward many of the 'new' mechanical-based industries of the twentieth century. But they were still few in number by comparison to the graduates of similar schools in Germany and the United States: in 1913, there were 40,000 students of science and technology in the United States, 17,000 in Germany and 5500 in Britain.[7] It is not surprising to find that Germany and the USA were soon to surpass Britain in industrial output.

Many men of science were aware of these issues. At a conference in May 1916, several leading British scientists gathered to lament the bias of the educational system against science. They expressed horror at the fact that so many educated politicians and civil servants were fluent in Greek and Latin but knew nothing of scientific method or of new developments in physics, chemistry and engineering. It was pointed out that the headmasters of thirty-four of the top thirty-five public schools were classicists; that only four Cambridge colleges were presided over by men with scientific training and that there were none at Oxford. Lord Rayleigh described this as 'truly deplorable'. The division within the class that ruled the country between those with a classical education and those with a scientific background was to exercise many people for decades to come.[8]

The great technological changes that transformed the Edwardian era were closely linked. Behind everything was electricity. New electric powered technologies were transforming streets and homes. The building of the electric underground railway in London was one of the great achievements of the age, even if much of the network was not completed until the 1920s. The spread of the motor car (there were 132,000 private cars and 51,000 buses, taxis and coaches in Britain in 1914) enabled cities to develop vast suburbs and to grow into huge conurbations in which much of the change was concentrated. By 1914, four out of five people in England and Wales lived in cities and Greater London had a population of seven and a quarter million. In Manchester, there was one cinema seat for every eight inhabitants. Rising prosperity, particularly among the lower middle classes if not the working classes, helped fuel the growth of new retail chains, the spread of a mass market and the development of national brands of food and drink. The explosion of the tabloid press selling at low, accessible prices relied on advertising which helped the spread of these new brands. Everyone agreed that technology was changing the patterns of life in an extraordinary way. All of this was bound to have an effect upon the military and upon thinking of how to fight the next war, if and when it came.

The Edwardian Army and the Royal Navy, run by two great departments of state at the War Office and the Admiralty at the north end of Whitehall, were largely conservative-minded operations whose senior figures shared the attitudes and values of the rest of the British elite. In Britain's forces as in most armies and navies around the world, military work involved the repetition of the same processes over and over again. Tradition was given much greater credibility than innovation. At the Royal Academy at Sandhurst there was no science on the curriculum. New ideas and new approaches were not welcome. Senior army and naval officers, almost exclusively the products of the public schools, had a suspicion of industry and science just like the rest of their class. They were professionals who believed they knew what they were doing, had been doing it for some time and saw no reason to change the way they were doing it. Socially, the army was deeply conservative, and the life of the officer class revolved around fixed rituals that were based on loyalty to a regiment or battalion. H.G. Wells, the great science fiction writer and one of the reformers of the era, saw the army as 'a thing aloof', an institution that 'had developed all the characteristics of a caste' and was 'inadaptable and conservative'.[9]

The biggest division within every army and navy in the world in the early twentieth century was that between officers and 'other ranks'. This again closely mirrored and sustained the fundamental division in society between owners and managers on the one part, and labourers or workers on the other – a division even extended to many sports of the day, like Edwardian cricket, where it was expressed as the difference between 'gentlemen' (who took part for the love of the sport) and 'players' (who played for money). The world of the late Victorian or Edwardian officer was one of luxury rarely seen today. Even the most junior officers had their own servants who would unpack their bags, lay out their clothes and probably clean their boots. Life in the officers' mess involved much ritual but was, like life in a country house or a wealthy city dweller's home, entirely reliant upon a division between upstairs and downstairs and the existence of a vast army of servants and domestic workers. With some infamous exceptions, there was very little familiarity across this divide, and while an officer would have responsibility for the care and welfare of his men, there were few occasions when an officer would mix or socialise with the men in his battalion. The activities of the two groups were kept rigidly separate. To many in Edwardian Britain, especially to those in the army or the navy, it probably looked as though these divisions would continue for ever.

However, the outpouring of inventions in the late nineteenth and early twentieth centuries – electricity, new chemical industries, the internal combustion engine, radio, cinema, powered flight – all began to challenge the nature of military thinking. The industrialisation of war and the introduction of new military technologies is one of the most striking features of the last decades of the nineteenth century. The British Army that went to war in the Crimea in 1854 was armed with muzzle-loading rifles designed at the end of the seventeenth century. These fired smooth-bore pellets and were reliable only up to a range of about eighty yards. The artillery fired spherical cannonballs to a range of about two miles. Sixty years later, Lee Enfield rifles firing conical bullets from a magazine were accurate up to 400 yards and a well-trained soldier could fire at the rate of fifteen rounds a minute. The newly invented machine gun, with a rate of fire of up to 600 bullets per minute, would soon prove to be a mass killing machine. Meanwhile, the artillery had been transformed. Breech loading and steel barrels, rifled to provide greater accuracy, enabled field artillery to hit targets at a range of up to seven miles. Heavy mortars could throw incredibly powerful ordnance over a considerable distance.

The American Civil War of 1861–5 has been called 'the first Great War of the Industrial Revolution'.[10] Railways played a key part in transporting troops to the front. Trenches were dug and barbed wire used extensively. The employment of armoured trains, land mines, balloons, the field telegraph, ironclad ships, submarines and sea mines began to transform the nature of war. Countries at the forefront of these technological changes, like Germany with its huge Krupp steel works, or Japan, had a great advantage. The Prussians defeated Austria in a ten-week campaign in 1866 and France in six months in 1870–1. Japan overcame Imperial Russia in a campaign that stunned the western world in the war of 1904–5. Britain's prowess was in its navy and the fifteen-inch guns installed in the latest of the huge steel-armoured dreadnought battleships had a range of nearly nineteen miles. And the Royal Navy was as large as the next two biggest navies combined.

What was different about the First World War was not only the level of industrialisation but also the scale of the conflict. The mass armies that assembled in 1914 were equipped with the latest scientific and industrial offerings but in addition, after several decades of rapid population growth, were of an unprecedented size. The population of Germany increased by more than one half between 1870 and 1910, from 41 to 65 million. Germany could assemble an army of 5,170,000 men including reserves in 1914. The French mobilised 4 million men at the outbreak of war. Russia, with a population of 161 million, also had 4 million trained men available. But of course every army also had to be fed, watered, uniformed, provided with boots, communications and transport as well as guns and equipment. So, when war came in 1914 it would be the first ever fully industrialised, scientific war. The big question was, were the military classes ready for this? And would it be the soldiers or the scientists who would discover what was needed to win a twentieth-century war?

The first application of science to the new technology that would go to war in 1914 came in what contemporaries saw as the extraordinary new science of aviation. Although the first dramatic developments in powered flight had happened elsewhere, it would not be long before the excitement of the first pioneers would spread to the British shores.

Part One

Aviators

2

The Pioneers

On Friday 30 April 1909, an event of great historic importance took place in a field near Leysdown on the Isle of Sheppey, where the Thames widens into a broad estuary. A young racing driver and mechanical enthusiast by the name of John Theodore Cuthbert Moore-Brabazon climbed into a fragile canvas-and-wire structure. With huge biplane wings and parallel vertical panels on either end, it looked rather like a large box kite. The structure had an elevator in the front and an enormous biplane-like tail. In between the two wings was a 10 hp engine with a crude three-ply wooden propeller. To fly this contraption, Moore-Brabazon had to lie flat on the lower wing and operate the limited controls with his arms stretched out in front of him. The wind was blowing very gently from the south when he climbed in, checked the rudder controls, started the engine and, once happy that it was running smoothly, instructed his friends who were clinging on to the wings to let go. He moved gently forwards. The engine pulled well and after a few yards a sudden gust of wind hit the craft head on. With so much lift from the large wings, the contraption took off and ascended perfectly into the air, to a height of about fifty feet.

Moore-Brabazon was just beginning to enjoy the experience when another gust of wind caught the aircraft from the side. It tipped suddenly off balance and the right wing rose higher in the air. Moore-Brabazon heaved the rudder control as strongly as he could to try to rebalance the aircraft, but to his dismay the pressure he exerted broke the control line. He was now in the air with no control over his machine. All he could do was to tilt the front elevator and glide slowly back down to earth. But the field was full of ditches and dykes and he now had to worry about finding a smooth spot on which to land. Only a few seconds later, still out of balance from the gust of wind, the left wingtip hit the ground heavily. The aircraft shuddered under the impact. Wires and struts snapped viciously. As the machine crashed into the ground, the shock of the impact forced the engine from its moorings and it shot forwards, missing Moore-Brabazon by inches, and buried itself in the earth. The pilot, bumped and bruised and a little dazed but otherwise unhurt, was trapped on the wing, tied down by the wires that had come loose and wrapped around him. The first sensation he felt was being licked by his two dogs who had run after the aircraft. The previous few moments had been neither impressive nor heroic. The aircraft had only travelled 150 yards and had never attained a height of more than about fifty feet. But Moore-Brabazon had made history. For this was the first powered flight by a Briton in Britain. It was, as he later said, 'an adventure into the unknown'.[1] For his efforts, Moore-Brabazon was awarded Certificate No. 1 from the English Aero Club, the first ever pilot's licence issued in Britain.

Although this was a British first, fliers had been taking to the air for some years in fragile canvas devices powered by one of the new generation of motor engines. After the Wright brothers' first, famous heavier-than-air flight on 17 December 1903 at Kitty Hawk in North Carolina, it had taken three years for flying to come to Europe when on 12 November 1906, in France, the Franco-Brazilian pilot Alberto Santos-Dumont flew, or some said hopped, a little over 200 yards in his own aircraft. And France was to be the centre of European aviation for some years to come. Gifted engineers like Gabriel Voisin and Henri Farman began to build and fly several new designs of aircraft. Indeed, John Moore-Brabazon was flying a French Voisin-Farman aircraft when he first took to the air on the Isle of Sheppey. He had lived in France and made several flights there beforehand, watched by crowds who cheered him on. He described these early flights as like 'sitting on a jelly in a strong draught'.[2]

Moore-Brabazon, better known simply as Brabazon or 'Brab', was typical of the pioneers who characterised the first chapter in the history of flight. His family were members of the Anglo-Irish aristocracy and he had attended Harrow, one of the top public schools. But while there he fell head over heels in love with the motor car. Like many of his age group with a mechanical bent, he became obsessed by the new engineering triumph and gave a lecture to his school scientific society on the workings of the internal combustion engine. His father, who had served in the Indian army for thirty years, thought his son should be getting a fine classical education and like many of his generation, as Brabazon put it, 'did not "hold" with motor-cars'.[3] Family pressure persuaded Brabazon to apply to Cambridge to study for an Applied Sciences degree. Bizarrely, he had to spend a term at a crammer to improve his Greek, which he had neglected at school, to get through his matriculation exams even to study sciences. But he only spent a year at Cambridge before dropping out.

While there, he met another young man from a similar social background who shared his obsession with motor cars: Charles Rolls was about to form a company to design and build automobiles. The two young men used to race cars together and won several speed trials in the Gordon Bennett races (named after the proprietor of the *New York Herald*) in Ireland in July 1903. After dropping out of Cambridge, Brabazon stayed in close touch with Rolls but moved to Paris, where he served a form of apprenticeship in the workshop of the Darracq Motor Company. Here he did a bit of everything, from assembling engines to fitting tyres on to new vehicles.

Before long, the young Englishman had an opportunity to try his hand once again at motor racing, and very good at it he turned out to be. The thrill of driving at speed seems to have captivated him, and soon Brabazon was winning motor races – and with them decent sums of prize money. He raced in Britain, in Ireland and on the continent for several different companies including Austin, Mors and Minerva. It was in one of the Minerva cars that he made an international name for himself by winning the celebrated Circuit des Ardennes in 1907. His car reached the then incredible speed of 70 mph.

Brabazon, like many of the bright young things of his era, enjoyed another enthusiasm. Having started in France in the 1780s, by the early twentieth century ballooning had become established as a form of reconnaissance for the military. However, it was now also a fashionable recreational sport for the wealthy elite; to fill a balloon with the required 45,000 cubic feet of gas cost five pounds (equivalent to about £500 today). While at Cambridge, Brabazon discovered the joy of sailing silently and majestically over the countryside suspended in a basket below a large spherical balloon. He loved the peace and quiet of ballooning, and in a way that perfectly summed up his class he used to say that 'to go up in a balloon is the only way to go into the air like a gentleman'.[4]

It was perhaps inevitable in 1906–7, when interest began to grow in the new science of aviation, that Brabazon and his chums like Charles Rolls would take up this latest fad. It offered the opportunity to combine a fascination for solving mechanical problems with the thrill of speed and the excitement of flight. After Brabazon's victory at the Circuit des Ardennes, several motor manufacturers approached him with requests to drive for them. But instead Brabazon spent more time mixing with the young men who were beginning to pioneer the new sport of flying.

One group Brabazon got to know well was the three Short brothers, Horace, Eustace and Oswald. They came from a family of north country mining engineers. The eldest, Horace, had designed an early form of amplifier in the late 1890s. The brothers initially set up a business in Hove, near Brighton, to construct balloons. After a few years they each put up £200 capital to form a company to design and build aircraft and moved to a larger site, erecting a set of hangars on an area of marshland at Leysdown on the Isle of Sheppey. When Brabazon returned from France he based himself there.

Most of the early pioneers emerged from the tradition of mechanical engineering whose champions had been part of the revolution in transport over the previous decades. The Wright brothers were bicycle manufacturers, while in France many of the pioneers were technicians who enjoyed trying to make machines work successfully. In Britain the first generation of aviation pioneers quickly established themselves, building their own aircraft or adapting the designs of others. Broadly speaking, they fell into two types. First there were the gentlemen pioneers, of whom Brabazon is a perfect example; another was his friend, the son of a peer who was correctly known as The Honourable Charles Stewart Rolls.

Rolls had been to Eton and Trinity College, Cambridge, and was one of the first people in Britain to acquire a motor car when in October 1896 he imported a Peugeot from France. He attracted an immense amount of attention when he drove it up from London to Cambridge, a journey that took eleven and a half hours. Rolls was tall, thin and looked every inch the aristocrat, but unusually for his class he had an instinctive understanding of things mechanical. Flamboyant, wealthy and still in his twenties, he cut a dashing figure on the early motor racing circuit and later was one of the first to join the RAC, which he helped to run for some years. He was a great publicist for the motor car.

In 1902, Rolls formed his own motor company in Fulham, dealing largely in French Panhard vehicles. But it was when, in 1906, he merged this concern with Frederick Henry Royce's Manchester engineering parts supply company that one of the greatest brands in motoring and aviation was created. The resultant manufacturing company, based in Derby, was set up with capital of £60,000 and the first Rolls-Royce was a four-cylinder, 20 hp luxury vehicle named the Grey Ghost. Although by the time of the merger Rolls had become obsessed with aviation, the Rolls-Royce company devoted its early years to the manufacture of luxury motor vehicles. It was only during the war that it produced its first aircraft engines. The very name Rolls-Royce of course still stands as a by-word for excellence.

Thomas Sopwith was another member of the wealthy classes who enjoyed pursuing his passions of yachting, motorcycling and motor racing and first came to aviation as a sport. Having bought a single-seat monoplane and taught himself to fly, in 1910 he established a British record by flying 107 miles in 3 hours 10 minutes. After this he opened a flying school at Brooklands. Among those who learned to fly at the school was Harry Hawker (whose company twenty-five years later built the Hurricane fighter). Sopwith went on to design and build his own aircraft through the Sopwith Aviation Company, where his models soon acquired a reputation for combining manoeuvrability with stability; his name would become a household word in the course of the war for the famous series of fighters he produced.

Some of the other British aviation pioneers, however, were not sons of the gentry, and – like the Short brothers – they represented a different tradition that also fed into the early aviation business. Alliott Verdon Roe was born into a middle-class family in the suburbs of Manchester. His family moved to Clapham Common in south London and he attended St Paul's School, but he was not happy there and left at fourteen to be apprenticed at a railway locomotive works outside Manchester. During his apprenticeship he developed an enthusiasm for cycling and a reputation as a fine bicycle mechanic. He then went to sea as an engineer for the British and South African Royal Mail Company, and while he was watching albatrosses glide in the wake of his ships he developed an all-consuming passion for flight.[5] He left the merchant navy in 1902 and became a draughtsman in the new automobile industry, where he came up with an improved gear-changing mechanism. But Roe spent as much time as possible building gliders and furthering his interest in the possibilities of aviation. In January 1906 an engineering supplement of *The Times* published a letter from Roe in which he dismissed ballooning and said the future of aviation lay with 'the aeroplane system'. He predicted that motor-driven machines would be flying over England by the following summer.[6]

But A.V. Roe, as he was known, had no wealthy friends to back him, and during the next couple of years he struggled to find funding to develop his ideas. In August 1907 he won a small amount of money from a competition flying aircraft models. Then, in the mews of a house in Wandsworth where an elder brother lived, he built his first aircraft. Its wooden frame was stiffened by wire bracing, and it had an upper wingspan of 36ft and a lower wingspan of 30ft. A local cycle shop mechanic helped with the welding work. There was no rudder, while the pilot's place was at the very front, by the forward elevator. The big advance was that Roe designed the first all-purpose control column, effectively a prototype joystick, with which the pilot could adjust the elevator and hence affect the aircraft's vertical movement.

Having built the aircraft, Roe had to find somewhere to try to fly it. He chose the motor racing centre at Brooklands near Weybridge in Surrey, as the club there was offering prize money of £2500 to anyone who could achieve a circular flight around their brand new track. A sum like this would set him up properly in business. Roe moved there in September 1907 and erected a shed on the infield to assemble his aircraft, but in part, no doubt, because he did not share the gentlemanly background of most of the club's members, he found the local track officials downright hostile to the presence of an aviator in their midst. For a long period he lived a hand-to-mouth existence, trying to get his craft to fly. However, despite facing a variety of difficulties and obstructions, he learned many lessons from every attempt to fly his aircraft. He started with a 9 hp twin-cylinder motorcycle engine, but in May 1908 he obtained a far more powerful 24 hp Antoinette engine from France, one of the best aero engines of its day. Roe later claimed that he managed to get airborne the following month in a modified and improved version of his aircraft, although at the time he made no announcement about what would have been a notable pioneering achievement. (In a dispute with Brabazon in the late 1920s a committee investigated his belated claims and decided not to accept them.[7])

Despite the setbacks, A.V. Roe went on to be a prime mover in the founding of the British aircraft industry. His company Avro produced one of the finest planes in the early part of the war, the Avro 504. And in the Second World War it would be Avro that produced such legendary British aircraft as the Lancaster bomber.

Frederick Handley Page was another of the new generation of engineer-mechanics. Trained at the Finsbury Technical College in north London, he was always interested in the latest trends in technology and it was while he was working as chief electrical designer at an engineering firm in south-east London that be became fascinated by flight. Unhappy with the time he was devoting to flying experiments the firm sacked him and, in 1909, in his mid-twenties, he decided to set up his own company, with capital of £500. For a few years he built one-off aircraft and supplemented his meagre income from aviation by teaching evening classes at Finsbury. His company, based first at Barking, moved to a bigger site in Hendon where he won a contract from the War Office to produce versions of an official design. He began to design passenger-carrying aircraft, the first of which flew across London with a single volunteer passenger in 1911. Later he went on to build larger twin-engined biplanes with 70ft wingspans that became the prototype for the first bombers. By the end of the war the name Handley Page had become synonymous with big bombers, and although the company went on to design and produce several civil airliners it was best known for the first of the Second World War 'heavies', the Halifax, and later for the sleek nuclear bomber, the Victor. Within years of the first flights, many of the names who would come to shape British aviation for the next half century were beginning to form small companies and become players in the aviation game.

A great deal of popular interest in early aviation was aroused by the biggest newspaper baron of the day. Lord Northcliffe was keen to promote new sports and technologies. A motoring enthusiast from the beginning, he had a collection of cars of which he was inordinately proud. He founded the Harmsworth Cup for motor-boat racing in the early twentieth century and was a founder member of the Aero Club. Seeing the massive interest aroused by the early pioneers of flight in France, and fearing that Britain was falling behind in this new science, Northcliffe launched the first of what would be a set of prizes intended to encourage advances in aviation. In November 1906 – just a week after Santos-Dumont's pioneering 200-yard flight in France – the *Daily Mail* offered an award of £10,000 to the first aviator to fly from London to Manchester. No one in Britain had yet even been able to get airborne, and the idea of flying from London to Manchester in one of the early machines seemed so utterly fantastic that the award provoked widespread ridicule. *Punch* mockingly responded by offering £10,000 to the first aeronaut who succeeded in flying to Mars and back within a week.[8]

But Northcliffe had his finger on the pulse. His interest in and encouragement of the early pioneers provided an immense boost to aviation, although it was a long time before anyone was able to fly from London to Manchester. And his readers devoured stories of the magnificent young men in their flying machines. So Northcliffe came up with a series of more modest and achievable awards. He offered £1000 to the first pilot in Britain to fly a circuit of a mile. This posed several problems, because the first generation of aircraft tended simply to take off, fly in a roughly straight line and then come down. Banking and returning to the point of take-off came later. But the award was nevertheless won when Moore-Brabazon managed to fly his craft at little more than twenty feet above the ground for half a mile to a marker, then used the rudder for almost the first time to turn slowly in a perfectly flat semi-circle. He felt the wind in his face change direction and after he completed the turn it blew at him from behind. This was a strange and eerie experience.

The pioneers were literally making it up as they went along. There was no instruction manual and certainly no rule book. They simply tried something new and if it worked they repeated it the next time. If it didn't work they either crashed to the ground, or at least tried to learn from what had gone wrong. And in the case of Brabazon's mile circuit, the *Daily Mail* correspondent, Charlie Hands, was there to write it all up for Northcliffe's eager readers.[9]

Northcliffe's next prize was to revolutionise the early flying industry. In 1908 he offered £500 to the first pilot who could fly a powered aircraft across the English Channel. By the end of the year no one had made the attempt, so Northcliffe doubled the prize money, and in July 1909 two pilots announced they were going to make an attempt. The first, Hubert Latham, set up camp at Sangatte near Calais and on 19 July headed off across the Channel in his Antoinette aircraft. He got to within six miles of the English coast when his engine failed and he crashed into the sea – to be rescued by a French naval destroyer. A few days later Louis Blériot arrived in Calais and prepared to make his own attempt.

Blériot had made his name and his fortune as a manufacturer of lamps for the early motor car industry. This gave him the time and opportunity to develop an interest in early flying machines. When he witnessed Santos-Dumont's first European flight in November 1906 he decided he would design his own aircraft. Several models followed, including a successful monoplane, and despite many mishaps and a crash in which he was badly burned, Blériot carried on designing new aircraft and flying them himself. One of the problems with most aero engines at the time was that they needed to run at maximum power, and few could run for more than five minutes before they overheated or seized up. Better engineering slowly improved engine durability and Blériot was able to make flights of fifteen, then thirty, then fifty minutes duration.

At dawn on Sunday 25 July 1909, Blériot set off from Calais in his model XI aircraft. The monoplane design was recognisably modern with a 25ft wingspan, a partially covered box-girder fuselage and a small rudder in the tail. The 25 hp Anzani three-cylinder engine with a two-bladed walnut wood propeller was mounted in front of the leading edge of the wing. A triangular system of warping wires braced the wings to a support in front of the pilot, who sat just behind the wing. There was an undercarriage with wheels that could slide up or down a steel tube. Flying at 45 mph, at an altitude of about 250 feet, Blériot crossed the Channel and made a bumpy landing in strong winds on a field on the cliffs above Dover Castle. He had been in the air for 36 minutes 30 seconds. The *Daily Mail* correspondent raced off in a motor car, picked him up and brought him down to the harbour where a huge crowd assembled.

Blériot became an instant celebrity around the world. But in Britain the reaction was particularly intense. The press were up in arms at this Frenchman's ability to breach Britain's historic insularity. The *Daily Express* headline was 'Britain is No Longer an Island'. The *Daily Mail* also noted that 'British insularity has vanished' but turned its sense of outrage more clearly into an attack on British lethargy for failing to develop a more advanced aviation industry. H.G. Wells blamed the British educational system for a national failure in this new science.[10] No one in any part of British life could now fail to recognise the huge potential and strategic significance of powered flight.

The critical issue was that although powered flight had been invented and mastered, who was it really for? What was the market for these early box kites? There were a few wealthy adventurers, like Brabazon, Rolls and Sopwith, for whom a new pursuit like flight offered a great excitement. And the sporting side of aviation and the breaking of early records was a sensation that would attract large crowds and sell a lot of newspapers: tens of thousands queued to see Blériot's aeroplane when it was displayed at Selfridge's brand new department store in Oxford Street. But beyond this, everyone recognised that there was really only one market for these early flying machines – the military.

This had been appreciated from the very beginning. The Wright brothers are often presented, especially in the United States, as a homely family of simple, enthusiastic, able mechanics – humble bicycle makers who became accidental heroes. In fact they were keen not only to pioneer the new science, but also to exploit it in order to make money. They were soon offering their knowledge to armies and navies in Britain and France, and by 1906 they had added the armies of Germany, Italy, Russia, Austria and Japan to their list of potential clients. They employed a professional arms dealer, Charles R. Flint, as their agent to negotiate with foreign governments. And his proposal was not just to sell single aircraft to foreign armies here and there but to sell entire air fleets; fifty aircraft at £2000 each.[11] Flint predicted that the military potential for aviation was big.

Historians have often claimed that the British armed forces were abysmally slow to recognise the potential of powered flight.[12] In reality, this was not the case. During the nineteenth century the British Army had certainly been slow to take up ballooning as a form of reconnaissance, to use an observer in a balloon to spy behind enemy lines. However, by the end of the century, the Royal Engineers – the most technologically inventive and forward-looking section of the army – had eagerly taken up and advanced the craft of military ballooning. Balloonists were at last taught the skills of aerial reconnaissance and of photography and signalling. Tethered army balloons could usually operate in winds of up to 20 mph and at a height of 1000 feet, from which an observer could often see at least two miles behind enemy lines. An army manual of 1896 declared that 'no modern army would be considered complete without balloon equipment.'[13] The Royal Engineers established the Balloon Factory, initially at Aldershot and then at Farnborough in Hampshire, with a depot containing a hangar, a machine shop, a foundry, a carpenter's shop and several gasometers to fill up the balloons. From this primitive start, Farnborough would go on to play a central role in the development of British military aviation for almost one hundred years.

In the early years of the new century the leading figure at

Farnborough was Lieutenant-Colonel John Edward Capper. He was educated at Wellington College and at the Royal Military Academy in Woolwich, from where he joined the School of Military Engineering at Chatham in 1880. As a trained army engineer he was involved in road, railway and bridge construction in India and in South Africa during the Boer War. In the autumn of 1904, Capper was sent to the United States, notionally to visit the St Louis World's Fair. But while there he called on the Wright brothers at their workshop in Dayton, Ohio. He was fascinated by what he saw and wrote a memo on his return stating: 'We may shortly have as accessories of warfare scouting machines which will go at great pace ... whilst offering from their elevated position unrivalled opportunities of ascertaining what is occurring in the heart of the enemy's country.' He concluded, 'America is leading the way, whilst in England practically nothing is being done ... [What was needed] was a proper experimental school.'[14] Less than a year after the first powered flight, the British Army's senior aeronautical expert had already identified the aeroplane as the future for military reconnaissance.

Capper continued to hammer home his new message at every opportunity. But military staffs are notoriously slow to respond to new ideas, and it took a few years for the momentum of aviation to build up. It was certainly hard to imagine the potential of powered flight when no one had yet managed to fly in Europe. And those who were talking up flight seemed to be the cranks and experimenters, of whom the military had an inbuilt suspicion. The army view remained that they themselves knew best how to soldier. The amateurs could have their fun, but at this stage there was nothing they could teach the professionals.

However, in 1905, a new player arrived on the scene who was to have a profound effect not just on the future of aviation but upon the impact of science and scientists across the military. In that year the Conservative government of Arthur Balfour collapsed and the Liberals formed a new government under Prime Minister Henry Campbell-Bannerman. He appointed as Secretary of State for War, the civilian politician with responsibility for the army, Richard Burdon Haldane, who went on to be one of the greatest reformers of the British military establishment.

Born in Edinburgh into a leading professional family, Haldane studied at Edinburgh Academy and then went to Göttingen University. There he fell under the influence of the German philosophical school of Hegel, which left him with a lifelong belief that rational, scientific principles should guide all action. After completing his education at Edinburgh University he began a legal career. Having been a great success chiefly as an appeals lawyer, becoming a Queen's Counsel aged only thirty-four, he entered politics as a Liberal Imperialist supporting Herbert Asquith and Edward Grey. A great believer in education who helped to found the London School of Economics in 1895 and Imperial College, London in 1907, Haldane was rather high-minded, a philosopher by training and a lawyer by profession, an intellectual who never tried to court popular opinion or play to the crowd. In his first government appointment as Secretary for War he began to implement new ideas inspired by his passionate belief that science could help to improve all aspects of life. The period of dramatic reform that he ushered in was to shape the military force with which Britain went to war in 1914.[15]

First, Haldane decided there was a need to organise the army into a mobile expeditionary force that could be trained and prepared to go abroad should war break out on the continent of Europe. The force would consist of six divisions with cavalry support. With this central decision in place, other changes could be made to fit the army for its new role. The various yeomanry and militia organisations around the country were abolished and reformed into a new territorial force that would provide homeland defence when the expeditionary force went overseas. Haldane appointed Douglas Haig as the new Director of Staff Duties at the War Office and Haig began to reform the training of the army. In 1909 Haig formalised the changes, and his Field Service Regulations became the core training manual for the British Army. At the top Haldane created a new Imperial General Staff to govern the strategic direction not only of the British Army but also of the armies of the several Dominions and the Empire. The new Chief of the Imperial General Staff would be the head of the army, reporting to the government. With the rapid growth of aviation it was inevitable that before long Haldane would apply his reforming zeal to this aspect of military activity as well.

While Haldane and his colleagues discussed their first reforms, changes were taking place in the embryonic world of powered flight. The infant motor industry showed the way forward. Engineer mechanics had founded several small companies by raising some cash and turning entrepreneurs. The nascent aviation industry went the same way. The pioneers were often brilliant and inspired engineers who rose to the challenges they faced. But they were not always good businessmen, nor did many of them have access to sufficient capital to establish their operations on the necessary scale. A.V. Roe, struggling to finance the production of his first aircraft in a London suburban mews garage, is a classic case in point. But the companies set up by the Short brothers, Frederick Handley Page, Thomas Sopwith and Harry Hawker were also held back by being too small. And as owner-managers, the founders were inevitably focused on the massive technological changes that were taking place rather than on plans for business growth. So, gradually, just as the aeroplane began to look as though it might have a serious military future ahead of it, bigger players came into the game.

Not surprisingly, some of the big arms manufacturers took a considerable interest in this new technology. Vickers-Maxim, one of the biggest industrial combines in the country, had started as a steel producer, grew with the development of the railways, merged with the company set up to produce machine guns and at the end of the nineteenth century diversified into shipbuilding by acquiring the shipyards at Barrow-in-Furness. In 1901 it bought the Wolseley car company and within a few years of the start of serious flying in Britain it was ready to become involved in aviation too. But, of course, its interest would be directed entirely towards the production of warplanes.

Another company that decided to expand into the aviation business was British & Colonial. Founded by one of the great Victorian entrepreneurs, George White, this had grown into the biggest producer of electric trams in Britain; White was known as the 'tramway king'. When he decided to form the British & Colonial Aeroplane Company he had ambitions for it to become the biggest producer of aircraft for Britain and the Empire. The new company was based at the Filton tram depot near Bristol and would make aircraft under the name 'Bristol'.

Meanwhile, the army's interest in aviation, directed and encouraged by Lieutenant-Colonel Capper at the Balloon Factory in Farnborough, had also grown considerably. Since the beginning of the century the Germans had been building the large, rigid airships that would become popularly known as Zeppelins. Wanting to keep his options for the future open, Capper therefore organised the construction of a rigid cylinder-shaped airship. Eager to promote the value of this new British craft, Capper had the *Nulli Secundus*, as it was called, flown from Farnborough to London in October 1907. Here it flew symbolically across Buckingham Palace, made a circuit around the War Office and then flew past St Paul's Cathedral. Thousands of spectators watched, cheering enthusiastically from below in the belief that Britain was catching up with the Germans. Unfortunately, as the *Nulli* tried to return home it met fierce westerly headwinds and consequently had to put down in the grounds of Crystal Palace in south London. Here, the airship thrashed around in the wind and had to be dismantled. It was returned to Farnborough by road, ingloriously, in pieces. Despite the public acclaim, the airship's arrival had not marked the start of a heroic new chapter in British aviation technology.

With a goatee beard, long flowing hair and a Stetson hat, Samuel F. Cody was an unlikely figure to come to the aid of the British Army. Modelling himself on Buffalo Bill, Cody had arrived in England in

1890 to stage Wild West cowboy stunt shows. They proved very popular and made him rich. In the early years of the new century he became fascinated with man-lifting kites and spent many years trying to persuade the military authorities of their potential. Although the army was initially sceptical, Cody was not one to take 'no' for an answer and he set up shop at the Balloon Factory, working with Colonel Capper. Soon his interest shifted to powered craft. By the autumn of 1908 he had built an aircraft based on the Wright biplane but powered by a 50 hp Antoinette engine. It was in this machine that he achieved the first powered flight in Britain at Farnborough on 16 October. Cody was practising his taxiing when to his surprise the wind caught the aircraft and he took off, flying for about 450 yards before crashing into a tree. Cody was lucky to walk away with only cuts and bruises.[16] As he was an American citizen, however, the Aero Club did not regard Cody's short hop as the first British powered flight.

Another aircraft designer, closer to the military establishment, was John William Dunne. As an officer serving during the Boer War, Dunne had become convinced of the need for effective aerial reconnaissance of the enemy's gun positions. Invalided out of the army, he started designing aircraft with swept-back, V-shaped wings, convinced that these would bring the stability necessary for high-quality aerial observation. Like many pioneers he was able to raise neither interest in nor funding for his designs until his father, a general, pulled some strings, enabling Dunne to start working with Capper at Farnborough in June 1906. Capper became a great supporter of Dunne's. For three years he worked on designs for the British Army; historian Hugh Driver has described this as the first 'uninterrupted, state-aided flight research' in Britain.[17]

In 1908, the reforming Secretary of State for War, Richard Haldane, finally turned his attention to military aviation. Looking at the pioneers, however, he did not like what he saw. He perceived them as a group of cranks, pranksters and hopeless dreamers coming up with impractical designs and offering thrills more akin to a fairground sideshow than a solid military operation. To Haldane, they added up to little more than a bunch of bicycle mechanics, wealthy sportsmen and motor engineers, accompanied by a flying cowboy. Everywhere, their experiments appeared to be based on simple trial and error. There was no scientific theory behind it all. Haldane was a great believer that science should provide the rational principles on which all human endeavours were to be based. This was the 'faith', as he called it, which drove his army reforms. But he could see important work going on elsewhere, particularly in Germany, a country of which he was a great admirer. No one had yet flown more than a few hundred yards and Haldane put that down to the fact that the experimenters were amateurs who lacked any proper theoretical understanding of the technology they were attempting to develop.

Lord Northcliffe was extremely critical of government policy on aviation and argued in a personal letter to Haldane that the country was way behind others in Europe. But aircraft at this very early stage were just too fragile and unreliable to impress the military. Haldane replied that not only was Britain behind Europe in scientific knowledge, but that 'dirigibles and still more aeroplanes are a very long way off indeed being the slightest practical use in war.'[18] He decided that science must be applied to aviation in order to give it value to the British military establishment.

Accordingly, Haldane turned to the all-powerful Committee of Imperial Defence, the body that looked at the broad strategic objectives for the defence of Britain and its empire. A week after Cody's first near-fatal flight, Haldane created a specialist sub-committee chaired by Lord Esher on 'Aerial Navigation'. He invited several leading figures of the day to give evidence. Following the Secretary of State's lead, the sub-committee took a very dim view of the experiments taking place at the Balloon Factory. To the committee members it always seemed that Capper was on the verge of achieving something significant but never quite getting there. Their report noted that despite all recent progress, the pioneers 'can scarcely yet be considered to have emerged from the experimental stage', and that the machines they were working on lacked 'practical value'.[19] The sub-committee therefore came to the remarkable conclusion that the army should abandon its aircraft trials, justifying this in part on the grounds of economy. They argued that private aircraft companies should carry out research and development; when the technology had improved and something of practical value was available, then the army could step in and buy it. In the meantime, Capper, Dunne and Cody could be dispensed with. Instead, Haldane called for a new start in military aviation.

On 30 April 1909, by a bizarre coincidence on the very same day that Brabazon made his first historic powered flight in Britain, Haldane announced the next and most important reform he was to make in the world of military aviation with the creation of an Advisory Committee for Aeronautics. In announcing the committee Herbert Asquith, who had taken over as Prime Minister of the Liberal government the year before, said it would ensure 'that the highest scientific talent shall be brought to bear on the problems [of aeronautics]'.[20] Its president was one of the most prominent men of science in Britain, the distinguished Cambridge physicist Lord Rayleigh, a Nobel Prize winner and former president of the Royal Society, while its chair was to be Dr Richard Glazebrook, the director of the National Physical Laboratory, a Cambridge mathematician who had immense energy and a reputation for getting things done. Several committee members were also members of the Royal Society: William Napier Shaw, another Cambridge mathematician, was director of the Meteorological Office and always keen to explore new ideas; Horace Darwin, the fifth son of Charles Darwin the evolutionist, was an engineer who later set up the Cambridge Scientific Instrument Company; and Sir Joseph Petavel was Professor of Engineering at Manchester University. There were also military representatives, but only one member came from outside the worlds of soldiering and academic science. Lord Rayleigh asked Frederick Lanchester, the motor car designer, to join. Not only had Lanchester been a prime mover in establishing the British car industry, but he had a technical mind that ranged over a variety of subjects. In addition to improving the design of motor engines he studied and wrote about the principles of aviation, before designing his own aircraft. Despite his lack of formal education, Lanchester was a popular committee member and he had a great sense of humour, while as the only one who had actually struggled with the aerodynamics of aircraft design, he would bring essential practical experience to the committee's deliberations.

With such leading scientists now applying their minds to the challenges of aviation, Haldane had what he wanted, a group that could provide the 'foundation of science' on which to base military aviation. No more trial and error. Now leading men of science would lay down the theoretical foundations of aeronautics. It was a breakthrough for science, for aviation and for the armed services.[21] Britain had taken the first step towards creating a form of laboratory to prepare the military for war in the air. But how quickly would their academic and theoretical deliberations take effect if war should come sooner than anticipated?

3

The New Science

The Advisory Committee on Aeronautics immediately set to work. Its distinguished members turned their minds to such subjects as the 'mathematical investigation of stability', 'the resistance of stabilising planes, both horizontal and vertical', the 'effect of rudder action', the 'effect of gusts of wind' and 'materials for aeroplane construction'. They studied the efficiency of different types of propeller blades, they investigated the general principles of aerodynamics and explored the reliability, steadiness, efficiency and design of motor engines. They looked at suitable fabrics for the covering of airships. They built a wind tunnel at Teddington in which to conduct aerodynamic experiments on models and a spinning arm with which to try out different types of propeller. They built a 60ft-high wind tower in a remote spot in Bushy Park in north London to observe both wind velocities and the vertical and rotary motion of air currents. They researched different types of aluminium alloy and assessed their strengths. At Imperial College, London, the committee set up two research scholarships in aeronautics for advanced engineering students.[1] They were certainly trying hard, as Haldane wished, to establish the foundations of a science of aviation.

The government was regularly criticised for making the committee too academic and for not accepting that the aviation companies were also doing valuable practical work. The question was asked time and again whether aviation should develop 'practically' or 'scientifically'. One critic, Arthur Lee, put it to Haldane in a debate in the House of Commons: 'While pure science is very well in its way, I think this is a case where it is of more value when diluted with a great deal of practical experience.'[2] Haldane continued to argue that 'Science must come first' and insisted that the purpose of the committee was not 'to construct or to invent' but to solve the many problems that arose in making aircraft stable, reliable and efficient 'by the application of both theoretical and experimental methods of research'.[3] Many, however, including Lord Northcliffe, remained highly critical of the government's interest in aviation, preferring to continue offering prizes for the practical-minded young pioneers. By 1914, the *Daily Mail* had handed out more than £24,000 (equivalent to at least £2.4 million today) in prize money for aeronautical contests.[4]

Advances in aviation were now happening at a dizzying rate. Britain's first international air show was held at Doncaster in October 1909 and attracted thirty thousand eager visitors. In June 1910, Charles Rolls, seeking to publicise advances in flight, made the first double crossing of the English Channel in a French version of the Wright Flyer. The Short brothers on the Isle of Sheppey made six Wright brothers Flyers under licence. Rolls bought one and donated it to the army. The Admiralty, too, began a relationship with the Short company that would continue well into the war. And at last investors were coming forward. John Dunne, sacked from working for the army, formed his own syndicate with backing from the Marquess of Tullibardine, Lord Rothschild and the Duke of Westminster. The development of aviation was moving into its next phase.[5]

But there was a negative side. Only a month after his double Channel crossing, Rolls was flying at an air show in Bournemouth. He came in to land at a very steep angle when his tailplane broke off and his aircraft plunged vertically to the ground from about eighty

feet. Rolls was killed instantly in the crash. One of the great proponents of the nascent motor car and aeroplane industries was lost.
John Moore-Brabazon was 'sickened' by the death of his friend and
did not fly again until the war.[6] In 1910, roughly one flight in every
500 resulted in a fatal accident. For the early pioneers, powered flight
was not just daring, it was downright dangerous.

While Haldane's Advisory Committee on Aeronautics continued
with its researches to try to reduce the risks, John Capper was
replaced at the Balloon Factory by a dour Dubliner, Mervyn
O'Gorman, a senior consulting engineer who shared Haldane's
scepticism of the amateur pioneers and his belief in the need for a
scientific base on which to establish aviation. It fell to O'Gorman to
try to marry the committee's theoretical work with the next phase of
design work for the army. Having brought in from Daimler a new
chief engineer, Frederick Green, O'Gorman decided to seek out a
new chief designer, perhaps recognising that theory could go so far,
but to get an aircraft flying for the army he would need someone
with practical skills. The person he chose on Green's advice was a
man who would become a legend in the British aviation industry,
Geoffrey de Havilland.

De Havilland was the son of a Buckinghamshire clergyman. His
father wanted Geoffrey to follow him into the church, but like many
young men of his generation he became fascinated by the mechanics and engineering of the new motor cars. He showed his technical
abilities when he was only a teenager by installing a generator in the
family home to provide electricity. In 1900 he started at the Crystal
Palace Engineering School, where as a student he designed and produced a highly efficient motorcycle engine. De Havilland had an
inquisitive mind but with this came a sort of restlessness and he
tended to move on rapidly from one thing to another. Bored with
motorcycles, he sold the patent to his design and it became the principal engine that powered the machines of the immensely successful
Blackburne company. He went on to an apprenticeship with a steam
engine manufacturer in Rugby, worked for the Wolseley motor car

company in Birmingham and then became designer for a motor omnibus company in Walthamstow. While there he came across Wilbur Wright's aeroplane displays in France. He later wrote that he realised immediately that 'this was the machine to which I was prepared to give my life.'[7]

De Havilland persuaded his grandfather to invest £1000 in his new venture and he left his employment to begin making drawings for his own aircraft. He opened a workshop in Fulham and started construction with whatever materials he could lay his hands on, using a 45 hp engine built for him by a company in Willesden. Like other pioneer designers, he then had to find somewhere to try to fly his aircraft, and he eventually gained permission from Lord Carnarvon to use a part of his Highclere Park estate in Hampshire called Seven Barrows. Here he got his machine flying, but he had so little experience of how to control the aircraft that he soon crashed it. He retrieved the engine and built a second, lighter aircraft which he again succeeded in getting airborne. So confident was he in this new aircraft that he took his wife and eight-week-old baby for a joyride in it. De Havilland then had the good fortune to meet Frederick Green, who knew that things were developing at the Balloon Factory. Green introduced him to O'Gorman, who agreed to buy de Havilland's aircraft for £400 and to take him on as chief designer. It was a lucky break both for de Havilland and the British Army.

De Havilland was now effectively the sole War Office aircraft designer, operating a sort of monopoly of aircraft manufacturing for the army. His first aircraft was called the Farman Experimental 1 or FE1. He soon designed and built an improved version with a 70 hp Renault engine, while a variant, fitted out with a Vickers-Maxim machine gun mounted in the nacelle, was called the FE2b. De Havilland must have been like a kid in a candy store, with facilities that he could barely have dreamed of a few years before in his Fulham workshop. He began a highly productive period and designed a series of 'pusher' planes (with the propeller facing backwards, 'pushing' the aircraft forwards) and a series of 'tractor' planes (with the propeller facing forward and slicing through the air). The tractor planes were named as Blériot Experimentals and were given a BE delineation, although unlike Blériot's monoplane, all of the army's BEs were biplanes. The BE2 proved to be a winning model and went on to be one of Britain's main fighter aircraft in the early part of the war. It was constantly modified and improved, and with its several variants more than 2000 BE2s would be manufactured over the next few years.

Meanwhile, within the army itself, there was still little interest at senior levels in the development of aviation. Most leading figures just could not see what value these fragile, unreliable contraptions, which only flew when there was no wind, could possibly be to the armed services. General Sir Douglas Haig is supposed to have said that 'flying can never be of any use to the army', and another commander commented after seeing a flying demonstration, 'These playthings will never be of use in war'.[8] In the context of such attitudes, the only officers with aviation experience were young enthusiasts who had learned to fly as a hobby at their own expense and who tried to persuade their peers of the military value of their aircraft.

Many were artillery officers who recognised the need for aerial observation. With guns now capable of firing over distances of up to seven miles, there was a great need for some form of forward observation, in order to know if a target had been hit or if shells were landing short or wide. Aircraft playing a reconnaissance role could potentially provide this function with far greater flexibility than tethered balloons had done, or so some private fliers argued. It seems extraordinary today that a professional body like the army should rely upon hobbyists to make the case for the introduction of new technology, but five years before the First World War this was the case. Moreover, with the acceptance of the recommendation that the army should discontinue its flying trials until the science of aviation was better understood, the little experimentation that still went on took place in a purely informal capacity.

Despite the lack of interest at the top, an old firing range at Larkhill, on the fringes of the huge military exercise area of Salisbury Plain and near the artillery camp at Burford, was turned over in 1910 to a new army flying field. Artillery officers like Captain John Fulton, the first army officer to be given a pilot's certificate by the Aero Club, and Captain Bertram Dickson, who had won several prizes for his flying feats in France, began to fly regularly there. The engineering giant British & Colonial, keen to ingratiate itself with the War Office in the hope of winning major contracts to build aircraft, offered to set up a flying school at Larkhill and was allowed to build three hangars on the site. Curious officers used to wander across and watch the flying, and gradually more and more of the technically minded became interested in the potential of flight. In September, two British & Colonial Bristol box kites actually took part in the army's annual manoeuvres. However, thanks to a combination of bad weather, engine failure and inexperience, they performed without much distinction. The prevailing mood was still that there was a long way to go before the dangerous sport of flying could contribute towards fighting a real war.

The French annual manoeuvres in the autumn of 1910 in Picardy also featured aircraft for the first time. Here the Farman biplanes and Blériot monoplanes performed far more effectively than their British counterparts. When *The Times'* war correspondent, Colonel Charles à Court Repington, wrote glowingly of the role the French aircraft had played it began to dawn on the War Office that they would have to do something to avoid being left behind. The Chief of the Imperial General Staff, General Sir William Nicholson, who had opposed aeronautical experiments by the army two years before, announced his conversion. He wrote in a memorandum in February 1911 that the army should 'push on with the practical study of the military use of aircraft in the field', making clear the reason for his conversion by writing that 'other nations have already made considerable progress in this training and in view of the fact that aircraft will undoubtedly be used in the next war, whenever it may come, we cannot afford to delay in the matter.'[9]

With support from the most senior officer in the British Army, things began to change. An Air Battalion was established in April consisting of an Airship Company based at Farnborough and an Aeroplane Company at Larkhill to be commanded by Captain Fulton. Conditions were still primitive. The Larkhill aviators lived in tents alongside the hangars and there were never more than four or five serviceable aircraft available at any one time, but the formation of the Air Battalion marked a very significant step. Flying in the army was once again 'official' and no longer the preserve of enthusiastic amateurs.

Meanwhile, naval aviation was beginning to develop in its own direction. The Short brothers at their base on the Isle of Sheppey, where many of Britain's early experiments in flight took place, developed a close link with the Royal Navy. They moved a few miles to Eastchurch and loaned two aircraft to the navy on the understanding that they would send four officers there for pilot training. The Admiralty were quick to accept the offer and called for volunteers. Two hundred officers eagerly came forward, and from these four were finally selected and began their training in March 1911. Led by Lieutenant Charles Samson, these officers were delighted with the experience and Samson wrote in his report, 'The rapid progress in the science of aviation is apparent to all ... Few people now deny that the aeroplane has come to stay.'[10] The pioneer fliers saw a great role for naval aircraft as scouts, extending the 'eyes' of ships at sea, and recommended that the Admiralty should take over Eastchurch as a base for naval flying. It was at this point that a new and dynamic figure was appointed as the Admiralty's political head.

In a Cabinet reshuffle in October 1911, Prime Minister Asquith asked Winston Churchill, then aged thirty-six, to take control of the Royal Navy as First Lord of the Admiralty. Churchill, who was at the time Home Secretary and one of the leading figures in the great reforming Liberal government of the day, 'accepted with alacrity'.[11] He brought formidable energy to the task and became one of the great naval reformers, doing for the Royal Navy what Haldane had done for the army five years before. He introduced a newly formed Naval War Staff, based on the example of Haldane's Imperial General Staff, to focus on strategic planning. He transformed service conditions and opportunities for promotion of the ratings below decks. He introduced a new class of faster and more heavily armed battleship, the Queen Elizabeth class. And in perhaps the biggest long-term revolution of all, he made the decision to shift the navy from the use of home-produced coal to oil in its ships. As a consequence of this hugely important change the government took part-ownership of the Anglo-Persian Oil Company (which later became British Petroleum, BP) in order to guarantee oil supplies, giving a new geopolitical importance to the entire Middle Eastern region.[12]

Throughout his life Churchill was interested in new technologies and new ideas, and he threw his support behind the development of naval flying. He visited Eastchurch and flew in a variety of aircraft and airships, while a stream of memos flowed from his desk in all manner of encouragement and support. So enthusiastic did he become that at one point he asked a group of naval officers at Eastchurch to teach him to fly. Having decided he wanted to qualify for his own pilot's licence, he took several lessons before the aircraft he had been using for training crashed in the sea, killing both instructor and trainee. Churchill's friends pleaded with him to give up the lessons and to think, if not of himself, then at least of his wife Clemmie, who was five months pregnant with their third child at the time. Reluctantly, Churchill gave up trying to obtain his flying certificate.

However, he continued to encourage naval flying, promoting attempts to enable aircraft to take off from ships at sea. Experiments were conducted from a wooden platform erected over the deck of the battleship HMS *Africa*, and in May 1912 Commander Samson became the first naval flier to take off from the deck of a moving warship, HMS *Hibernia*, during the Portland Naval Review.

Advances were also made in the use of aircraft with floats that could land or take off from water, craft to which Churchill proudly gave the name 'seaplane'. With the active encouragement of the First Lord of the Admiralty it looked as though naval aviation might even eclipse the army's efforts.

Soon after his appointment, Churchill made a major contribution to the development of military aviation. When the Committee of Imperial Defence debated the future of military flying, its general feeling was that the Air Battalion formed six months earlier was insufficient and that a larger unit was needed, better resourced and with more aircraft and more technical support. Churchill wrote a memo arguing for a new 'Corps of Airmen' in order 'to make aviation for war purposes the most honourable, as it is the most dangerous, profession a young Englishman can adopt'. He claimed there should be no rivalry between the two services to 'prevent the real young and capable men [from either army or navy] ... from being placed effectively at the head of the corps of airmen'.[13] Churchill's intervention was decisive. It was agreed to form a new unit and in April 1912 the Royal Flying Corps (RFC) was established to bring together and supervise naval and army flying. At last, military aviation was guaranteed an independent role in the armed services of Britain.

When it was formed, the Royal Flying Corps consisted of a Military Wing, a Naval Wing, a Flying School and a Reserve. Major Frederick Sykes was put in charge of the Military Wing, Commander Samson in charge of the Naval Wing. The Commander of the Corps was General Sir David Henderson. The son of a Clydeside shipyard owner, Henderson had studied engineering at Glasgow University and then went on to Sandhurst. He was commissioned into the Argyll and Sutherland Highlanders and saw active service in the Boer War, where he was wounded in the siege of Ladysmith. For some years he acted as Director of Military Intelligence, in which role he laid down the principles for the process of collecting, interpreting and using field intelligence, and in 1907 he wrote the army manual on *The Art of Reconnaissance*. At about this time he became interested in aviation and, aged forty-nine, he learned to fly. He was the oldest person at the time to be awarded a pilot's certificate.

Henderson and Sykes between them now came up with new terminology and a new plan for the Military Wing. Flying would be organised around 'squadrons', consisting of eighteen aircraft in three 'flights' of four, each with two extra in reserve. Every squadron would have its own stores and trained mechanics and would operate as an independent entity. Eight squadrons were planned – although only five had come into existence by the time war was declared in August 1914. Henderson brought one other feature to British military aviation. He sported a wide, bushy moustache that was enthusiastically copied and became characteristic of hundreds of fliers for decades to come.

The year 1912 turned out to be pivotal in the history of British military aviation, and not only with the creation of the Royal Flying Corps. In August, a formal competition was held at Larkhill to find suitable aircraft for the RFC. The terms were announced well in advance to ensure that the different companies had time to prepare their models. Aircraft had to fly at a speed of 55 mph, and must be able to take off from grass in no more than 100 yards and carry at least one passenger or observer. They had to climb at a rate of at least 200 feet per minute to a height of 1500 feet, carrying a load of 350 lb, and stay at that height for an hour. They must be able to land on rough ground and come to a halt within 75 yards. Not many aircraft at the time could meet these specifications. But thirty-two aircraft were entered, although with breakdowns and engine failures only twenty-four actually participated. They included machines built by British & Commonwealth (the Bristols), by Short Brothers, Avro, Vickers and Sopwith, while the French models competing included Blériots, Farmans, Deperdussins and Hanriots. Some were biplanes and others monoplanes. Some had pusher engines; others had tractor engines. The competition was controversially won by the flying cowboy, Samuel Cody, in one of his own designs, known as the 'flying cathedral' due to its large size and powered by a huge Austro-Daimler 120 hp engine.[14] The RFC bought two models of this aircraft, the Cody V, but after the pilot was killed when one of them crashed, the other was discarded and never saw active service.

One of the judges in the competition was Mervyn O'Gorman, superintendent of the Balloon Factory, which in 1912 was renamed the Royal Aircraft Factory. In order to avoid a clash of interests, aircraft designed and produced at the Factory were not allowed to compete. However, the overall reaction was that the standard of the entrants had been disappointing. This confirmed to O'Gorman and others that the best work was being done in their own factory. Although de Havilland had not been able formally to enter the competition, he had raised the bar at the start by flying his own BE2 at speeds of up to 72 mph and to a record height of 10,560 feet. This was a truly impressive achievement for 1912, and in some ways the competition did nothing more than prove the superiority of the factory's designs. O'Gorman put this down to the work of the Advisory Committee and to the application of scientific principles to the design of its aircraft. The RFC placed an initial order for twelve BE2s after the competition, the order being sub-contracted out to other manufacturers who were issued with detailed blueprints by the Royal Aircraft Factory.

The army's summer manoeuvres of September 1912, when two sides competed in a war game carried out in Norfolk, also demonstrated the advancing status of aircraft. The 'Blue' Army under the command of General Sir James Grierson used their aircraft for reconnaissance very smartly. On the first day 'Blue' aircraft spotted the 'Red' army within an hour, enabling the 'Blue' cavalry to attack them. And after three days the 'Blue' forces, who had consistently made effective use of their reconnaissance aircraft, won a complete victory. Grierson wrote with extraordinary foresight after the exercises, 'Personally, I think there is no doubt that before land fighting takes place, we shall have to fight and destroy the enemy's aircraft ... warfare will be impossible until we have mastery of the air.'[15] The commander of the defeated 'Red' forces was General Sir Douglas Haig.[16]

The concept behind the formation of the RFC had been that both army and naval aviation would come under one joint command. However, army flying increasingly came to be focused on providing reconnaissance for the expeditionary force that would travel to the continent of Europe to fight its battles. The RFC consequently developed plans for its squadrons to travel abroad and base themselves in Europe, with a squadron assigned to support each army corps. Scouting aircraft were not intended to act as fighters, they were to carry out reconnaissance and get their intelligence back as quickly and efficiently as possible.

But this was not what Winston Churchill had in mind for naval aviation. He wanted aircraft that could be used to defend home bases, ports and naval installations from attack and so must be able to shoot at enemy aircraft. And he wanted to continue to develop flying from ships, not only for long-distance reconnaissance but also for offensive purposes. In 1913 a naval aircraft dropped a torpedo for the first time, while other aircraft experimented with all sorts of bombs and machine guns as ways of attacking warships at sea.

The two wings of the RFC gradually drifted further and further apart. In February 1914 Churchill wrote a memorandum outlining his views, stating with remarkable foresight: 'The objectives of land aeroplanes can never be so definite or important as the objectives of seaplanes, which, when they carry torpedoes, may prove capable of playing a decisive part in operations against capital ships.'[17] It was clear that Churchill wanted a separate division for naval aviation and he did much to support what in July 1914 formally became recognised as the Royal Naval Air Service (RNAS). A new uniform was designed for naval fliers to make them stand apart from their army brothers in the RFC. Eastchurch was the central air station for naval flying, and the alliance with Short Brothers developed further. The company opened a new factory at nearby Rochester to concentrate on producing aircraft for the navy. Sopwith too turned his attention to the design and building of seaplanes, as did the Avro company. At the time of its formation, the Royal Naval Air Service possessed 39 aeroplanes, 52 seaplanes and a few small airships, along with 120 trained pilots. Morale was high, the fliers were highly motivated and in the summer of 1914 they were as ready for war as a small force like this could be.

In the two years after the military flying competition, the Air Advisory Committee and the Royal Aircraft Factory continued to make real progress. In the light of the 1912 manoeuvres a new scouting aircraft was produced drawing extensively on the advances made in the science of aeronautics. Henry Phillip Folland, a former apprentice at the Lanchester Motor Company who had gone on to work at Daimler, designed the new aeroplane. Green had brought Folland to Farnborough as a draughtsman for Geoffrey de Havilland but he soon proved to be an inventive and talented aircraft designer in his own right. His new scout, known as the Reconnaissance Experimental One (RE1), was a variant of the super-successful BE2c designed particularly to provide a more stable platform for observation. A two-seater biplane with the observer sitting in front of the pilot, the RE1 was powered by a 70 hp Renault engine and could fly at a maximum speed of 83 mph. Like the BE2, several successful variants of the RE series were to fly throughout the war.

Some historians have criticised the work of the Royal Aircraft Factory. They have argued that by imposing a rigid state monopoly on aircraft design, Haldane severely limited the procurement of alternative designs and so denied the army access to inventive work that could have been done elsewhere in the privately owned aviation companies.[18] There is some truth in this: several gifted designers ended up as subcontractors duplicating models that had been designed at Farnborough. However, the other side of the argument is that it had taken some time to understand the military potential of flight. Aircraft had been too unreliable and too dangerous for many experienced people to appreciate what help they could be to the military. As it was, on the very same day that a British pilot first flew in Britain, the government established a committee of some of the finest brains in the land to help create a scientific foundation on which flying could build. There was nothing lax or lethargic about that. When planes were still hopping just a few hundred yards over the ground in a straight line, scientists were exploring the principles that might establish aviation as a new arm of the military. The Royal Aircraft Factory provided real encouragement to an extremely able aircraft design team who developed a range of ideas. During the war, new technologies and aircraft design were to advance even faster than in the years of peace, and some of the aircraft the RFC went to war with in 1914 were later shown to be inadequate. Nevertheless, by the time of war, Britain had a firmly established aircraft industry that was manufacturing about ten aircraft each month, making sales to continental states as far afield as Greece, Romania and Imperial Russia. Things had come a long way from the industry's humble origins only eight years before. But the greatest challenge was still ahead. Would the application of scientific ideas help when the biggest test of all came in August 1914?

4

Observing the War

Europe went to war very suddenly in the summer of 1914. Few people had anticipated that a flare-up in the Balkans would ignite a European conflagration. War came, almost literally, out of a clear blue summer sky. And virtually everyone expected it would be over as quickly as it had started, probably 'by Christmas'. The popular press had whipped up patriotic fervour in all the European nations in the preceding years. In France and Germany, children learned patriotic songs and poems at school, while the French longed to be avenged for their defeat by Germany in 1870. Citizens of both nations grew up knowing of their long-term rivalry with the other. In both countries, as well as in the Austro-Hungarian and Imperial Russian empires, there was conscription and millions of young men left families, villages and towns to train with the army.

A military culture was dominant in much of continental Europe. In Britain, where conscription did not exist, there was instead immense pride in the Royal Navy as the largest in the world and ruler of the waves, although there was a growing fear that Germany was a rival to this supremacy. There was tremendous enthusiasm for the Empire, a belief in the superiority of the Anglo-Saxon races, and throughout the country people regularly gathered to celebrate royal birthdays and the anniversaries of military triumphs. Military chiefs such as Kitchener were probably more respected than most political leaders. Books of imperial derring-do and stories of wartime national glories were immensely popular.

As a result, across Europe, millions cheered the coming of war. In Vienna, Berlin, Moscow, Paris and London crowds assembled to hail the declarations of war. Many young people saw war not only as an adventure but as some sort of liberating force, part of their destiny. Throughout history it had been necessary to fight wars and everyone was convinced that victory would be theirs. In Vienna, a young man wrote, 'everywhere one saw excited faces. Each individual ... was part of the people, and his person, his hitherto unnoticed person, had been given meaning.' In Britain, the young Rupert Brooke wrote, 'Now God be thanked who has matched us with this hour.'[1]

In retrospect it was bizarre that young men should cheer with such immense enthusiasm the news that many of them would be sent to their deaths, and that women, likewise, should encourage their husbands, sons and brothers to head off into the carnage. The fact was, of course, that apart from a tiny number of visionaries who predicted the holocaust that would follow, most people had no sense of what a modern war would involve. It was assumed that huge armies would fight a couple of battles, the fleets would engage each other at sea, some ground would be occupied and then everything would be over. The diplomats would put together a peace settlement and a few territories would be exchanged. With so little awareness of the destructive capacity of modern artillery, machine guns, aircraft and bombs, there was barely any dread of war, only a senseless enthusiasm for it and a desire among the young to be part of it before it was all over.

Much has been written about how Europe went to war in August 1914 according to railway timetables.[2] When the mobilisation of great armies began and the troop trains started to move, the diplomats could no longer control events. Tied together by alliances, the states and empires of Europe slid inexorably into conflict. Austria-Hungary was the first. After a Serbian terrorist assassinated the heir to its empire in Sarajevo at the end of June, Austria-Hungary issued an ultimatum, declared war on Serbia and bombarded Belgrade. Russia then announced a general mobilisation to defend its Serbian ally.

Germany feared war on two fronts against both Russia and France, who were united in an alliance. The only way to prevent a two-fronted war was for Germany to strike and knock out France – who the Germans knew would be quicker to mobilise – while the great Russian bear was slowly limbering up. According to the Schlieffen Plan, prepared in detail a decade in advance by the German General Staff, Germany must immediately invade France before attacking Russia. And to avoid the heavily defended French border they would cross through Belgium. But Britain was pledged to defend Belgium's neutrality. So, as German troop trains moved into Belgium to attack France instead of Russia, Britain declared war on Germany.

There was nothing inevitable about this rapid progression to war. Previous crises in the Balkans had not led to a general European conflict. Recently historians have stressed how 'improbable' was the sequence of events that led to war.[3] Everywhere, general staffs had resolved that to defend their countries they must mobilise their armies to attack. Diplomats assumed this would deter the others, and in earlier crises the tactic had succeeded in defusing the situation. But this time deterrence did not work, and everything moved so fast. On 28 July the Great Powers of Europe were at peace. By midnight on 4 August every major European nation, except Spain and Italy, was at war.

According to the plans laid down by Haldane in his army reforms, Britain now prepared to send its expeditionary force, the BEF, not to protect Belgium but to serve on the left flank of the French army. In the first weeks of August, the small British Army mobilised. Regiments requisitioned entire trains and moved south. Men were recalled from summer holidays and rushed by train to the Channel ports. Here in mid-August the army began to cross to northern France, where it seized yet more trains in order to take up its position.

In some ways the armies that marched off to war in the hot summer days of 1914 were similar to the armies that had gone to war in Napoleon's day. All European armies moved with the support of thousands of horse-drawn carts and wagons. The infantry still marched on foot. Horses hauled the guns of the artillery. The cavalry were armed with sabre and lance. And the French infantry in their blue coats, bright red trousers, képis and white puttees could easily have belonged to a previous era. But that was as far as it went. In other ways the armies that went to war in 1914 were very much of the twentieth century. The men had been transported towards the front by railway. The rifles that the infantrymen carried were now breech loaded and capable of a rapid fire impossible a hundred years earlier. The field guns of the artillery were also more accurate and able to fire far more rapidly than in earlier conflicts. The invention of cellulose nitrate (at the time called gun cotton) had launched the development of a new generation of high explosives. Furthermore, the Germans in their field grey and the British in khaki looked far more modern than had their counterparts in the brightly coloured uniforms of previous eras. And the British and German armies went to war with huge supplies of canned food. Tins kept the food fresh and made it possible for armies to range over long distances without having to resort to foraging as in previous wars. Alongside each troop of foot soldiers was a curious mix of slow-moving vehicles, some with solid tyres, bumping along over tracks and roads. Then, in addition, there were the strange-looking aeroplanes of the Royal Flying Corps that accompanied them.

The Royal Flying Corps was mobilised along with the rest of the British Army. On 13 August, in a variety of aircraft, the first pilots flew across the Channel, landing in Amiens. This was the first time British military pilots had accompanied an army marching off to war. When the squadrons had assembled in France they joined a fast-moving, mobile army and followed the BEF as it took up its position alongside the French forces. Each army corps was allocated a squadron of aircraft.

The RFC had virtually no ground transport and rapidly assembled a set of unusual vehicles that made it look more like a travelling circus than a military convoy. There were one or two 30 cwt army Leyland trucks, but the other vehicles were all commercial lorries that had been quickly requisitioned. These included two Maples furniture vans, a Peek Frean Biscuits delivery truck and a large red van used to deliver HP Sauce, which had the words 'The World's Appetiser' emblazoned in gold letters on its side.[4] There were large vehicles carrying fuel, while a range of private limousines followed, lent by officers and civilians 'for the duration', included Sunbeams, Renaults, a Mercedes Tourer and, most bizarrely, a Rolls-Royce two-seater coupé. At their head was a Daimler in which Major Robert Henry Brooke-Popham carried a supply of gold, which he used for buying spares for the squadron and provisions for the officers' mess. Sometimes the convoy stopped overnight at a chateau, sometimes at a farmhouse where the men slept in barns; sometimes they pulled in to a hotel. On other occasions they just stopped at the side of the road, where the men slept under hedgerows.

Airfields were known as 'landing grounds' and often consisted of nothing more than a flat, open field. Preparing a new landing ground might involve simply carting the sheaves of harvested corn to one side of the field. The aircraft were still of course remarkably primitive, wooden frames covered with canvas and held together by wire, while pilots still flew largely by instinct. A naval aviator, Captain Ivon Courtney, one of the officers who had taught Winston Churchill to fly, wrote: 'We had no headphones in those days. Once airborne we would shout at one another and hope the wind carried something approximating to what we said across to the other fellow. One didn't normally use instruments; they were all in a box but we "old" fliers scorned them, we liked to fly on "ear" as it were.'[5] Navigation was even more rudimentary and relied upon recognising distinctive features in the landscape below. If a pilot got lost he would simply land and ask where he was. Despite advances in the science of aeronautics, flying was still very much a seat-of-the-pants type activity.

That year a beautiful summer turned into a glorious autumn, and many young soldiers reflected on the incongruity of going to war on days of such warm, delightful weather. Maurice Baring of the RFC remembered:

> The time passed in a golden haze. Never was there a finer autumn ... I remember the heat on the stubble on the Saponay Aerodrome; pilots lying about on the straw; some just back from a reconnaissance, some just starting ... And then the beauty of the Henry Farmans sailing through the clear evening ... the moonlight rising over the stubble of the Aerodrome, and a few camp fires glowing in the mist and the noise of men singing songs of home.[6]

In fact the hot weather offered less than ideal conditions for flying in the flimsy, light, underpowered contraptions that the RFC took to war. The heat of summer created sudden thermal updrafts that tossed the aircraft around in the sky. Sometimes the heat generated thunderstorms and flying became impossible. On one occasion, a summer storm wrecked many aircraft that were parked on a landing ground. They were literally picked up and thrown about in the wind. However, despite such adversities the RFC began to fly reconnaissance missions within a week of arriving in France.

The first ever reconnaissance flight was flown by Captain Philip Joubert de la Ferté on 19 August in a Blériot XI. He had no observer but was asked to check if enemy troops were present in an area to the west of Brussels. Struggling with cloudy weather and with very little knowledge of the countryside, Joubert soon found himself lost but was reluctant to come down in order to discover his location. However, when he spotted what seemed to be a parade ground in a military garrison in a large town he decided to land and ask where he was. He discovered he was in the town of Tournai, and after an 'excellent lunch' with the commandant of the garrison he flew on in the afternoon. Once more he quickly became lost and when short of fuel had to land again. But this time the local gendarmes came out and threatened to arrest him, as he had not been issued with identification papers and they thought he was a spy. He was helped by a linen manufacturer from Belfast who happened to be visiting the town. The salesman draped a Union Jack over the aircraft. Once the locals understood who Joubert was the mood changed. He was able to buy petrol and eventually get on his way. He returned to his landing ground near Mauberge to deliver his report at 5.30 that evening.[7]

Despite the haphazard nature of some of this early flying the reconnaissance pilots of the RFC soon began to establish their worth. On 22 August, aerial scouts flying ahead of the advancing BEF spotted what they correctly identified as the massed troops of the German First Army marching south. On the following morning, the British Army had its first major encounter with German troops at what became known as the Battle of Mons. A furious engagement took place along the Mons-Condé canal, where the well-drilled British soldiers fired their Lee Enfield rifles so rapidly into the phalanxes of enemy troops that the Germans believed the British were firing machine guns at them. As the battle raged, the reconnaissance pilots flew beyond the battlefield and spotted large German troop movements attempting to outflank the British line. They also reported on a fierce engagement with the French army and observed that the French, on the British right, had started to withdraw.

In the evening this vital information was brought to Sir John French, the commander-in-chief of the BEF. Most military reconnaissance up to this point had come from light cavalry. Although Sir John was a cavalry man he quickly realised the value of the intelligence that his airmen were bringing him. His troops had performed well and had held up the German advance, but French saw that he was now in an isolated position, with the French army withdrawing

on one side and the Germans about to outflank him on the other. He
ordered an immediate withdrawal. This at least led to the survival
of the British Army. Had the BEF been surrounded and wiped out
in its first engagement it would have been a catastrophe for the
Allied cause. Although their intelligence resulted in the famous
'retreat from Mons', the reconnaissance pilots had saved the day for
the British Army. Sir John French acknowledged that 'their reports
proved of the greatest value.'[8]

One of the many problems that faced these brave early fliers was
rifle fire from the ground. Unused to seeing aircraft and lacking
understanding of their role, most soldiers assumed they must be
enemy craft of some sort and opened fire on them. It rapidly became
essential for aircraft to carry a form of national marking to prevent
such friendly fire incidents. The French put a blue spot on the under-
side of the wings of their aircraft, surrounded by a red ring; the
Germans marked their machines with a black cross. The RFC's initial response was to paint a union jack on each wing, but these were
too small and so they adopted the symbol of a red spot, surrounded
by a white ring and outside that a blue ring. Thus was born the
roundel that would identify all British military aircraft for the next
hundred years.

Over the following weeks, RFC reconnaissance flights continued
to provide the Allied leaders with vital intelligence about the enemy.
The aerial observers scribbling on their notepads were the first to
spot the German army's change of direction at the end of August.
This resulted in the Allied decision to stand firm on the river Marne,
an action that eventually succeeded in halting the German advance
and forcing their troops to turn back. In the rapid manoeuvrings that
followed, the reconnaissance crews were able to keep their generals
well informed both of enemy movements and the location of the
Allied armies' advanced units. They could pass the information back
to headquarters staff more quickly than it could be carried over land.
They could report on the status of bridges, on the road network and
provide other vital information about the landscape through which

the armies were passing. Captain Joubert once described a report that his observer wrote up on landing as being as extensive as 'an early Victorian novel'.[9] In September, Sir John French paid tribute to the RFC pilots and observers when he wrote: 'They have furnished me with the most complete and accurate information, which has been of incalculable value in the conduct of operations. Fired at constantly both by friend and foe, and not hesitating to fly in every kind of weather, they have remained undaunted throughout.'[10] It was only five years since Blériot had crossed the Channel, but reconnaissance aircraft had already proved themselves vital agents of modern warfare.

In the last two months of 1914, the nature of the fighting was transformed from a mobile to a static war as both sides dug in. For four years they were to face each other in two lines of trenches that meandered from the Belgian coast, across north-east France and down to the Swiss border. Between the two opposing front lines was the barren landscape of no man's land, thick with barbed wire and at times only a few dozen yards wide but providing a gulf that was increasingly difficult to cross. It was soon apparent that troops well dug in and protected by machine guns and a formidable mass of wire were able to beat off almost every form of attack that could then be thrown against them. The Germans from the first saw the Western Front as a defensive line and took the high ground, or whatever territory was easiest to defend. As a consequence, from ground level, the Allies could rarely see beyond the German front line. So the aerial observers soon became vital in trying to see behind the German lines, to study their trench layouts, to identify where their artillery was positioned and to watch for reinforcements of men and material being brought up to the front line. Since this information was too detailed for a pilot or observer to record during a patrol, the time was ripe for the next major advance in aerial reconnaissance.

Photography was well established by 1914. It had grown from a means of portraiture to a popular and accessible hobby for millions. George Eastman had revolutionised amateur photography by introducing the box brownie camera in conjunction with flexible roll film. Lenses had grown enormously in quality with improvements in optics in the late nineteenth century, but to take advantage the professionals still relied on cumbersome glass plates that had to be exposed one at a time. In the latter part of 1914 the French began to make extensive use of aerial photography. They had good cameras and lenses and were soon able to take excellent photographs of the enemy trenches, labelling the photos that resulted as 'aerial maps' and forwarding them as a courtesy to British headquarters. With the cavalry unable to carry out reconnaissance patrols across no man's land, the British commanders soon realised that aerial photographs could provide invaluable information about what was going on in the enemy trenches and behind his lines. But hardly anyone within the RFC had any knowledge of photography.

In November 1914, John Moore-Brabazon, now aged thirty, who had given up flying after the fatal accident of his friend Charles Rolls in the summer of 1910, decided to join up to do his bit for the war effort. Still at heart an aviator, he was commissioned as a second lieutenant in the RFC despite his lack of recent practice, and because of the extent of his flying experience, which pre-dated even the formation of the corps, the War Office sent him out immediately to France. He joined 9 Squadron at St Omer, which was not an operational squadron but was tasked with servicing aircraft and exploring new technologies to aid military aviation. Though he had undergone no basic training of any sort, he was told on his first morning, as the newest officer present, to march the men from their billets to the aerodrome. With no knowledge of drill he pretended to the sergeant-major that he had lost his voice and asked him to carry on with the men on his behalf. As the sergeant-major got the men in line, yelled out the appropriate orders and marched them to the right place, Brabazon realised he had a lot to learn to adapt to life in the army.[11]

Brabazon did however have some knowledge of photography and, with another officer, Lieutenant Charles Campbell from the Intelligence Corps, was soon selected to find out what the French were up to and how the British Army could best copy them and keep up. Joining their small team was Sergeant-Major Victor Laws, who had experimented with aerial cameras before the war and was the only person within the RFC who had any understanding of how such equipment functioned. Along with its use for mapping, Laws also noticed that the movement of men or vehicles crushed grass or soft ground in a way that was still noticeable as shading on a black and white aerial photograph taken a couple of days later. This marked the beginning of photo interpretation and the science of reading movements across land by the study of aerial photographs. Together Brabazon and his team were to lay the foundations for photo intelligence within the British Army.

It seems incredible now, when the use of aerial imagery is so widespread, that the army should have been reluctant to take it up. But some officers thought there was something 'unfair' or 'ungentlemanly' about photographing the enemy behind his lines. Brabazon noticed that officers always wanted to 'play by the rules', feeling that the use of such photography to expose what one's opponent was up to seemed to 'invade a privacy that had always been accorded to the enemy'. Others simply objected to aerial photography because it was new and they had never used it before. Brabazon recalled that when he and his two colleagues were allocated to the recently created No. 1 Wing to develop new techniques for aerial photography, they were 'about as welcome as the measles'.[12]

Nevertheless there were some senior officers who saw the potential, and one of these was Colonel Hugh Trenchard, the commander of No. 1 Wing. Trenchard had learned to fly, at his own expense, at Brooklands before the war., One of the principal pioneers of military aviation in Britain, Trenchard was to become leader of the RFC and later of the RAF. At the start of the war he was forty-one years old, a giant of a man, very tall, broad shouldered and with an enormously loud voice from which he acquired the appropriate nickname of 'Boom'. He was however not an easy communicator and could be taciturn to the point of appearing rude, especially to junior officers who were often met with total silence when reporting to him. Like many senior figures in the RFC, he had an almost crusading passion for the air arm and regarded it as his task to preach the gospel of aviation to the often unbelieving generals. A great enthusiast for aerial photography, he used always to carry a few photos in his pocket, taking any opportunity to try to persuade a sceptical general of their value.

In January 1915 the RFC established its first experimental aerial photographic unit. In the earliest cameras they used, the lens was extended by a form of bellows, but these devices proved hopelessly impractical in the air. Other, more portable cameras were tried although there was always the risk that the vibration of the aircraft would blur the image, or that the intense cold at altitude would fog the glass plate and spoil the picture. One of the principal problems was that Zeiss, a German firm based at Jena in Bavaria, had produced most European optical lenses before the war. There was naturally now a huge increase in demand from the army for specialist products such as binoculars and range finders as well as for aerial cameras, yet bizarrely, this had not been anticipated. And with the coming of war, all exports from Zeiss to Allied countries had ceased. So, on 6 January 1915, the War Office wrote to the Royal Society to ask if any of the scientists on their War Committee could advise on how to improve the optical effectiveness of British-made lenses. Noting ruefully, 'It appears that the best British optical glass is inferior to the German,' a War Office official asked the distinguished scientists of the War Committee, 'Can any means be suggested for improving the process of manufacture?'[13] It was one of the first instances in which the British Army appealed directly to the scientific community to help solve a military problem. An historic watershed had been crossed.

The consequence of the War Office appeal was that one company, Woods in Derby, began work on improving British optics and developing the glass coating that was so vital to the production of a high-quality lens. But this would take time and in the interim a

French company, Para Mantois, started to produce the priceless lenses. Using these optics initially, Brabazon and his team went on to design and introduce a new generation of cameras. The first, known as the 'A-type', was a fragile device that the pilot or observer had to hold by hand over the side of the plane. The A-type used 5 in x 4 in glass plate negatives, and the observer had to separately load and unload each plate. It was a cumbersome process that required skill and patience, but the quality of the photographs taken was excellent.

The British commanders began to plan an offensive for March 1915 in the area around Neuve-Chapelle, a town in the Artois region of north-east France. General Sir Douglas Haig was to command the assault and his chief intelligence officer was Brigadier John Charteris. In the weeks before the attack, aerial observers comprehensively photographed the German trench defences for up to a mile behind their front line; when pieced together, the photographs formed composite photo-mosaics covering a wide area. From these, the planners were able to make a detailed map of the German trench positions. Charteris wrote in his diary on 24 February, 'My table is covered with photographs taken from aeroplanes. We have just started this method of reconnaissance, which will I think develop into something very important ... Photographs cannot lie.'[14] The photographic maps helped Charteris and Haig understand the extent of the German defences and on the first day of the offensive, British soldiers succeeded in breaking through the German line.

However, there were insufficient reserves to follow up this initial success, and the Germans were quick to reinforce their positions. After three days there was stalemate once again. Nevertheless, the photographs had greatly aided those planning the attack, and it became standard practice for trench maps marking the German dispositions based on aerial photos to be made and distributed to the assaulting infantry. Tens of thousands of photo maps would be produced during the war and sometimes hundreds of copies of each map were needed. As soon as a new unit took up a stretch of front line, the intelligence officers were given detailed maps of the enemy positions in front of them. And infantry officers in every unit from battalion upwards would study these maps before making any sort of raid or assault on the enemy trenches.

The technology and the procedures advanced rapidly under the demands of war, which as ever proved to be a 'mother of invention'. Most aerial photographs were taken vertically downwards, but pilots soon learned the value of taking photos at an oblique angle which allowed the 'eyes in the air', as the reconnaissance aircraft were called, to see further behind German lines.[15] New techniques emerged to improve the role of aerial observers in spotting for the artillery. Gunners experienced a major shortage of shells in 1915 and therefore could not afford to waste ammunition, but firing over such long range made it impossible for even forward ground observers to get a clear fix on where shells were landing. Aerial observation from 3000–4000 feet had a huge advantage over spotting from the ground.

At first aerial observers marked German artillery positions on a map before dropping it to the ground near the gunners in a leather bag. Soon lightweight telegraphic radios were fitted to aircraft, enabling far more efficient communication with ground controllers by Morse code. Technicians devised a 'clock-code' system whereby concentric circles, each given a letter and representing the distance from an enemy target, were drawn on a celluloid disc radiating out from the point on a map where the target was located. Outside the circles the numbers of a clock face were drawn, with 12 o'clock representing due north. The observer plotted where each of his own side's shells were landing in relation to the target, and communicated the letter and number back to the gunners on the ground. So, with a scale of say, 100 yards for each concentric circle, if the observer sent back the signal 'B3' it would signify that their shells were landing approximately 300 yards due east of the target. The gunners were then able to correct their aim and hopefully score a bullseye.

To replace the awkward hand-held models, Brabazon and his team came up with the 'C-type', a new camera that was fixed to the aircraft fuselage. The pilot activated the C-type as he flew across enemy lines. This further improved the quality and range of aerial photography. With warfare achieving a new intensity, there were by now no longer any officers left who felt it was 'un-gentlemanly' to photograph from the air what the other side were up to, and photo intelligence developed rapidly.

In August 1915, Hugh Trenchard became commander of the RFC. Immediately he began to plan for a massive expansion in the size of the corps from just a few squadrons to several dozen. He pushed for the development of faster aircraft with more powerful engines, equipped with better weapons along with the widespread use of radios. And, in an expansion of aerial warfare, he promoted the development of larger aircraft with an improved capability to carry bombs. All of this was in addition to the continued development of aerial reconnaissance both for the artillery and with the use of photography. By now Trenchard was pushing on an open door as there was no longer any serious opposition to the growth of the air corps. When Lord Kitchener, the Minister for War, was presented with a plan to add fifty more squadrons to the RFC, he simply wrote on the memo, 'Double this.'[16]

The Germans of course were also rapidly developing their own aerial observation techniques. *Drachen* or 'sausage balloons' usually flew at a height of about 6000 feet and were placed two or three miles behind the German lines. From this height a well-trained observer equipped with a good telescope could report back in detail on movements behind the Allied lines. In the gaps between the *Drachen* were two-seater observation aircraft flying figures-of-eight and reporting on any activity in the Allied trenches. As the RFC and the French aviators tried to prevent German observation flights from crossing Allied lines, so an air war developed with each side trying to protect its own airspace.

In the first examples of aerial combat, pilots would lean out the side of their cockpits and fire at enemy planes with revolvers or rifles. From the spring of 1915 the observer in the standard British observation aircraft, the BE2c biplane, was issued with a portable Lewis machine gun, but in firing it he had to avoid hitting his own propeller, wings or rigging – no easy task when trying to shoot at a fast-moving enemy target. The Vickers FB5, a 'pusher' aircraft with the propeller behind the cockpit, allowed the observer a clearer field of fire from the front seat. But the real breakthrough came in late 1915 when the German Fokker E2 *Eindecker* monoplane appeared. Its forward-firing machine gun was synchronised to fire through the plane's propellers without hitting them. This ingenious invention enabled the E2, cruising at high altitude, to dive steeply down on Allied aircraft, aiming and firing from the front. Lieutenant Oswald Boelcke soon mastered the tactic of diving with the sun behind him to launch a surprise attack on the slower British or French observation aircraft, darting away quickly before other Allied planes could respond. The Fokker *Eindecker* was the first modern fighter aircraft, and thus a new branch of warfare – aerial combat – was born.

During 1915, Boelcke's number of 'kills' grew dramatically. So, in early 1916, RFC headquarters ruled that every reconnaissance aircraft must be accompanied by at least three fighter escorts. The BE2c had a maximum speed of about 72 mph, and soon picked up the unfortunate nickname of 'Fokker fodder' as it could not match the performance of the faster German fighters diving out of the sky to attack. This revived earlier criticisms about the limitation of British military aircraft design. In a debate in the House of Commons in March 1916, Noel Pemberton Billing accused the RFC of 'murdering' their own airmen by sending them into combat in inferior aircraft to those of the Germans. This accusation profoundly shocked many people, coming as it did from a man who had served very ably in the Royal Naval Air Service before leaving to enter politics. A judicial inquiry following Pemberton Billing's accusations exonerated the RFC but blamed the Royal Aircraft Factory at Farnborough where the BE2 range, which made up about 70–80 per cent of all aircraft acquired by the RFC until the middle of 1916, had been designed and produced.

The result of this public row was to open the market to manufacturers like the Sopwith company that had new ideas about design and performance. The fact was that the air war had progressed rapidly in directions no one had anticipated in 1912. When de Havilland had designed the BE2 series, the principal demand was for aircraft to provide a stable platform for observation. No one had foreseen aerial combat. By 1916, new types of aircraft with far greater manoeuvrability were needed, and their design and production began.[17] And in the following year the Ministry of Munitions took over all aircraft procurement.[18]

By the summer of 1916 a new generation of better armed, faster British reconnaissance and fighter aircraft had appeared, restoring the balance and keeping German attackers away from the slower observation aircraft. The Nieuport Scout, the FE2b, the DH2 and the Sopwith Strutter were able at times to outfly the German fighters. As the RFC grew dramatically in size, 'dog fights' between groups of Allied and German fighters became a common sight in the skies above the trenches. In the see-saw air war of 1914–18 aerial supremacy was to tip from one side to the other as new aircraft were produced, and better aircraft and weapons were then designed to defeat them. But by the middle of 1916 the pendulum had swung back from the Fokkers so that aerial supremacy over much of the Western Front lay with the British and the French.

As fighter planes proved their superiority over observation aircraft, so the reconnaissance planes flew higher to avoid interception. And the higher the observers flew, the greater the need for better cameras with longer lenses. Moreover, the lower temperatures experienced at higher altitudes caused moisture to condense in the cameras, fogging lenses and cracking glass plates.[19] The need to train dozens and then hundreds of support staff for the aerial observers was a top priority. Promoted to second lieutenant, Victor Laws became head of the new RFC School of Photography at

Farnborough. Officers and NCOs were trained here in all aspects of developing and printing glass plates, making enlargements, maintaining aerial cameras, and preparing maps from photographs. Speed was of the essence in developing and distributing the photographs. Before long every front-line squadron had its own small unit of one officer, an NCO and five men to process aerial photos, a task often performed in army trucks fitted out as mobile darkrooms. The men were known as 'stickybacks', as the photos were often still wet and sticky when they were handed out.

Developing and processing aerial photos was only half of the operation. The photos remained useless unless effectively analysed by skilled interpreters. A new class of photo interpreters learned techniques of using shadows to measure the scale of objects on the ground, and studied how to find and identify machine gun positions, artillery posts, the location of unit headquarters and to analyse the tracks and movement of troops. They were soon able to assess enemy strongpoints and to recognise passages through the wire as well as trench dugouts.[20] First the French and then the British produced manuals on photographic interpretation, listing the key German targets to be identified in the trenches and methods for measuring the size of objects on the ground. Building up a reference library of aerial photos was a key element of photo interpretation; by comparing present cover with previous coverage of the same area, the interpreters could easily identify the construction of new defensive positions or the location of new gun emplacements. So the method of labelling and storing aerial photographs had to be standardised to make retrieval of previous aerial imagery possible.[21]

Meanwhile Brabazon and his team continued to advance the development of aerial cameras. He applied a strictly scientific approach to the solving of problems, like finding better mounts to reduce vibration and blurring, and the use of flares to open up the prospect of night photography. A new wide-angle camera was produced with a 10in lens that could photograph an area of three miles by two from up to 20,000 feet with good, clear results. In early 1916 the team came up with a new 'E-type' camera made of metal instead of wood, with an improved remote control facility for the pilot. Later, as anti-aircraft defences forced aircraft to fly at ever greater heights, Brabazon developed the 'L-type', which had a long 20in lens and improved mechanisms for feeding the plates through the camera gate.

By the time the Battle of the Somme began in July 1916, the RFC had grown to 27 squadrons with a front-line strength of 421 aircraft. And new aircraft were continuously rolling off the production lines in Britain. Observers were able to photograph the German trenches extensively from the air and produce detailed, accurate aerial maps. Photo interpreters identified a new, third German defensive line from the photographs, spotted major clusters of barbed wire and gun positions and marked them up on the maps. The battle raged throughout the summer and autumn. When it was finally called off in the mud of November the Allied lines had advanced only a few miles despite the months of killing. Reconnaissance aircraft had taken some 19,000 photos and 430,000 prints had been made of these.[22] Along with the terrible casualties on the ground, almost 800 aircraft had been shot down, including many photo reconnaissance planes, and 252 RFC pilots had been killed.[23]

However, by now it was the dog fights between Allied and German fighters that captured the public imagination. In the popular press at least, aerial reconnaissance had taken a back seat in the new air war. As the war on the ground became mired in mud and barbed wire, the press lauded the new breed of war heroes, depicting them as modern knights of the air, brave, daring and chivalrous. The French press was the first to use the term 'ace' to describe a fighter pilot who had shot down five enemy aircraft. Later the Germans, the British and the Americans used the same term, although each nation defined it slightly differently.

Captain Albert Ball was the first British ace to find fame. A skilled pilot with fast reflexes, good eyesight and excellent marksmanship, he shot down three enemy planes in a day on five separate occasions.

His total of 'kills' stood at forty-four when in May 1917, during a routine patrol, he flew into a cloud and appears to have become disoriented before crashing to his death. Among the many other British aces, both Edward Mannock (sixty-one kills) and James McCudden (fifty-seven) were awarded the VC. Becoming a flying ace involved more than just mastering the science of aeronautics. As well as being a supremely good pilot who was totally at home in his aircraft, an ace had to have several special qualities. He usually only attacked when he knew he had the upper hand, either through a superiority in numbers or in approaching the enemy unseen. And he needed to have an obsession with attacking enemy aircraft, a quality that was more than an aggressive impulse but verged on a sort of addictive compulsion. One pilot, Cecil Lewis, later reflected, 'The fighting aces of World War One as I remember them were young men of high fettle and great energy. They seemed to burn. Perhaps their metabolic rate was greater.'[24]

The most legendary flyer of the war was the German pilot Manfred von Richthofen, a Prussian aristocrat and huntsman who commanded a squadron of bright red Fokker triplanes. His unit became known as the 'Flying Circus' and the British press dubbed von Richthofen the 'Red Baron'. Richthofen had trained as a cavalry officer before the war, and fought on both the Eastern and Western Fronts in 1914. But after static trench warfare set in he transferred to a flying unit in 1915. With Boelcke's death in an accident in October 1916, Richthofen became the leading German ace and took command of his own squadron. Like Boelcke and other aces that had preceded him, Richthofen was calculated and ruthless about tracking down enemy aircraft and going in for the kill when he and his squadron felt they had an advantage. Having clocked up a tally of eighty kills, he was finally shot down in April 1918 far behind British lines. A British pilot flew over his home airbase to drop a message informing the Germans of his death, and British officers gave him a funeral with full military honours.

After the Somme, the Germans introduced the Albatros DI, and the pendulum swung again, this time in their favour. Powered by a 160 hp Mercedes engine, the Albatros could fly at up to 110 mph and with twin Spandau machine guns could fire 1600 rounds a minute. When the ground fighting resumed at the battle of Arras in the spring of 1917, the RFC no longer enjoyed aerial supremacy but still had to carry on performing the daily rounds of reconnaissance and artillery spotting. During April 1917 the RFC lost 250 aircraft shot down and 400 airmen killed or wounded. The month became known as 'Bloody April'.[25]

A new set of aircraft names now became the fighting champions of their day. The Germans had their Fokker biplanes and triplanes, the British responded with Bristols, Vickers and Nieuports. The Germans then employed Albatroses, while the RFC introduced first the Sopwith Pup and then the Camel, which once again brought them mastery of the skies. The new generation of aerial warriors attracted much popular attention, but for most pilots on both sides, their flying careers on the Western Front proved tragically short – the average life expectancy of a young flyer in the RFC in early 1917 was just two weeks.[26]

On 1 April 1918 the RFC and the Royal Naval Air Service were merged to form a new military service, the Royal Air Force. But its creation came at a time of crisis on the Western Front, to which the Germans had transferred huge numbers of troops from the Eastern Front after Russia dropped out of the war following its revolution in 1917. The Germans, though, were now wary of giving too much away to aerial observation, a concern which percolated right up to the German commander. General Ludendorff sent a memo warning of the 'constant observation of our organisation by the enemy's aeroplanes in trench warfare and the technical improvements introduced in aerial photography'. Ludendorff instructed his commanders to 'hide our artillery positions from air observation' and to make 'greater endeavours than in the past' with 'camouflage work'. He went on, 'The plates employed for photography from great heights are extraordinarily sensitive to colours and show up clearly the slightest differences in shades, even those imperceptible on the ground.'[27]

Ludendorff's instructions were a great tribute to the success of Allied photo intelligence. Despite the German camouflage, throughout February 1918 British reconnaissance patrols had picked up the movement of large numbers of troops, the building of new supply dumps and the construction of eighteen new airfields. The Allied flyers were able to predict that a major offensive was imminent, even if they did not know where and when it would come.

On 21 March 1918, the Germans launched their anticipated attack and threw the French and British armies reeling back. The offensive began during a period of foggy weather when aerial reconnaissance was impossible and so the location of the attack, along the Somme, came as a complete surprise to the Allies. Within a few weeks the British Army had lost more than 100,000 men, surrendering much territory captured so painfully over the last two years, and by the second week of April the British commander-in-chief, Sir Douglas Haig, viewed the situation as highly perilous. His famous Order of the Day on 11 April concluded with the words, 'With our backs to the wall and believing in the justice of our cause each one of us must fight to the end.'[28] Then, on the following day, the weather cleared and RAF reconnaissance aircraft brought back for the first time detailed and accurate information about the German positions. More photographs were taken that day than on any other since the beginning of the war. Using the intelligence they picked up from these photos, the Allies could again bomb and shell German positions with accuracy. Haig noted in his diary the 'satisfactory results of today's work in the air'.[29]

The German offensive rolled forwards for another three months but slowly began to run out of steam. And when the Allies launched their counter-offensive in July, aerial photographs had already delivered to the Allied commanders detailed outlines of the German positions. Furthermore, having destroyed so many German aircraft the Allies now enjoyed total air supremacy, and British and French aircraft were able to fly up to eighty miles behind the German front line to report on the movement of reinforcements by train. American aircraft too were now arriving in large numbers; by the summer of 1918 there were forty-five American squadrons and twenty-one Photographic Sections on the Western Front. When it came to aerial observation and photo interpretation, the Americans lined up to learn the new science from the French, who had led many of the innovations in the war years.

The Allied offensive combined rolling artillery bombardments ahead of advancing infantry, with huge numbers of tanks supported by ground attacks from RAF aircraft that were closely integrated into the battle plan. The static war of the last four years became a mobile war once again and aerial photography was used to record the German withdrawal across France and Belgium until the German high command finally agreed to an armistice in November. By the final campaigns of the war, photo intelligence had unquestionably come of age. In the course of 1918 over ten million aerial photographs were delivered to the armies in Belgium and France.[30]

The First World War had not entirely been fought on the Western Front. A hugely successful campaign was conducted in Palestine during 1917–18, when General Allenby led a combined army from Egypt across the Sinai to defeat the Ottoman Turks at Gaza before advancing north to capture Jerusalem. He seized all of Palestine from the Turks and by the end of the war had advanced as far as Damascus. Allenby's armies had received support on the east of the Jordan river from the Arabs, who had risen in revolt against the Turks and, encouraged by Colonel T.E. Lawrence 'of Arabia', had advanced along the right flank of Allenby's army.

Much of the area over which these Allied armies marched had never been properly mapped, but Lieutenant Hugh Hamshaw-Thomas of the RFC was to change all that. Hamshaw-Thomas had been a leading palaeo-botanist before the war, well known for his study of fossils, and was a Fellow of Downing College, Cambridge. He was a quiet, unassuming and studious young man whose painstaking approach to the study of Jurassic fossils was now brought to bear on aerial photography. From late 1916 he led an aerial reconnaissance unit in Egypt that produced huge photo mosaics. The detailed maps produced from these mosaics covered more than 500 square miles of Egypt, Palestine and Syria and made a significant contribution to the success of the advancing armies moving into territory that had been controlled by the Ottomans for centuries.[31] Hamshaw-Thomas's work demonstrated yet again the importance of aerial photography to military intelligence.

By the end of the war, enormous strides had been made in the science of aerial reconnaissance. Cameras with long lenses manufactured in the newly created British optical industry could produce detailed images of vast areas from as high as 20,000 feet, while flexible roll film had begun to replace fragile glass plates. At the same time, the process of interpreting information from these photographs and of producing detailed maps of up to about 1:20,000 scale had been perfected. Photo intelligence had become the principal source of understanding where the enemy was located, how he was moving his reserves and what he was likely to do next. As the artillery's long-range guns had become the major aggressive weapon of the war, so they too had used aerial reconnaissance to sustain their weapons' accuracy and deadly effect.

It's fair to say that the twentieth-century battlefield had been reinvented by the use of photo intelligence. The foundations had moreover been laid for aerial mapping to chart great swathes of the globe in the 1920s and 1930s. In the Middle East aerial photographs would be used to map geological patterns and to decide where best to dig for oil. In both North and South America and in Australia it became possible to survey huge tracts of land from the air. An entirely new science had thus come out of the war. But aerial reconnaissance and photo interpretation were not the only forms of intelligence to be transformed in the laboratory of war from 1914 to 1918.

Part Two

Code Breakers

5

Room 40

On the first day of war, 5 August 1914, only a few hours after the cable ship *Alert* had severed the telegraph cables linking Germany with the outside world, the director of the Intelligence Division of the Admiralty, Rear Admiral Henry Francis Oliver, was walking across Horse Guards Parade towards his club in Pall Mall for lunch when he met his friend and colleague Sir Alfred Ewing, the director of Naval Education. Oliver asked Ewing to join him, as he had a problem he thought Ewing might be able to help him with. For some days he had been receiving in his capacity as head of intelligence intercepted cables that appeared to be German naval signals. But they were in some sort of code and meant nothing to Oliver. However, he knew that Ewing had an interest in ciphers, since the pair had a few months before discussed 'a rather futile ciphering mechanism'. Over lunch, Oliver said he was looking for someone to try to decipher the signals, which would no doubt grow in number over the next few months. As naval education was likely to be put on hold until the war had been won, Oliver suggested that Ewing might like to run this small section for him. Ewing promptly agreed.[1]

Sir Alfred Ewing was at the time fifty-nine years of age. A short, thickset Scot with bright blue eyes beneath shaggy eyebrows, he had a quiet voice with a mild Scottish accent and always dressed immaculately regardless of the fashion in a grey suit with a striped waistcoat, a mauve shirt and a dark blue bow tie with white spots.[2] The son of a Scottish clergyman, he had won an engineering scholarship to Edinburgh University and spent some years with cable firms working on engineering research, including such varied subjects as magnetism and Japanese earthquakes. He returned to the academic world at Dundee University in his early thirties and was appointed Professor of Mechanical Engineering at Cambridge in 1890. Five years later he received the Gold Medal of the Royal Society for his work on magnetic induction. In the reforms of the Royal Navy in the early years of the twentieth century, the First Sea Lord, Jacky Fisher, persuaded him to take up a new position in the navy and become Director of Education. He made a great success of this and was knighted in 1911. His hugely successful career had made him one of the nation's leading scientists and he would go on to become President of the British Association for the Advancement of Science.

Ewing now turned to the world of codes and ciphers, a subject in which he was interested but about which he freely admitted he was grossly ill-informed. Taking a scholarly approach, he first visited Lloyds the insurers and the General Post Office to study their code books. He then spent some time in the stacks of the British Museum examining old, dusty books on code making. With more and more intercepted signals flooding in to the Admiralty, up to one hundred a day, it was evident that Ewing would need assistance, so he turned to what could best be described as the Royal Navy's 'old boy network'. Knowing they would have the necessary qualities of discretion for such top secret work, he asked some of the teachers in the naval colleges at Osborne and Dartmouth, men he knew and trusted, to join him.

One of these was Alexander Denniston. Thirty-three years old, Denniston was another quietly spoken Scotsman. A great sportsman who had played hockey in the Olympic Games held in London in 1908, he had one of the great skills now needed in Ewing's team. He was a brilliant linguist, had studied at the Sorbonne and the University of Bonn and was a fluent German speaker. Thinking the assignment to the Admiralty would be a short one as the war was likely to be over in a couple of months, Denniston agreed to join Ewing. He was eventually to retire from the world of code breaking only in 1942.

Ewing, Denniston and a few others recruited at the same time were the first official code breakers specifically employed to crack coded messages sent by the German navy and by German diplomats abroad. Given the prevailing strong sense that studying the enemy's signals was a rather 'un-gentlemanly kind of activity', they were all sworn to secrecy.[3] They gathered in Ewing's cramped office and had to hide their papers whenever he had a visitor. And as they stared at the rows of meaningless letters and numbers that were arriving daily at the Admiralty, they had very little idea either of what the signals meant, or where to start in attempting to make sense of them. Meanwhile, however, developments were also taking place elsewhere.

Less than a week after work had begun, General Sir George Macdonogh, the Director of Military Intelligence in the War Office, approached Ewing. The army, too, were starting to receive intercepts of German messages but could make nothing of them. He suggested that they pool resources and try jointly to decrypt some of the messages. Accordingly, in a passing spirit of collaboration, Denniston and a couple of others spent some time at the War Office, but the rivalries and suspicions between the army and the senior service were too deeply entrenched for anything much to come out of this period of cooperation.

By contrast, the Germans made a good start in the intelligence war. In the first major battle of the conflict, at Tannenberg on the Eastern Front at the end of August 1914, they were able to listen in to all the Russian radio communications, including precise orders as to where specific units were to locate. Because the Russians had very few trained radio operators and lacked any encryption system, nearly all of this information was sent unencrypted or 'in the clear'. All the Germans had to do was tune in to the signals and translate them from Russian, and the entire Russian battle plan was revealed. With this intelligence the Germans were able to marshal their forces correspondingly. They moved several divisions very efficiently by railway from one side of the vast battle zone to the other, and the result was a massive German victory. For relatively small German losses, the Russians lost 78,000 killed or wounded, and 92,000 men were taken prisoner. After a second victory in early September, the Russian Second Army was almost completely wiped out. It took the Russians six months to recover from the disaster. It was clear that this was going to be a war in which listening in to the other side's radio communications would play an important part.

Although the Royal Navy had begun the war by taking the dramatic and pre-emptive step of destroying the German undersea cables, they had no plan as to how to intercept the radio signals that the enemy would then be forced to send. Since 1908, when a leading scientist from the Marconi company, H.A. Madge, joined the Admiralty as an expert on long-range wireless telegraphy, most naval vessels had been equipped with on-board radios. This led to a revolution in the way the world's navies operated. Instead of admirals being in charge of their fleets and captains their vessels once they were at sea, they were now subject to control from the Admiralty in London. Only at the Admiralty was a full strategic and tactical overview available. As Churchill, First Lord in 1914, wrote, 'from its wireless masts or by cable it [the Admiralty] issued information often of a vital character to ships in many instances actually in contact with the enemy. It was the only place where the supreme view of the naval scene could be obtained. It was the intelligence centre where all information was received, where alone it could be digested, and whence it was transmitted wherever required.'[4] The German navy had an equivalent in the *Admiralstab* in Berlin and used a powerful transmitting station at Nauen outside the capital to send long-wave radio signals around the world. It was these signals that naval intelligence was intercepting in abundance, though so far it had no way to interpret or understand them.

Although the science of radio was still in its early days it had grown well beyond infancy. Better tuning of equipment, finer calibration of transmitters and receivers had all greatly improved radio signalling and reception, and a network of amateur radio enthusiasts had grown up. In the years before the war, there had been a public scare about German plans for an invasion of Britain, and popular novels had helped to fan the flames. Erskine Childers' classic 1903 thriller *The Riddle of the Sands* had depicted two young men discovering a plan for a German invasion of Britain while on a sailing holiday in the Baltic. In 1909, another popular novelist, William Le Queux, wrote *Spies of the Kaiser*, in which the fiendish Germans used wireless telegraphy to send information back to Germany about the movement of Royal Navy vessels in the North Sea. The idea of secret agents sending intelligence reports to Germany from underground radio stations took hold in the public imagination. And so, on the declaration of war, the Defence of the Realm Act closed down *all* amateur radio stations. This closedown provided an unexpected boost for the navy.

At the outbreak of war there was only a single Admiralty long-wave listening station, sited at Stockton-on-Tees near the Durham coast. This station was not going to be up to the task of listening in to a wide range of enemy signals sent on many different frequencies. However, a saviour appeared on the scene. Russell Clarke was what used to be called a 'radio ham'. A barrister and an amateur radio enthusiast who, with an eccentric friend, Colonel Hippisley, a Somerset landowner, had before the war tuned in to German naval signals for fun. He went to see Ewing and told him that the Germans were transmitting many more signals, and at shorter wavelengths, than the navy realised. Clarke said he would be prepared to intercept these signals if granted official facilities to do so. Ewing gained permission to set Clarke up with his own listening post, and Clarke chose the coastguard station at Hunstanton on the coast of the Wash near King's Lynn in Norfolk. This was ideally suited to pick up German naval radio signals sent out across the North Sea. The expanded station was staffed by GPO engineers and under Clarke's instruction soon began to intercept a growing number of signals. Ewing also approached the Marconi company, who agreed to set up three separate listening stations to intercept naval signals. Before long there were eleven stations around England receiving a range of messages, all of which were duly transcribed and dispatched to 'Ewing, Admiralty'. The nation's first 'Y Service' or wireless listening service had been created.

But, of course, unlike the Russian messages, German messages (like those of the British and French) were encrypted, and the German navy employed at least three major codes. Ewing and his small team at the Admiralty had still not worked out how to decode the mass of signals that now began to arrive on their desk. For several weeks they continued with their research, and with the team growing in size they moved upstairs out of Ewing's small office to a larger space on the first floor. Located at the end of a corridor and looking out over one of the Admiralty's inner courtyards, adjoined by a small side room in which a camp bed was assembled for anyone working long hours who needed a break, the new office was quiet and out of the way, and the whole corridor was marked with large 'No Admittance' signs so that the top secret work could carry on uninterrupted and unobserved. Known innocuously by the number the Admiralty had given the office, Room 40, it would remain at the heart of naval code breaking for the rest of the war.

There followed a series of extraordinary strokes of good fortune. In mid-August the German merchant steamship *Hobart* moored off the coast at Melbourne. It was possible that the captain did not even know that war had been declared in Europe. A particularly resourceful officer of the Royal Australian Navy, Captain Richardson, went on board in the civilian disguise of a quarantine officer and requested to search the ship. Imagining that the captain would attempt to destroy any confidential papers, Richardson kept a watch on him and sure enough, in the early hours of the following morning, the captain slid back a secret panel in his inner cabin and took out a set of documents. Richardson pounced and seized the documents at pistol point . They included the *Handelsverkehrbuch* (HVB) codebook used by the *Admiralstab* to communicate both with merchant ships and at times with vessels of the German High Seas Fleet.

It was clear that no one in Australia realised how important this code book was, as it took nearly a month for them to inform the Admiralty in London of what they had seized. The Admiralty immediately instructed the Australian Navy to send the code book to London, but even the fastest steamer in those days took about five weeks to make the voyage and the precious document did not arrive until the end of October. By that time, though, a chance incident had already offered up another prize.

On 25 August, two German light cruisers had gone on patrol near the entrance to the Gulf of Finland, only 200 miles from Kronstadt, the base of the Imperial Russian Navy. A thick fog descended and one of the cruisers, the *Magdeburg*, ran aground on the Estonian coast. As an accompanying destroyer tried to pull it free the fog lifted, and two Russian cruisers spotted and closed in on the German ships. However, the *Magdeburg* was stuck firm on the sands. In the scramble to evacuate, the Russians captured the ship's captain and several of his men. In the confusion, the German master failed to follow his orders to destroy all the top secret code books. The Russians found in the main chartroom of the ship the *Signalbuch der Kaiserlichen Marine* (SKM), the code book used by German warships to communicate with each other and with the *Admiralstab* in Berlin, and alongside it details of the current cipher key and a German map of the Baltic and the North Sea, marked up with the grid used by the German navy to locate their vessels. It was a spectacular find. In a gesture of considerable generosity, recognising that Britain was the greatest sea power of the day, the Russians offered to hand over the code book to the Royal Navy if they sent a vessel to collect it. HMS *Theseus* immediately set off from Scapa Flow and after various delays brought back Captain Smirnoff of the Imperial Russian Navy, who personally presented Winston Churchill with the priceless code book on 13 October.

Only four days later a saga began that resulted in yet another stroke of luck. Four German destroyers were out laying mines off the Dutch coast when they were spotted by a British patrol led by the light cruiser HMS *Undaunted*. A brief engagement followed in which all four German ships were sunk. Following his orders this time, the captain of the only destroyer in the squadron that held secret code books gathered all his papers into a lead-lined chest before throwing it overboard. It immediately sank to the bottom of the sea. But this was not the end of the story. In the last week of November a British trawler was fishing some miles away from the incident when it dragged up the chest in its nests. It was in due course handed over to the Admiralty, who on opening it found that the chest contained the *Verkehrsbuch* (VB), the third major code book used by the German navy. Within a thirteen-week period the Admiralty had acquired three remarkable documents.

The acquisition of the German code books was an intelligence coup of unique proportions. The code breakers in Room 40 now had the means to follow the daily orders to every ship in the German navy. Never again would any military service have so much vital deciphering information dropped into its lap.

However, even with the code books in their hands Ewing's team found it difficult to decipher the signals. Some of the messages sent in SKM appeared to be no more than weather reports sent daily to and from Berlin, while the rest were still unreadable. Luckily another figure near the centre of events at the Admiralty was able to step into the breach. The Paymaster of the Fleet, Charles Rotter, was a German expert and when he looked at the messages he realised that they had been superenciphered: once encoded, the letters had been switched according to a key to yet another set of letters. Rotter began searching the messages for the most common sets of letters in the German language and for the words that were likely to be most frequently repeated. He could make a start, for instance, by looking for regularly used call signs. Once he had decoded a few letters it was like doing a crossword – with some of the clues solved the others were easier to fill in. Working with the code book and with his knowledge of German and of naval affairs (knowledge which none of the civilian code breakers possessed) Rotter was able to find the key to the SKM signals within about a week, providing a substitution table which finally enabled the decoding of the messages.[5] It was the crucial breakthrough. Denniston and his colleagues were recalled from the War Office and told to work exclusively on the naval signals. From November 1914 the code breakers of Room 40 started to read the German signals and to build up a detailed understanding of the operations of the German navy and the movements of the High Seas Fleet.

In that same month a final change set the direction of naval intelligence for the rest of the war. When Oliver was promoted to Chief of the War Staff, Captain Reginald William Hall became Director of Intelligence at the Admiralty. Hall, known as 'Blinker' because of a nervous tendency to blink his eyes when talking, would play a decisive role in the operations of Room 40. Bald, with a long aquiline nose, Hall was the sort of person who, through the strong sense of energy, confidence and authority he carried with him, created an instant impact when he entered a room. He had come from a naval family, joined the Royal Navy at the age of fourteen and worked his way up, becoming a captain in his mid-thirties. He soon acquired a reputation for being strict but fair and enlightened, willing to overturn tradition when necessary and always concerned for the welfare of his men. As captain of the brand new battle cruiser HMS *Queen Mary* just before the war, he had pushed the ship's crew to the highest levels of efficiency and they had the top gunnery record in the navy. At the same time he upgraded the status of petty officers and introduced the first ever ship's cinema, a book shop and a chapel. He even installed a laundry on board, challenging the age-old tradition that sailors should wash their own clothes. Senior naval men feared this was the end of all that was highly valued in the navy. But Hall wanted to make a ship feel like the crew's home as well as their place of work, and the men responded positively. His second-in-command later wrote that 'there has seldom been such a loyal, hard working and happy ship's company as the one that manned the *Queen Mary'*.[6] Within a few years all the major warships in the Royal Navy had adopted Hall's reforms.

Hall led his ship to war, but his physical condition was already poor and within a few months he was forced to resign his command due to ill health. It was a bitter blow to an ambitious captain still only in his mid-forties. It looked to Hall as if his naval career might be over just as the war was getting interesting. However, the Admiralty was aware of Hall's special qualities and, needing to replace Oliver, offered him the position of Director of Naval Intelligence. On arriving, he was amazed to hear that a small group at Room 40 were beginning to decipher German naval messages – the secret was so tightly guarded that no one outside the Admiralty knew of this work. But he brought to his new task immense enthusiasm, great organising ability and the capacity both to lead and to manage a team, pursuing every facet of the work with relish. Tracking down spies, sending back to Germany messages with misinformation to deceive the enemy as to British intentions, and interviewing enemy prisoners were all activities he enjoyed immensely. Intercepting diplomatic cables would become a major feature of naval intelligence as well. Having rather surprisingly found himself in charge, Hall seemed to thrive in this cloak-and-dagger world. He would play a major role in the war of intelligence.

Winston Churchill soon became obsessed with the high-grade intelligence that Room 40 was beginning to supply; he had always been interested in espionage and got a thrill from reading enemy signals in the raw. But, failing to realise the need for the material to be properly analysed and put in context by trained intelligence experts, he believed that simply having the decrypts was good enough in itself. As a consequence, when he drew up a set of guidelines as to how the record books should be kept and who should have access to them, total secrecy was the rule and as few people as possible should be allowed to see the decrypts. Churchill told the Prime Minister of the existence of the source, but kept from the rest of the Cabinet the fact that the Admiralty was listening in to and reading all the major naval and diplomatic signals sent in and out of Berlin. This limited release of information and lack of awareness of the need for field commanders to be kept fully informed, was to have fateful consequences.

Churchill did ask Hall to bring in an expert who could extract from the German signals any matters of naval significance that the civilian code breakers might have overlooked. And Hall's appointment for the task, Commander Herbert Hope, was a good one. Hope, like his boss, would have a major impact on the work of Room 40 for the rest of the war. But at first, in line with Churchill's rule that secrecy must prevail, Hope was set up in a room all by himself on the other side of the Admiralty from Room 40. Every day the code breakers brought him five or six decrypted messages, his role being to assess their significance and pass this on to Hall. At first Hope felt he had been thrown in at the deep end and his 'remarks were very amateurish'.[7] He soon realised that he would be far more effective if he was alongside the code breakers and able to discuss the work with them. But the suggestion was refused, on the grounds that no one was to be admitted to Room 40. It was only after a chance meeting with Jacky Fisher, who had returned to the Admiralty as First Sea Lord, that Hope was allowed at last to build up a relationship with the code breakers, whom he came to regard 'as fine a set of fellows as it would be possible to meet'.[8] Extracting details and information from the decrypts, Hope became the first proper intelligence officer in the Admiralty and effective director of Room 40.

It was only a few weeks after Hope joined the code breakers that the first spectacular opportunity came to make use of the decrypts. Admiral Jellicoe, who led the Grand Fleet, was stationed at Scapa Flow, the giant anchorage in the Orkneys, from where it would take time to deploy his warships to the North Sea if the German High Seas Fleet were to emerge from its harbours at Wilhelmshaven and Kiel. So his vessels were instructed to maintain long and exhausting patrols, sweeping the North Sea for thousands of miles in a fruitless search for German warships. Only some unique piece of intelligence could give him enough forewarning to place his ships in the right spot to find the enemy vessels if they ventured out of harbour.

On the afternoon of 14 December the code breakers in Room 40 decrypted a signal sent earlier that morning in SKM code by the commander of the German fleet. It stated that on the next day a raiding party of battle cruisers led by Rear Admiral Franz von Hipper would make an excursion into the North Sea. Its objective was not clear, but the mission was possibly to shell the English coast and lay a new minefield, with the intention of drawing British ships in to their destruction. Admiral Wilson, who had come out of retirement to act as a senior operations officer, alerted Churchill and the First Sea Lord. After a brief consultation, the triumvirate sent an urgent cable to Jellicoe in Scapa Flow just before midnight, saying 'Good information just received shows that German 1st Cruiser Squadron with destroyers leave Jade River [the estuary at Wilhelmshaven] on Tuesday morning early [15 December] and return on Wednesday night.'[9] Taking advantage of the early warning, they instructed Jellicoe to send a squadron of battle cruisers, led by Vice Admiral Beatty, to a position off the Dogger Bank; they were not to attempt to intercept the German raiders, as it was not known exactly where they were going to be, but were ordered to cut them off and destroy them as they returned to Wilhelmshaven on the Wednesday afternoon.

Churchill was in his bath in his rooms at the Admiralty the following morning when he was told that German ships were shelling Scarborough and Hartlepool on the north-east coast. He jumped out of the bath, dried himself quickly and ran down to the War Room. There he and Fisher realised that Jellicoe was perfectly positioned between the raiders and their home ports, 'cutting mathematically their line of retreat'. Churchill was hugely excited that his ships now had in their grasp 'this tremendous prize – the German battle cruiser squadron whose loss would fatally mutilate the whole German Navy and could never be repaired'.[10] However, nothing is simple when it comes to the sea. A mist suddenly developed, dramatically reducing visibility. Hipper's German ships withdrew on the predicted course, but Beatty and his battle cruisers could see little more than a mile through the mist. There was a minor engagement but the German ships sailed right past the British vessels. 'They seem to be getting away from us,' Wilson commented in the Admiralty War Room.

It was true. Not a single German ship was sunk. A great opportunity to fatally weaken the German Navy had been missed. Instead of his moment of glory, prompted by brilliant intelligence work, Churchill had to face the howls of outrage from a furious public who demanded to know why the navy was incapable even of protecting Britain's own coast. 'Where was the Navy?' screamed the newspaper headlines. As a result of the shelling, one hundred civilians had been killed and five hundred wounded in their homes. Were the Admiralty asleep, the newspapers demanded to know? Churchill reflected, 'We had to bear in silence the censures of our countrymen. We could never admit for fear of compromising our secret information where our squadrons were, or how near the German raiding cruisers had been to their destruction.'[11] He just had to hope there would soon be an opportunity to avenge this failure. He did not have to wait long.

Churchill was in his office at the Admiralty at midday on 23 January 1915 when Admiral Wilson marched in unannounced. 'First Lord,' he said with a glow in his eye, 'these fellows are coming out again.' It appeared that German radio signals had been intercepted and deciphered giving instructions that Admiral Hipper was to lead another major raiding force of four battleships and six cruisers, accompanied by support ships, to the Dogger Bank to try to intercept British patrols. It might be that they intended another attack on British coastal cities. Churchill and Wilson (Fisher was at home suffering from a cold) discussed what their response to the intelligence should be. After two hours they sent a series of orders to Admiral Jellicoe at Scapa Flow, telling him to dispatch the Grand Fleet, and to Vice Admiral Beatty at Rosyth giving him a precise location off Dogger Bank to which he should sail with his battle cruisers. The naval commanders responded to their instructions within hours and the two sets of dreadnought battleships closed on each other in the middle of the North Sea. The British were convinced they had set a trap, and that the German fleet would sail right into it.

As dawn broke the following morning the first reports of sightings of enemy ships arrived in the Admiralty War Room. Then came reports of ships opening fire, and next the news of battle. 'There can be few purely mental experiences more charged with cold excitement than to follow, almost from minute to minute, the phases of a great naval action from the silent rooms of the Admiralty,' wrote Churchill later. With the battle raging in a deafening crescendo of shells hundreds of miles away, he remembered that 'in Whitehall only the clock ticks, and quiet men enter with quick steps laying slips of pencilled paper before other men equally silent who draw lines and scribble calculations, and point with the finger or make brief subdued comments ... a picture always flickering and changing rises in the mind, and imagination strikes out around it at every stage flashes of hope or fear.'[12] But, once again, out in the middle of the North Sea things did not go as those back in London had hoped.

Jellicoe claimed that he had been informed too late of the German excursion and that when battle commenced he was still 140 miles from the scene. On the other hand, Beatty, in his flagship HMS *Lion*, was on station and ready to pursue the German warships with his battle cruisers, which were faster than the German vessels. But the German fire was more accurate than that of the British ships. *Lion* was badly hit and had to slow up. As the British cruisers caught up with the Germans they poured fire at the *Blücher* and the *Derflinger*, but both ships refused to sink. Then, in the late morning, the British sighted what were thought to be U-boat periscopes off the starboard bow of the *Lion*. The British force turned ninety degrees and in so doing lost pursuit of the enemy ships. In a final hurrah the remaining British battle cruisers sank the *Blücher* and severely damaged the *Seydlitz*, which managed to limp back to its base with the *Derflinger*, both ships ablaze and their decks strewn with the dead and wounded. The British claimed a victory and certainly the German navy had been given a fright. But they had once again missed the possibility of completely destroying a German naval squadron.[13]

In the post mortem that followed it was clear that there were problems with gunnery control on some of the British ships and it seemed that the side armour of Beatty's battle cruisers was not as protective as it was believed to be. But when it came to the intelligence it was evident that the Admiralty were keeping too much control over their precious decrypts. As a result of Churchill's insistence on keeping the intelligence to as few people as possible, when the decrypted German signals first became available only four men outside Room 40 were allowed to know what the German signals were saying: Lord Fisher, the First Sea Lord; Admiral Oliver, the Chief of the War Staff; Admiral Sir Arthur Wilson, chief of operations, and Churchill himself. They alone decided how to act on the intelligence they received. 'Blinker' Hall, as Director of Intelligence, was not consulted, nor was Hope, the chief intelligence officer in Room 40, who might well have picked up a broader picture of what the Germans were up to. In the case of the Battle of Dogger Bank, as it came to be known, not only had there been a delay in getting the information to Jellicoe which meant that he did not take part in the battle, but, even worse, it had been known all along, from intercepted signals sent to Berlin, that the German U-boats were forty miles south of the action. But this was never conveyed to the commanders in the battle. So Beatty had pulled away from the pursuit at the critical moment and allowed the German ships to escape for no reason at all. It was one thing to intercept the German signals and to decipher them, but if the intelligence discovered was not then used effectively the whole advantage was lost. Churchill and the old men of the Admiralty (two of the four were in their seventies and

Wilson had fought in the Crimean War) had not understood this. The battle had not shown up a failure in the gathering of intelligence, it had revealed a failure to *process* intelligence.

There was however an amusing postscript to the Battle of Dogger Bank. HMS *Lion*, Beatty's flagship, was forced in to dock on the Tyne for repairs. German signals indicated that a U-boat had been sent to wait outside the port to torpedo the ship and finish it off when it left. At this point of the war, Hall and an MI5 officer were using the identity of 'D', a German spy who had been secretly arrested and tried, and was awaiting execution, to feed misinformation to the Germans. Hall remembered seeing a Japanese album containing photos of Russian ships damaged in the Russo-Japanese war of 1904–5. Here he found a photo of a badly damaged Russian ship that looked very much like *Lion*. He sent the photo to Berlin as if from 'D', saying it showed *Lion* in for repairs and so badly hit that it would not come out for at least two months. The *Admiralstab* decided this was far too long to wait and the U-boat was quietly withdrawn.

Hall continued to send signals from 'D'. In April 1915, as a decoy for the attack on Gallipoli, he passed on information that the Allies were planning to invade Schleswig-Holstein. Observers reported that German troops from another front suddenly arrived in the district and frantically began to dig trenches. As a final payoff to this saga, it was known that 'D' was one of the spies who only supplied information for money. So Hall continued to charge large sums for each piece of misinformation he supplied to the Germans. When the money came through the MI5 officer bought a new car and Hall redecorated his office with the proceeds.[14]

The *Admiralstab* in Berlin was still unaware that the British were deciphering most of their signals. An inquiry after the capture of the *Magdeburg* had concluded that the Russians had probably not found the SKM code books, and when Hipper's fleet found the British waiting for them off the Dogger Bank, instead of becoming suspicious they concluded that a spy must have reported their departure from Wilhelmshaven. They believed that their code books and superenciphering system were impregnable; it was utterly impossible that an enemy could decipher their signals. They did however tighten up their procedures. It was a complex business to issue new code books, which might take months to reach ships on the other side of the world. It was much easier to change the ciphering key which reset the substitution table. Earlier in the war the key had been changed only about once every three months. This was increased to once a week and by 1916 to once every twenty-four hours, at midnight. However, the cryptanalysts in Room 40 grew so sophisticated that it soon took only a few hours to find the new key. This would be the task of the night shift; it was thought a very poor show if by the time the morning shift arrived the new day's ciphers were not completely readable. Still the Germans went on transmitting in their codes and ciphers, believing that no one could possibly read them, while the staff of Room 40 grew continuously until by 1916 there were about fifty code breakers working throughout the day and night.

In May 1915 *The Times* had revealed that the British assault at Neuve Chapelle in March had been effectively strangled by the lack of shells. Because no one in the War Office had anticipated the huge numbers of shells that the artillery would need in order to sustain trench warfare, it was widely felt that Asquith's Liberal government were not doing enough to lead the war effort, and the ensuing political crisis led to the government's collapse. A new coalition government of Liberal and Conservative ministers was formed, with Asquith continuing as Prime Minister, but the price of Conservative support was Churchill's removal from the Admiralty. He was blamed for what was called the 'fiasco of the Dardanelles' and the failure of his plan to knock Turkey out of the war by sending a naval expedition to capture Constantinople. For months, whenever he got up to speak in the House of Commons, Conservative members would interrupt him with shouts of 'Remember the Dardanelles'. Devastated, and feeling that he had been made a political scapegoat, Churchill was out of office and influence for the first time in nearly ten years. Many thought his political career was finished. He went into a profound depression, what he called his 'black dog'.[15] The Conservative politician Sir Arthur Balfour became First Lord of the Admiralty, while, for separate reasons, Fisher resigned as First Sea Lord. But the system the two men had created with regard to keeping the spoils of intelligence within a tiny clique, survived them. No significant change in the processing of intelligence was made at the Admiralty.

After the Battle of Dogger Bank, the German High Seas Fleet remained holed up in its harbour at Wilhelmshaven for the next fifteen months. But in May 1916, Admiral Reinhard Scheer decided to make another venture into the North Sea. The plan was for Rear Admiral Hipper once again to take out his battle cruiser squadron in the vanguard and when spotted by the British fleet, to turn and lure the British ships into the guns of Scheer's waiting High Seas Fleet. In addition, several U-boats were withdrawn from the Atlantic and stationed off the British coast, from Aberdeen southwards, to alert the German admirals to the British fleet's movements and to be on hand to close in for the kill during the engagement. The Germans never thought they could totally defeat the Royal Navy, but they hoped to cause sufficient losses for its fleet to be reduced in size to something more equal to that of the Germans, and to weaken or even to break the ever-tightening British naval blockade of Germany.

The first signals relating to their intentions were picked up and deciphered in Room 40 in the last week of May. As usual, the coded messages provided only fragments of evidence and needed interpretation. Admiral Oliver insisted on keeping control of the intelligence and asserted that only he could send out orders to the British fleet. Assisting him now was the Director of Operations, Captain Sir Thomas Jackson, an officer steeped in the traditions of the Royal Navy and deeply suspicious of new-fangled developments. According to William Clarke, one of the new code breakers who joined Room 40 in March 1916, Jackson had only ever visited the intelligence room twice, the first time to complain that he had cut his finger on one of the red boxes containing the decrypts and the second, when a delay in finding the new cipher key had caused a temporary suspension in forwarding the decrypts, to say how pleased he was not to be bothered with all that 'damned nonsense'.[16]

By the late afternoon of 30 May, Oliver believed there was enough evidence to order Jellicoe to depart Scapa Flow and Beatty to leave Rosyth, and for them to rendezvous off the Scottish coast. The Grand Fleet, the pride of Britain, now assembled in the North Sea with a mighty force of twenty-eight dreadnought battleships, nine battle cruisers, thirty-four cruisers and seventy-nine destroyers. Jellicoe hoped for the chance of another Trafalgar, to destroy the German fleet and to establish British naval supremacy for the next hundred years. Scheer and Hipper sailed out towards them with a fleet of twenty-two battleships, five battle cruisers, eleven light cruisers and sixty-one destroyers.[17] It would be the biggest confrontation of the giant dreadnought battleships in history.

Soon after midday on 31 May, a single small incident fundamentally turned the outcome of events. Again, it involved the mis-transmission of intelligence received in Room 40. As Scheer left Wilhelmshaven on the afternoon of 30 May, he had issued a spate of orders by radio. One of these was an instruction to switch call signs and to inform all vessels that his usual sign 'DK' would now be used by the wireless telegraph station in Wilhelmshaven harbour. Late on the following morning, while the Grand Fleet was awaiting orders, one hundred miles east of the Scottish coast, Captain Jackson made one of his extremely rare visits to Room 40. Knowing Scheer's usual call sign, he asked for 'DK's current location. Wilhelmshaven, he was told. Asking no further questions and not waiting to be told that the 'DK' call sign had been switched, he turned and marched out, reporting to Oliver that Scheer and the High Seas Fleet were still in their base port. Oliver immediately sent a message to Jellicoe telling him that it seemed the German fleet was still in harbour. Jellicoe therefore thought he had stolen a march on the Germans and so made no attempt to speed his warships to the rendezvous. The resulting delay was to prove fatal.

Ninety minutes after Jellicoe received Oliver's signal, Beatty's battle cruisers, now seventy miles ahead of Jellicoe and the rest of the Grand Fleet, sighted Hipper and his battle cruiser force. They immediately gave chase and Hipper turned and ran, heading south. His orders were after all to draw Beatty and the rest of the British ships into the guns of the High Seas Fleet, about fifty miles behind him. Hipper opened fire on the warships chasing him and once again, as at Dogger Bank, the German gunnery proved lethal. Two of Beatty's fastest ships, HMS *Indefatigable* and HMS *Queen Mary*, were hit and the ammunition magazines on both ships blew up, causing massive loss of life. Beatty is supposed to have turned to his flag captain and uttered the legendary words, 'Chatfield, there seems to be something wrong with our bloody ships today.'

Further south, the captain of HMS *Southampton*, who was leading a light squadron to search out enemy vessels, now spotted the main German fleet and passed this information on to Jellicoe and Beatty. Amazed, both men began to question the accuracy of the intelligence they were being sent from the Admiralty, who had told them that the German fleet was still in harbour.

Beatty halted his pursuit of Hipper's ships and turned north to head back towards the rest of the Grand Fleet. Hipper turned his force around and joined Scheer in pursuit of them. The two giant fleets were now heading directly towards each other, and at around 6.30 p.m. they made contact. The Germans succeeded in hitting another battle cruiser, HMS *Invincible*, proving again that the British armour was not good enough. The ship exploded and went down so quickly that only six men out of a crew of more than 2000 survived. And Beatty's flagship, the recently repaired HMS *Lion*, was badly hit once more. But the German ships themselves were now taking an immense hammering. Hipper's battle cruisers received a particularly heavy mauling, and the *Lutzow* was seriously damaged. Scheer ordered his fleet to do an emergency 'battle turn away' and, laying down a smoke screen, started to withdraw. However, British ships had come between him and his direct route back to

Wilhelmshaven. As the light began to fade there were a confusion of ships scattered across the sea off the point at Jutland, and the British sank one of the pre-dreadnought battleships, the *Pommern*. Finally, at 8.30 p.m., as darkness fell, the last shots were fired.

Back in the Admiralty, Commodore Hope and the team in Room 40 were working flat out decoding and passing on intercepts of messages between Scheer, Hipper and their ships within thirty to forty minutes – fast enough for the information gleaned to be of major value to the commanders at sea. Jellicoe, who had slowed his approach to Jutland, still had enemy ships within his sights and could have caused a vast amount of damage to the German navy. But it seemed that the fog of battle had descended. He hoped to regroup his forces and to continue to press his attack on the German fleet at dawn. At 9 p.m., Room 40 received an intercept from a German destroyer giving the detailed position and course of the rear cruiser in Scheer's fleet, the *Regensburg*. This was deciphered and Oliver sent it on to Jellicoe less than an hour later. When Jellicoe checked this, it turned out that he was in the exact position given for the *Regensburg*. It was an unfortunate piece of luck. The German destroyer had got its sighting wrong by ten miles. Room 40 correctly passed on the location they were given, but as far as Jellicoe was concerned this was the last straw. What with the earlier message saying that the German fleet was still in harbour when they were already at sea heading straight for him, it appeared that the Admiralty seemed to be sending him nothing but duff intelligence.

The problem was in reality the same old issue of the failure to pass on effective intelligence. As the British admiral regrouped his ships and waited for action in the morning, Scheer was conducting a night manoeuvre to escape south and head for home. Between 10 p.m. that night and 3 a.m. on the following morning, Room 40 produced sixteen decrypts, every one of which would have provided help for Jellicoe in taking out Hipper's badly damaged battle cruisers. But only three were passed on.[18] In the Admiralty, Oliver continued to insist that he and Admiral Wilson alone were capable of drawing the right conclusions from the intelligence that Room 40 delivered to them almost minute by minute. It was yet another dismal failure of the system Churchill had created for processing intelligence.

When dawn came up, the German fleet had got away. They had suffered badly, but not as badly as the Grand Fleet. The British had lost more than 6000 men, the Germans 2500; the British had lost fourteen ships, the Germans eleven. Both sides claimed a victory – the Germans because they had sunk more ships and killed more men than they had lost themselves, the British because they had forced the German High Seas Fleet to flee back to its base port, from which it was rarely to emerge again for the rest of the war.

There was never another confrontation between the two powerful navies. The era of the great battleships was over. The British blockade of Germany continued slowly to strangle the country of food and other supplies, while the Germans concluded that they must now resort to an all-out U-boat war to try to starve Britain into seeking a truce. But the controversies about what became known as the Battle of Jutland raged for some time. The press and no doubt most of the British people were disappointed that there had been no outright victory. They wanted to know who to blame. Behind closed doors, Beatty blamed Jellicoe for being over-cautious and allowing Scheer to escape, and a feud between the two men lasted for many years.

As the dust settled, the Admiralty reassembled the players. Jackson, who had fatally misunderstood the position of the German fleet, was promoted to rear admiral and sent out to command the fleet in the Red Sea. Beatty was promoted to commander-in-chief of the Grand Fleet and Jellicoe to First Sea Lord. At the Admiralty over the next two years, he used to regularly visit Room 40 of an evening before dinner and read the log of all the major decrypts that had come in. He was 'polite and gracious' and would discuss the day's key issues with the code breakers themselves.[19] He, at least, appreciated the need to understand the whole situation revealed by the decrypts.

Room 40 had performed well. On three occasions it had provided an early warning to the British fleet that German warships were coming out to fight. The first time, the British response had been spoilt by the weather; the second, by failures at sea. And on this third occasion, when it really mattered, at the Battle of Jutland, it was a series of errors by the Admiralty that denied the Grand Fleet the victory it desperately hoped for. To have passed on the information that the Home Seas Fleet was still at Wilhelmshaven when in fact it was already in the North Sea heading north, without even checking with 'Blinker' Hall or Commodore Hope that such intelligence was accurate, was a blunder of massive proportions. It meant that Jellicoe slowed down and only reached the German fleet in the early evening. If his superior firepower had had more than just two hours to assault the German ships then victory would almost certainly have been his. Further blunders throughout that evening meant that Jellicoe was unaware of the location of the German ships. Hope later said he had a full list of every German ship that had been sent out and at any one time a good intelligence assessment could have provided an overview of their location. Instead, at the critical moment, Jellicoe had lost all confidence in the intelligence he was getting from London. It was a pathetic outcome from what had been a brilliant opportunity.

Britain could nevertheless be proud of its code-breaking record in the First World War. Room 40 had made extraordinary progress. It was a precursor to the far more celebrated work that would be carried out at Bletchley Park in the Second World War. By then, Churchill, who was this time in full command, and those around him had learned how to use the intelligence received far more effectively; whereas Room 40 had played an important part in the war at sea, Bletchley Park would perform a vital function across the entire war effort. However, the impact of the work carried out in Room 40 was itself to reverberate far more widely than in the naval war alone.

6

The Great Game

In March 1915 a group of German army officers and diplomats left Constantinople (Istanbul) disguised as a travelling circus, their secret mission to incite a jihad or holy war against the British in Persia (today's Iran). Persia was of great importance to the British; it was there that the Anglo-Persian Oil Company had the concession to drill for oil. And demand for oil was rapidly growing to fuel both the Royal Navy and the burgeoning mass of motor vehicles and aircraft powered by the internal combustion engine.

One of the German group was a former vice consul from the Persian Gulf, Wilhelm Wassmuss, later nicknamed 'Wassmuss of Persia' after Lawrence of Arabia.[1] Wassmuss, like Lawrence, loved the desert, its people and their customs. He was committed to raising the local tribes against Britain and seems to have left the main group as they headed into Persia and travelled inland with a small raiding party to blow up the Anglo-Persian oil pipeline where it crossed the mountains of Abadan, in the sort of guerrilla raid that would have appealed to Lawrence. The story then becomes confused. Either Wassmuss was captured by a Persian khan who planned to hand him over to the British, or he was surrounded by a British patrol while still in camp. Whatever the truth, he managed to flee at the last minute at night, wearing only his pyjamas and

leaving behind all his baggage. The small British force, frustrated
that their prey had escaped, collected up his bags, which were eventually taken back to London and stashed away in the basement of
the India Office.

When 'Blinker' Hall at the Admiralty heard about the incident he
became curious and sent a representative from Room 40 to search
the German diplomat's bags. There, wrapped in the vice consul's
long woollen underpants, was a code book. On examination back in
Room 40 it was found to contain the secret codes used by the
German diplomatic service for sending messages from Berlin to
Madrid and Constantinople. Most significantly, it was via Madrid
that the Germans sent all messages to their diplomats in north, south
and central America. Hall had sniffed out another remarkable find.
From the code book found in the vice consul's long johns, the British
could now read cables sent between Berlin and the Germans' diplomatic mission in the United States.[2]

It might seem strange today that the arm of government that was
now presented with the prize of reading German diplomatic messages to America was the Admiralty. But in those days the Foreign
Office would never have entertained the idea of intercepting the
diplomatic mail of neutral governments. That sort of thing just was
not what gentlemen did. In the years before the First World War, a
proper system for the collection of intelligence had been established
with the formation, in 1909, of the Secret Service Bureau, partly with
the encouragement of Winston Churchill while he was still Home
Secretary. This had two parts, one to handle domestic and imperial
intelligence in what became known as Military Intelligence 5 (or
MI5). The second dealt with foreign intelligence matters and became
known as the Secret Intelligence Service, or MI6. But the establishment of a state intelligence service did not bring with it a great
revolution in intelligence gathering. Neither MI5 nor the SIS had any
capability for code breaking and both were run on a shoestring
before the war. In 1914, MI5 only had a staff of fifteen, including the
office caretaker, and the SIS was similarly tiny.[3]

It needed someone like Hall, a complete outsider to the diplomatic and security establishment, to realise that, in wartime, code breaking could be a valuable source of potential intelligence. And of course, as the Royal Navy had interests in every corner of the globe, what the Germans were up to in far-flung places could be justified as being of immense potential interest to the Admiralty. Moreover, the only other cryptanalyst department in government was at the War Office, whose small group of code breakers were only concerned with intelligence of military significance. So despite the existence of a nascent security apparatus, the initiative in intercepting diplomatic messages was taken by 'Blinker' Hall at the Admiralty.

Already, the Admiralty had picked up several signals that were not of purely naval interest and had stored them away in a cupboard. Now they were in possession of the new code book, it was time to get them out again and have a look at what was going on. Hall thought they might contain useful information about the political and economic affairs of Germany.

To interpret this intelligence, Hall needed to recruit a new set of individuals. He formed another department to focus entirely on reading the diplomatic mails and responsible to him alone, not to Ewing as Director of Intelligence. Initially, it operated down the corridor in Room 45, but it was still generically known as part of Room 40. To head the new group Hall chose George Young, a man in his early forties with just the right international experience for the task. Educated at universities in France, Germany and Russia, Young had an excellent grounding in diplomacy, having served as an attaché in Washington, Athens, Constantinople and Madrid and as a more senior diplomat in Belgrade and Lisbon. He was in many ways typical of the new breed of pre-war professional diplomat, suave, sophisticated and mysterious – but one who was prepared to resort to any means to get one up on his enemy.

Hall and Young, probably with some advice from Ewing who was very well connected in the academic world, went on to recruit a remarkable team from the universities. Dilwyn 'Dilly' Knox had been educated at Eton and was a classics don at Kings College, Cambridge. He was to prove the most brilliant and intuitive crypt-analyst in Room 40. Frank Birch was another old Etonian and a history Fellow at Kings, Cambridge. His strength was in interpret-ing the intelligence from the decrypts. From London University there was Prof. L.A. Willoughby, a German specialist, while from Oxford came L.G. Wickham-Legg, a Fellow of New College, and Philip Baker-Wilbrahim, a Fellow of All Souls. Not all the new recruits were from the universities. There was George Prothero, editor of a political review; Thomas Inskip, a barrister and later Lord Chancellor; and Captain Loch, a former international correspondent of the *Daily Telegraph*. Several wounded naval officers were recruited to supply additional manual help, like the one-legged Lieutenant Haggard, and Lionel Fraser and Edward Molyneux, who later went on to become a famous dress designer. Then there was the Rev. William Montgomery from Westminster Presbyterian College, Cambridge, a specialist on the writings of St Augustine; and Nigel de Grey, another old Etonian who had failed his entrance exam to the Diplomatic Service because, although fluent in French and German, he had not reached the required standard in Italian. So he had joined the Royal Naval Voluntary Reserve as a volunteer while working full time for the London publishers, William Heinemann. Later in the war, Montgomery and de Grey would play a major role in the biggest intelligence coup of all.

Breaking yet another Admiralty tradition, Hall began to recruit women – much to the horror of the established senior naval staff. There being no 'old girls' network', Hall decided that all female recruits must, first, be totally secure and reliable. In practice, this meant that many of them were the daughters, sisters or wives of respected naval officers or other pillars of the establishment. Next, they must either be good linguists – and indeed there were many able female language specialists coming out of the universities of Britain, even if at this stage they were not able to take degrees in

many of the older universities – or they should be able to type. The
team of formidable women who now arrived to do their bit included
Mrs Denniston, the wife of the code breaker who already worked in
Room 40; Miss Henderson, the daughter of Admiral Wilfred
Henderson; Joan Harvey, the daughter of the Secretary of the Bank
of England; and Miss Robertson, the headmistress of Christ's
Hospital for Girls and a brilliant linguist. Chief of the secretarial staff
that supported Room 40 was a rather masculine woman nicknamed
'Big Ben' who smoked cigars. She was in fact Lady Hambro, the wife
of a city banker.[4] It seems there was only one lasting affair during the
war between these young men and women, when 'Dilly' Knox fell
in love with and later married his secretary, Miss Olive Roddam.

Many of these men and women were recruited because someone
already in Room 40 introduced them to Hall when a vacancy came
up. This was the way such recruitment was done, as a personal recommendation usually guaranteed the reliability of the recruit.

The list of individuals now at work in the Admiralty is in many
ways strikingly similar to the disparate but brilliant group that more
than twenty years later were recruited to Bletchley Park. And, as in
the following generation, this group worked well together and
clearly sparked each other off. But it was a sign of Hall's great skill
that he was able to recruit so many variously talented civilians to an
organisation, such as the Admiralty, that tended only to value the
sort of familiar service that it recognised in its own. It was another
illustration of how many remarkable men, and in this case women,
from a range of backgrounds, academic and otherwise, came to do
their bit in the laboratory of war. Hall himself later wrote, 'It must
have been the most heterogeneous staff that ever came together.
Men and women of every profession and class joined us. We had
few precedents to follow and worked to make our own rules as we
went along. We had our successes, but we also had our ignominious
failures. And ... we had our comic opera moments, when it was difficult to realise that we were in the midst of war.'[5]

One of the subjects of deepest political dispute in the United

Kingdom before the war was Ireland. At the time, the entire island of Ireland formed part of the Union with Britain, but since Gladstone's day there had been talk of Home Rule, and this was the great political ambition of the majority, Catholic nationalist population. But the minority, Protestant unionist population of Ulster refused to be part of what they predicted would be a Catholic state and demanded to remain within the United Kingdom, where their privileges would be preserved. And, in the spring of 1914, the Irish Question had threatened civil war. The country had drifted into two armed camps as tens of thousands of Irish Volunteers drilled openly in the south, and 100,000 men joined the Ulster Volunteer Force under Sir Edward Carson in the north. In April 1914, 20,000 German rifles were smuggled into the north of Ireland to equip the Ulster Volunteers. In the south, Sinn Fein, literally 'ourselves alone', was a political movement that went further than many nationalists wanted and called for an independent, free Irish republic. The declaration of war in August 1914 brought a temporary cessation to the threat of civil strife, and Irishmen of both hues rushed to join the British Army. But it did not mark any sort of resolution to the deep-seated divisions within Ireland itself.

The British government now feared that a nationalist uprising in Ireland would open up Britain to an attack by Germany through the back door. It was not surprising that the Irish nationalists, using the principle that 'my enemy's enemy is my friend', should turn to Germany for support, and indeed the German embassy in the United States was happy to encourage Irish-American hardline republicans. In late 1914, Hall picked up information that Sir Roger Casement, an ex-member of the British consular service and a keen Irish nationalist, had left New York and reached Germany. There Hall followed his progress as he appeared to try to form an Irish Legion from among the Irish prisoners of war. Next came information that he was about to land in a German ship equipped with arms supplied by Berlin, in a remote spot in south-west Ireland from where he would launch a Sinn Fein uprising against British rule.

Hall and Assistant Commissioner Sir Basil Thomson from Scotland Yard's Special Branch now conceived of an extraordinary ploy to test out the support for Sinn Fein in the west of Ireland. Chartering a grand 500-ton steam yacht by the name of the *Sayonara*, Hall made plans for it to appear to belong to an American who was loud and sympathetic to the Sinn Fein cause. He installed a supposed German-American owner, a skipper and a full crew who were to sail along the remote bays and inlets of the south and west of Ireland to observe where support for Sinn Fein was strongest and try to find where Casement was intending to land. When he came ashore they were to overpower him and any Germans accompanying him, and arrest him as a traitor.

Hall assembled a remarkable cast of characters for the venture. Lieutenant Simon, who had served under him in the *Queen Mary*, jumped at the opportunity of playing the part of the American skipper; as well as being a fine seaman, one of his great accomplishments was his ability to imitate a variety of American accents. Major Wilfred Howell was cast as the German-American owner of the yacht who was to display strong sympathies for the German cause. He had been educated in Austria and spoke German like a native, but was a British citizen who had served in various imperial forces in which he had won a DSO. A crew of fifty naval ratings were supplied from Portsmouth. They had to dispense with their Royal Navy uniforms, dress as American sailors, chew gum and generally behave in a way that would make observers think they were American. A radio set was hidden in the ship along with a stash of arms that would be pulled out to arrest Casement if he landed as expected.

A variety of adventures befell the *Sayonara* and its fake crew over the weeks of its travels around the south-west coast of Ireland. They were nearly arrested when pulled in by Admiral Sir Charles Coke, who was in command of the Irish coast. When Coke learned the real reason for the yacht's presence he agreed to play along. He invited a group of naval officers to lunch with the 'owner' and the 'skipper', and the British officers were outraged when the 'Americans' told them that Germany would win the war and had a far more impressive navy than the British.

At one point of their mission HMS *Cornwallis*, a British patrol vessel, intercepted the *Sayonara*. The captain interrogated the 'owner' and the 'skipper' and immediately grew suspicious. Reporting back to the Admiralty that the owner spoke mostly German and only broken English, he thought the yacht must be on some sort of spying mission and requested permission to detain the vessel and its crew. Hall intercepted the communications and sent back instructions that the Americans were to be allowed to carry on with their business. The captain of the *Cornwallis* was furious, and Hall was very amused when his private letters were later intercepted and he was found to have described the Admiralty as being full of 'silly old gentlemen'.

At another point in the *Sayonara*'s journey, Lord Sligo, a local Anglo-Irish landowner, rushed to London in person to demand a private meeting with Hall. Claiming that the *Sayonara* was undoubtedly up to no good and was clearly in the pay of Germany, he reported to Hall that he had seen with his own eyes the crew planting mines in Westport harbour. The actors on the *Sayonara* performed brilliantly during the comic opera of their adventure along the Irish coast, but in Berlin the Casement landing was postponed, so Hall had to recall them and return the yacht to its owner.[6] And in reality it is highly unlikely that the adventures of the *Sayonara* produced any intelligence of real value about the strength of Sinn Fein support in south-west Ireland.

Thanks to the newly discovered diplomatic code books, Hall was however able to follow the correspondence between Count Bernstorff, the German ambassador in Washington, and Berlin. He read reports that Sinn Fein had plans for an armed rising in Ireland but he could see that the German General Staff was wary about the strength of Sinn Fein support for Casement in the west. There were many months of delays but in April 1916, Room 40 deciphered a message in which the ambassador said that an armed uprising was planned for Easter Sunday, and requested the shipment of up to 50,000 rifles to support the rising. In the exchanges that followed, Room 40 discovered that a gun-running vessel called the *Libau* was to be sent from Germany, disguised as a Norwegian merchant ship and carrying 20,000 captured Russian Mauser rifles, ten machine guns and a million rounds of ammunition. Sir Roger Casement was to go to Ireland separately in a U-boat and to land at Tralee Bay in County Kerry. With all this detail, the Admiralty had no difficulty in apprehending the gun-running ship off the Irish coast. As it was being escorted in to Queenstown (now Cobh) the German crew scuttled the ship and surrendered as prisoners of war. The following day, Casement was arrested within hours of setting foot on Irish soil.

A delayed armed rising did take place in Ireland on Easter Monday, 23 April 1916. After Padraig Pearse, its romantic poet-leader, read out a proclamation declaring the existence of the Republic of Ireland, 150 revolutionaries occupied the General Post Office in O'Connell Street in Dublin. The British response was swift and decisive. Martial law was declared, troops were rushed in and HMS *Helga* shelled the city with high explosive shells from Dublin Bay. After six days the rising was crushed and much of the centre of the city was left in ruins. Most of the leaders were arrested and executed. Only Eamon de Valera, who claimed American citizenship, was pardoned. Moderate Irish nationalists were outraged by what was seen as British barbarity. Although few Irish men and women had been sympathetic to Sinn Fein at the beginning of the rising, many were converted to republicanism in its aftermath. The focal point of Irish politics shifted for ever as a consequence of Britain's response to the Easter Rising.

Sir Roger Casement was tried and found guilty of treason, despite claiming in his defence that he had realised the rising could not succeed and had come to try to stop it rather than foment it. His trial became something of a cause célèbre: America put pressure on the British government to commute his death sentence, while several

significant figures, including George Bernard Shaw and Sir Arthur
Conan Doyle, came out on his side. Using a set of diaries seized from
Casement, Hall then seems to have engaged in one of his most
unsavoury 'dirty tricks' campaigns. The so-called 'Black Diaries', in
which Casement described in some detail scenes of gay sex, revealed
him to be apparently a promiscuous homosexual with a penchant
for young men. This caused outrage in the more innocent atmosphere of the time, and Hall realised that if Casement's supporters
knew of the content of the diaries they would abandon their
demands for leniency. It seems that Hall and Commissioner
Thomson leaked extracts from the diaries to the British and
American press and to some MPs.[7] So discredited was Casement
that it put an abrupt end to the campaign to save him. He was
hanged at Pentonville Prison for high treason on 3 August.

Spain was another country known to be sympathetic to
Germany. Following the example of the *Sayonara* escapade, Hall
sent another luxury yacht, the *Vergemere*, on a cruise down the
Spanish coast to discover if Spaniards were doing anything to help
the German war effort, for example by preparing to refuel U-boats
before their long missions out into the Atlantic. This time he
decided not to use a fake American as skipper, but to select a sports-
loving British peer who must appear to be so wealthy that, despite
the war, he was still bent on taking a pleasure cruise down the
Mediterranean. The person he found to play the part of the stage
aristocrat was Sir Hercules Langrishe, a charismatic Irish baronet
and keen sportsman. Langrishe cruised from port to port, inviting
influential Spaniards on board the *Vergemere* for lavish champagne
parties, all sponsored by Hall. He seems to have picked up enough
intelligence for Hall to realise that Spain needed to be continually
watched. As an unexpected by-product of the operation, Langrishe
convinced many Spaniards that reports that the British were suffering great privations at home could not be true. The Germans
apparently mounted a similar exercise at the same time, taking a
steamer along the Spanish coast to sound out local opinion. But,

unlike Langrishe's extravagant champagne parties, the Germans on their boat served nothing more exciting than a glass or two of beer. Not surprisingly, the Spaniards much preferred British to German hospitality.

Hall had proved to be a ruthless and determined intelligence chief, willing to do anything to attain his ends. Hall and his Scotland Yard colleague Thomson made a formidable double-act. Thomson invited Hall to be present when he was interviewing several suspects and realised that Hall was extremely able at getting people to talk. But sometimes it is difficult to separate fact from fiction in the stories that grew up around Hall. At one point, he is supposed to have interviewed a naval figure who had been brought in from a steamer off Ireland. The man was thought to be a German but denied this, claiming to be an American who had never set foot in Germany in his life. Hall listened to what the man had to say and was apparently sympathetic to the suspect when he suddenly barked out in German the order for 'Attention'. The man automatically leapt upright before realising the game was up and admitting his true German identity.

On another occasion, Hall and Thomson together interviewed Margaretha Zelle, better known by her nickname of Mata Hari. Famous in Europe as an oriental exotic dancer, Zelle was Dutch and had learned to dance in a supposedly native Hindu style in the Dutch East Indies. Before the war she had become famous in Paris for her erotic dancing and for posing for nude photographs, clad only in a few discreetly placed jewels. She conducted well-known affairs with various senior French officers and officials. After the outbreak of war she carried on this lifestyle and was suspected of passing on to the Germans information she gathered by seducing French officers.

She regularly travelled between Paris and Holland by boat from Spain, and when she landed at Falmouth on one of these journeys she was arrested and taken to London for interrogation. Hall and Thomson were convinced she was spying for the Germans and was carrying information which she had committed to memory. But there was insufficient evidence to charge her and she was allowed to continue on her journey. It was said later that Hall and Thomson had freed her because in Britain there was no death penalty for female spies at the time, but had passed on details to the French who arrested her on arrival. In fact this was not the case; Zelle continued spying for the Germans for another year before she was arrested, found guilty and condemned to death in France. She was executed by firing squad in October 1917.

Hall's powers of persuasion became equally legendary. Early in the war, he had become aware that the Home Office was not censoring the overseas post effectively and suspicions reached him that enemy agents were simply sending their reports back via contacts in neutral countries through the regular post. When Hall sent in his own Admiralty team to the post office at Mount Pleasant in London where overseas mail was handled, they soon found detailed information as to how Germany was buying goods from neutral countries to beat the blockade imposed by Britain. This invaluable evidence helped the navy to tighten the blockade on Germany.

However, when Hall's unofficial censorship activities became known he was summoned to see the Home Secretary, Reginald McKenna. Hall arrived to find McKenna looking very stern, standing in his office by the fire, resting against the mantelpiece with his hands on his lapels. He said it had been brought to his attention that without his authority Hall had been 'tampering with the King's Mail', and threatened him with a two-year prison sentence for committing such a heinous crime. Hall later wrote that he felt like a naughty schoolboy being called in to see the headmaster for misbehaving. When, however, he asked permission to speak and explained to the Home Secretary what he had been doing, why he had been doing it and what he had found as a consequence, the mood changed completely. McKenna became fascinated by Hall's work, and the two men sat on the settee and talked for some time. By the end of the meeting, the Home Secretary was suggesting the establishment of an entirely new organisation, the War Trade Intelligence Department, to monitor German trade in goods prohibited under the blockade. By the time Hall left the room, McKenna had become a great fan and told Hall that the Admiralty should continue censoring the post until the new organisation could be formed.[8] The War Trade Intelligence Department was to do vital work in ensuring the effectiveness of the blockade. And Hall had talked his way out of one of the biggest crises of his intelligence career.

Hall was always keen to feed false information to the Germans, to deceive them into wasting precious time or resources on something completely useless, as he had done with the misinformation about an Allied invasion of Schleswig-Holstein. In the summer of 1915 he came up with the idea of planting on German intelligence a false code book; by sending messages using the code he could then feed misinformation into the German system. Hall and his colleagues spent some time planning how to get the book into German hands without giving the impression that it had been planted, and the scheme they came up with was worthy of any pre-war spy novel. There was a hotel in Rotterdam, in neutral Holland, where English businessmen used to stay. It was known that German agents operated there and that a certain 'blond lady' would always check in whenever any senior-looking British official arrived. It was also known that the hall porter was in the pay of the Germans. Hall persuaded Sir Guy Locock, who had been in the Diplomatic Service for twelve years, to agree to take the fake code book out to Rotterdam in a secret dispatch case with the apparent intention of passing it on to the British consul. Well versed in the ways of diplomats abroad and able, in Hall's phrase, to 'carry a despatch case in just the right way', Locock was bound to attract the attention of the German agents at the hotel.

Locock checked in and, after unpacking in his room, went out for a walk. He found a spot among some barrels in the nearby harbour where he could hide but still see the window of his room. Having found this observation spot he returned to the hotel where he noticed a blond woman checking in, alone, to a room further along his corridor. As it was a weekend, Locock could not deliver his important cargo for a couple of days and had to wait in the hotel. After dinner, the hall porter got into conversation with him and suggested it must be very dull for the visitor being by himself in a strange city. Locock agreed. In a confidential manner, the porter told him of a very fine club in Rotterdam where any gentleman from the hotel would receive a warm welcome. Locock put on a show of great interest and within minutes had rushed out of the hotel in full view of the porter. However, instead of going to the club, Locock went to his observation point in the harbour, where in a few minutes he saw the light in his room go on and a shadow cross the blind. He imagined the blond woman looking through his things and finding that, in his apparent haste to depart to the club, he had not locked the despatch case that he had hidden below his clothes. Before long the light in his room went out. But of course no agent worth their salt would simply steal the code book. That would be a giveaway; when the diplomat returned and found the book missing, its use would be instantly cancelled and no advantage would be gained. So Locock had to wait in hiding while the blond woman and any accomplice photographed the entire book and then returned it to his room. Shortly after 1 a.m. he saw the light in his room go on again briefly. Locock waited another half hour and then returned to the hotel in a very merry state, ensuring he spoke with the hall porter – or at least tried to speak to him, as by this point the Englishman was slurring his words so badly that he appeared to be barely capable of speech. The hall porter would think he had spent a delightful few hours at the club.[9]

Hall knew that if the Germans were to accept the code book as genuine, he would have to send some fairly meaningless coded messages to which a British ship somewhere would have to respond. However, after all the cloak-and-dagger play acting it seems that no substantial pieces of misinformation were ever sent or, if they were, they remain secret. But there is a fitting postscript to the story. A year after the blond woman had photographed the original code book, Hall judged it was about the right time for an appendix to the code to be issued, and found another agent to take the new document out to Rotterdam. Instead of repeating the same ruse, this time the agent approached the Germans in the Dutch city and, with a story of disaffection and hatred of Britain, offered to sell them the appendix. The Germans paid £500 for the appendix, a very substantial sum. It is not recorded what happened to the money.

From the moment he took charge of naval intelligence, Hall had started to collect and sift information from every area where British forces were or might be engaged. By the summer of 1915 he had an efficient network in place in the eastern Mediterranean, a region that was to become of enormous interest to Britain. German and Turkish radio signals regarding troop movements were intercepted and sent back to Room 40 for decipherment. Looking further afield, Hall found evidence that the Germans were trying to foment trouble in India by encouraging a revolt against British rule. It seemed that a group of revolutionary Indians living in America had approached the German embassy in Washington and asked for sums of money and supplies; indeed the Germans eventually paid out a small fortune. By following the German diplomatic cables, Hall knew precisely what was going on and who was involved. Nothing ever came of German attempts to overthrow British rule in India. But because Germany's undersea cables had been destroyed at the beginning of the war, and German diplomatic messages between Berlin and Washington were all sent by radio, Hall had the means to keep an eye on these communications that had the potential to stir up major trouble for Britain.

By the autumn of 1916, after the terrible loss of life suffered at both Verdun and on the Somme, and after the Battle of Jutland had shown they could not defeat the Royal Navy at sea, the German War Staff started to consider the possibility of unrestricted submarine warfare – that is, instructing its U-boats to sink on sight merchant ships of all nationalities. By October well over one hundred U-boats were ready to operate in the Atlantic. The Germans calculated that if they could sink about 600,000 tons of shipping per month, the Allies would be able to survive for only six months before Britain was brought to its knees and forced to seek a truce.

The sinking of American ships would risk causing outrage in the United States, of course, and the Germans had to consider the possibility that America would declare war on them as a result, but in Berlin it was decided that this was a risk worth taking. Stories of the USA's total unpreparedness for war convinced the Germans that American troops would make no difference to the struggle on the Western Front in the six months before the Allies threw in the towel. Moreover, President Woodrow Wilson was so committed to a policy of peace that the Germans thought there was a strong possibility he would not declare war even if American ships were being sunk. So, on 9 January 1917, the Kaiser gave his permission to launch unrestricted submarine warfare from 1 February.

A week after the decision was taken, on the morning of Wednesday 17 January, at about 10.30 a.m., Hall was at work in his office on the usual round of papers when one of his leading code breakers asked to see him urgently. Nigel de Grey and the Rev. William Montgomery had partially deciphered a message that they could see was of immense importance. De Grey entered Hall's office and asked his boss, 'Do you want to bring America into the war?' 'Yes my boy,' replied Hall. 'Why?'

De Grey explained that they had just deciphered a 'rather astonishing message' from the German Foreign Office to their ambassador in Washington.[10] In the first half of the message it was clear that Berlin was telling its ambassador that it proposed to begin unrestricted submarine warfare on 1 February. The second half was less complete but seemed to say that if this new strategy was likely to bring America into the war, then the Washington ambassador should contact the German ambassador in Mexico. There were additional words or phrases that did not make sense, like 'Japan', 'inform the President', 'war with the USA' and 'our submarines'.[11] The message, or telegram, came from none other than the German Foreign Secretary, Arthur Zimmermann.

Zimmermann's telegram had in fact been sent via three separate routes. First, there was the radio transmission, sent from Nauen on long wave using a new code referred to in its header as 15042. This was a variation of the code the cryptanalysts in Room 40 already possessed, known as 13042, and the difference might explain why de Grey and Montgomery had not been able to decipher the full message. Second, Berlin had started to send its diplomatic messages to the Americas via a neutral country, Sweden. The Swedish Foreign Ministry would forward a German message in the diplomatic bag to their ambassador in Buenos Aires, who would then act as a sort of consular postman and send it on to the relevant German embassy, in this case the one in Washington. Third, and most remarkably, Berlin had also sent the message in code, tagged on to another cable to the US State Department as part of an agreement the Germans had made to send confidential messages to their ambassador in Washington. To discuss war with America on a cable sent through the US State Department was a scandalous abuse of a diplomatic privilege given to Germany and a remarkable affront to the USA, but the German Foreign Office was confident that its codes were so secure that no one would ever decipher such messages.

Over the next couple of weeks Room 40, who had been monitoring these diplomatic communication routes for some time, picked up all three versions of the telegram and deciphered the full message. It was clear that in addition to warning his man in Washington of the decision to resort to unrestricted submarine warfare, Zimmermann was cooking up a new and more devious plan. He hoped the Mexicans could help to bring Japan into the war on the German side. Furthermore, he was asking his Mexican ambassador to discuss with President Carranza of Mexico the possibility of a German alliance. Mexican-American relationships were not good and, most audaciously of all, Zimmermann was offering German support for a Mexican invasion of Texas, New Mexico and Arizona to recover lost territory should America enter the war. News of German intentions to aid an invasion of their own country would doubtless spark such fury in the United States that it might in itself provoke America into war.

Hall was hugely excited to receive the deciphered telegram and instantly recognised that it was political dynamite. But it presented him with a major problem. How could he use the information without giving away the fact that Room 40 was listening in to and reading German diplomatic codes? If Berlin thought there was any possibility that British Intelligence was deciphering their radio messages they would immediately change all their code books and it might take years before Room 40 could catch up again. Alternatively, using the version that had gone through the neutral countries, either Sweden or the United States, would reveal that Britain had been reading the diplomatic messages of friendly neutrals. This too would be seen as totally unacceptable behaviour. Although there was nothing Hall wanted more than for America to join the war on the Allied side, he was not willing to compromise Room 40 for that end. He faced the eternal problem confronting intelligence chiefs, how to make use of the most secret information without revealing the source from which it had arrived.[12]

Hall decided the best thing to do for the moment was nothing at all. He instructed de Grey to destroy all versions of the decrypt except one, which was kept under lock and key in his office, and then he simply waited. Possibly the Germans' use of open war at sea might force the Americans into the conflict and it would be unnecessary to do anything further. However, he knew that President Wilson had just won an election on the basis of keeping America out of the war. Although some within the United States, mostly on the eastern seaboard, were sympathetic to Britain and France, many other important groups – like the Irish and German communities – were hostile. And in the west of America, most people saw the European war as remote and irrelevant. It was not at all clear how the Americans would be likely to respond.

It seems remarkable now that Hall shared this extraordinary piece of intelligence neither with his masters at the Admiralty nor with the Foreign Office. Matters relating to the relations between foreign powers were obviously their principal concern. Something on this scale would probably have been referred up even to the Prime Minister. Hall had great respect for Sir Arthur Balfour, who had left the Admiralty to become Foreign Secretary. But he had sufficient self-confidence, or what might better be described as arrogance, to feel he should try to control this story himself.

Hall therefore sat back and followed decrypts of the exchanges between the Berlin Foreign Office and their ambassadors that Montgomery and de Grey passed on to him almost daily. He could read that Bernstorff in Washington was hostile to the concept of unrestrained submarine warfare and tried to persuade the diplomats in Berlin to reconsider their policy. They refused, a fact which Hall was no doubt relieved to read. The German ambassador in Mexico, Heinrich von Eckhardt, reported back that the Mexican president was friendly to Germany and in return for support would consider allowing Germany to use Mexican naval bases to refuel and service its U-boats. This threat promised an even greater escalation of the war.

When the date for the declaration of open and unrestricted submarine warfare came on 1 February, the Americans broke off diplomatic relations with Germany and proclaimed a policy of 'armed neutrality'. But nothing further happened. As a man of peace, President Wilson was clearly not about to declare war. So, on 5 February, Hall decided to share his secret with the Foreign Office. He paid a visit to Lord Hardinge, the Permanent Secretary, and laid out what he knew, asking to enlist Balfour's support. Hall wrote that Hardinge was 'his usual cool self, interested but cautious'.[13] The remarkable intelligence discovered by Room 40 would now begin to work its way through the slow-moving bureaucracy of the Foreign Office.

While Hall was waiting for a decision from the Foreign Office, another deciphered exchange between Mexico and Berlin landed on his desk. The Americans had sent a military posse into Mexico to pursue a bandit who had been raiding across the American border. It seemed that President Carranza was deeply upset by American military action within his country and he now told Berlin that he was definitely interested in the possibility of German support for an invasion of the southern states of America and was favourable to Mexico's naval bases being used for German submarines. Hall decided he had to inform the Americans.

Calling in an intelligence officer from the US embassy in London, Ed Bell, Hall showed him the full text of the so-called Zimmermann Telegram. Bell exploded with rage when he read about covert German support for an invasion of his country, but after a moment's reflection he said it was so outrageous that it must be a hoax. Hall persuaded him that the telegram was authentic. 'This means war,' Bell concluded. 'Only *if* it is published,' replied Hall, who now explained to the American that he could do nothing without Foreign Office approval.

Rapidly, the American ambassador in London, Dr Walter Hines Page, became involved. A great Anglophile who for some time had tried to persuade the President to join the Allied war effort, Page must have been champing at the bit to get news of the secret cable back to Washington. But Hall's persuasive powers were so great that he convinced Page to hold off until the Foreign Secretary had decided how to proceed.

Events were beginning to move rapidly. Balfour resolved that the best way to proceed was for Hall to inform the US embassy of the existence of the Zimmerman Telegram himself. He knew and respected Hall from his days at the Admiralty and minuted, 'I think Captain Hall may be left to clinch this problem. He knows the ropes better than anyone.'[14] Of course, Hall was one step ahead of Balfour and had already informed the ambassador. Between them, Hall and the two Americans, Bell and Page, then concocted a masterly plan for the release of the information. They realised that if it was announced that the Americans had themselves discovered the intelligence then no suspicion would fall on Room 40. And of course Berlin had sent the Zimmermann Telegram to Washington as an extension of another message being sent to their ambassador. If the Americans found and deciphered that cable in their own State Department inbox, then they could take credit for the intelligence coup. The only problem was that the American secret service did not have the code books necessary to decipher the message.

Page met Balfour on 23 February. The Foreign Secretary formally gave the American ambassador a coded copy of the Zimmermann Telegram and Page returned to the American embassy with it. Nigel de Grey then went to the embassy and, with Bell, deciphered the telegram. As the American embassy was legally a part of the United States of America, the diplomats could now correctly claim that the telegram had been deciphered on American soil. Page then sent to the President and the Secretary of State a message that included a translation of the telegram, confirming exactly where in the message trail at the Washington cable office they could find the original.

There was a short delay in Washington while Wilson and his Secretary of State, Robert Lansing, decided what to do with this astonishing piece of intelligence. They asked for a copy of the code books to be sent to Washington so their own code breakers could go back over the diplomatic cables at the State Department. At Hall's insistence, the government refused this request. The Foreign Office explained to the Americans that the Germans never used only a single code, but jumbled several together in one message, and that only a few experts were able to decipher the completed message. Instead, the British government offered to decipher anything the Americans wished to read.

Within a few days, the Zimmermann Telegram was passed on to the American press. As anticipated, the American public were outraged. Many Americans believed the telegram to be a forgery, even perhaps that Britain had put it out in order to draw the United States into the war. President Wilson insisted it was authentic and had been deciphered by Americans. Then, on the following day, Zimmermann himself admitted at a press conference in Berlin that the telegram was genuine. The peace-loving President and the American people had to accept that Germany had been planning to assist a neighbouring government to invade their country. Within a month, the President recalled Congress and on 6 April 1917, the United States declared war on Germany.

In Germany, there was an investigation into how the telegram had been intercepted. The American press was full of wild stories. One told of how backwoodsmen from Arizona had smuggled themselves into a German communications centre in Brussels and stolen the crucial code book. Others told of Americans paying huge sums to buy the code books from renegade German officials. Yet more described how American secret servicemen had broken into a trunk owned by the Swedish ambassador in Washington and found the telegram there. While such accounts were probably dismissed in Berlin, the German Foreign Ministry interrogated its ambassador in Mexico, thinking his embassy might be the weak link in their communication chain. When this was ruled out, they concluded there must have been a leak in Washington. Meanwhile, Hall followed decrypts of the exchanges, noting to his delight that Berlin carried on using the same ciphers.

Not only did Room 40 continue deciphering messages from Berlin uninterrupted, but Hall himself came in for high-level praise from Washington. The President sent a letter to Page in London asking him 'at an early time ... to assure Admiral Hall of my very great appreciation of what he has done and of the spirit in which he has done it'.[15] In a cable to a British official, Hall proudly boasted with unusual directness and honesty, 'Alone I did it.'[16]

It was an outcome that must have astonished even the wily Hall. America had declared war on Germany. The Germans thought the leak had occurred in Washington and continued to use the same codes on their diplomatic cables. There had been no suspicion about the existence of Room 40 or of its role in the revelations.

Furthermore, the Americans had even secretly thanked Hall for everything he had done. The integrity of British intelligence had survived and the undercover work continued. Hall had done a brilliant job. At least one historian has described the story of the Zimmerman Telegram as 'the greatest intelligence coup in history'.[17]

Looking back on the affair, American ambassador Walter Page wrote to President Wilson at the end of war in praise of 'Blinker' Hall. 'Neither in fiction nor in fact can you find any such man to match him ... The man is a genius – a clear case of genius ... I do study these men here most diligently who have this vast and appalling War Job. There are uncommon creatures among them – men about whom our great-grand children will read in school histories; but of them all, the most extraordinary is this naval officer – of whom they'll probably never hear.'[18]

Page was wrong in one respect. In the late 1920s and 1930s stories began to come out about Room 40 and Reginald 'Blinker' Hall came to be known as Britain's 'great spy chief'.[19] Certainly, he and the extraordinary group he assembled around him had done a great deal to help the Allies win the war.

The United States of America was now in the war, but the Yanks were very slow in coming. It was just over a year before an American Expeditionary Force of a mere five divisions, led by General Pershing, engaged with German troops on the Western Front. Meanwhile, there began a catastrophic increase in the U-boats' sinking of Allied ships. In April 1917, the month America declared war on Germany, one ship out of four leaving British ports never returned. Nearly one million tons of shipping, two-thirds of it British, were sunk in that single month alone. It was clear that for the war to go on the Allies would have to continue what they had been doing for two years, to try every military means they could to win it.

Part Three

Engineers and Chemists

7

The Gunners' War

In the summer of 1914, William Lawrence Bragg had just been elected as one of the youngest Fellows of Trinity College, Cambridge. Bragg's family, originally from Cumberland (today Cumbria), had moved to Australia in the late nineteenth century, and he was the eldest son of a renowned mathematician and physicist, a professor at Adelaide University. Bragg grew up in Australia but his family returned to Britain when he was nineteen, after his father's appointment as Professor of Physics at Leeds University. Lawrence went up to Cambridge to read mathematics and entered an academic world that was alive with new ideas and the spirit of intellectual discovery. He became a member of a discussion group in which historians, classicists, engineers and scientists argued late into the night about the latest ideas of the time.

After his graduation, Bragg began collaborating with his father on a project to investigate the nature of X-rays. The two men were part of a circle, heavily influenced by scientists like Ernest Rutherford in Manchester and Albert Einstein in Berlin, who were doing pioneering work on atoms and electrons, and by the German physicist Wilhelm Röntgen, who had discovered X-rays. Bragg's research soon brought him fame as the joint formulator with his father of a new law that made it possible to calculate the position of atoms within a crystal by observing how an X-ray is diffracted by the crystal's surface. Their work gave birth to X-ray spectroscopy and the ability to study the detailed atomic nature not only of crystals but of every sort of chemical structure, and Bragg remembered it as 'a glorious time when we worked far into every night with new worlds unfolding before us in the silent laboratory'.[1] Bragg senior and junior jointly published a book entitled *X-rays and Crystal Structure*. For this book and the formulation of 'Bragg's Law', the father-and-son partnership were jointly to be awarded a Nobel Prize. It is the only time such a team has been awarded the prize, and Lawrence Bragg, at the age of just twenty-five, was and still is the youngest person ever to become a Nobel laureate.

At the beginning of August 1914, Trinity College, Cambridge was a lively place. Bertrand Russell, the great mathematician-philosopher and one of the most respected Fellows of the college, had only a few years before co-authored *Principia Mathematica*, laying down the foundations of modern mathematics. Russell was amazed at the sudden turn to war and wrote a letter to the *Manchester Guardian* arguing that Britain should remain neutral. Had Bragg looked out of the window of his rooms into Trinity's Great Court he would have seen Russell talking with the economist John Maynard Keynes, who was visiting his brother-in-law, another eminent scientist, Archibald Vivian Hill.[2] One of the most famous scientists in Britain, Lord Rayleigh, the Nobel Prize winner and ex-president of the Royal Society, was chancellor of Cambridge University and another Fellow of Trinity College. And although still young, Bragg already had working for him a research student, Edward Victor Appleton, whose subsequent work would lay the foundations for the development of radar and win yet another Nobel Prize. Bragg was part of a glittering scientific and intellectual community with a brilliant future ahead of him.

Bragg was planning a visit to Germany to give a paper on his new law when war was declared in August. The Edwardian world of science in which new ideas had so easily crossed borders and when scholars had travelled extensively around Europe came to an abrupt end and Bragg's trip was abandoned. Bragg had previously joined a university cavalry troop and had spent many summers at camp, training in marksmanship and horse riding, so within two weeks of the declaration of war, in a flurry of patriotism, he volunteered to join the Royal Horse Artillery. He was posted to a Leicestershire Territorial battery and spent a ghastly year in a very different world, training near Diss in Norfolk alongside the hunting and shooting set with whom he had little affinity. He wrote, 'I was very much out of my element as my knowledge of horses was not at all extensive, and my fellow officers and men were Leicestershire hunting enthusiasts.'[3]

However, things changed dramatically for Bragg in the summer of 1915. The French army had been making experimental attempts to locate the position of enemy artillery from the sound of the guns firing. They were not making much progress, but the War Office thought the British Army should also pursue the idea and looked around for someone suitable to carry out the research. They approached Bragg, who leapt at the opportunity of doing original research again. In September 1915, assisted by Harold Robinson, a member of Rutherford's team at Manchester University, Bragg was sent out to France to carry out experiments. General Headquarters of the British Army (GHQ) was based at St Omer, about thirty miles south-east of Calais.[4] Its role was to support the commander-in-chief, and it included all the military intelligence and mapping sections, as well as controlling the press and censorship. Bragg and Robinson now became part of the Maps Section of GHQ.

Once again, Bragg was in his element, performing cutting-edge scientific work. He soon established the principles of what became known as 'sound ranging'. A series of microphones would be set up in a long line over maybe six miles. When an enemy artillery piece fired, he could record the precise time at which the sound of the gun reached each microphone. From the difference in time it took for the sound to reach each microphone, it was possible using the velocity of sound to plot a series of arcs on a scale drawing. By linking these arcs a circle could be drawn, and the location of the enemy's artillery piece would lie at the centre of the circle. Although Bragg quickly worked out the mathematical formulae for these calculations, it proved far more challenging to find microphones sensitive enough and a recording system accurate enough to record the time intervals precisely. Standard microphones picked up too many other sounds, like rifle fire or dogs barking, which made it difficult to identify the low boom of distant gunfire. It was while Bragg was setting up the first sound ranging experiments behind the front line south of Ypres that news of his Nobel Prize came through. The village curé with whom he was billeted produced a bottle of Lachryma Christi to celebrate.

During 1916, Bragg and his team continued to experiment with different forms of low-frequency microphones, finally opting for a device with a heated platinum wire that cooled when hit by the sound wave of the gun's detonation. This acted as a reliable apparatus for filtering out all other sounds and recording the distant explosion of the enemy gun. By the end of the year, Bragg had also found a way to record the time interval between the arrival of the sound waves at several different low-frequency microphones with enough sensitivity to operate an efficient system. But the recordings were useless when the wind was blowing in the direction of the German lines, as it frequently did. At GHQ they began to lose patience with the young man whose ideas so few of the senior officers could even begin to understand. However, Colonel Edward Jack, the officer in charge of the Maps Section of GHQ, kept faith with Lieutenant Bragg and fought for him to get the resources he needed to continue his research. When, by the spring of 1917, Bragg had finally worked out how to account for weather conditions and wind direction, his sound ranging system could locate enemy guns to an accuracy of about fifty metres, good enough for the purpose of returning artillery fire.

Hundreds of sound rangers were eventually built before the end of the war, and sound ranging became the principal method for locating the enemy's artillery after aerial observation. At Passchendaele in the summer of 1917 the sound rangers located 190 German guns in the first three weeks of the battle; at Cambrai they again correctly identified the sites of all the major pieces of German artillery, and at Vimy Ridge they successfully located a heavy howitzer that was hidden in a wood eleven miles behind the front. By 1918 Bragg had moved on to teaching the principles of sound ranging technology to other young artillery officers many of whom were physicists and applied mathematicians, including Charles Darwin (grandson of the famous evolutionist), Edward Andrade and J.M. Nuttall. Bragg organised regular meetings for these officers to exchange ideas and reports on the use of the technology.

Bragg rose to the rank of major, and was awarded the MC for bravery and an OBE for his scientific work. He would go on to be one of the most distinguished scientists in Britain and was closely involved with the epoch-making discovery of DNA in the 1950s. Bragg, Robinson, Darwin and Nuttall were among the first British scientists in uniform, carrying out important scientific work on the battlefield, setting a precedent for much similar and vital work that would be done in the Second World War.[5] And Bragg's work in the artillery from 1915 to 1918 was a prime example of how academics came forward to bring their expertise to bear in the laboratory of war.

The Great War was dominated by the big guns of the artillery. However, when the British Army had gone to war in 1914 it was seriously under-gunned, both by comparison to its French allies and most especially to the German army. Every German army division at the start of the war possessed sixteen 150mm heavy howitzers; a British division possessed only four 60-pound guns by comparison.[6] The Allies were stunned by the unexpected use of German heavy howitzers on the battlefield. The French did not like such heavy guns but put their faith instead in the 75mm quick-firing mobile field gun. The '75' was light and highly manoeuvrable and was often described in 1914 as 'the best field gun in the world'. It was known to the French themselves in sacred terms as 'God the Father, God the Son and God the Holy Ghost'. Each French division had between forty and sixty of these guns.

In addition to the artillery, meanwhile, each German infantry division fought with twenty-eight machine guns, whereas the British and French had only twenty-four per division.[7] And in Britain in 1914, Vickers Maxim were only producing forty machine guns per month, by contrast to the vast output from the German armaments giant, Krupps, based in Essen on the Ruhr. Its 1914 workforce of nearly 80,000 produced 280 guns a month.[8]

With the digging of trenches and the building of fortified lines defended by the use of artillery came stalemate on the Western Front, and several inventive minds tried to come up with ideas on how to break the impasse. Many people contacted both the War Office and the Admiralty with ideas for weapons that they were convinced would win the war for the Allies. There were endless cranks who came up with crazy schemes. But there were also serious inventors who applied their minds to solving the problems of overcoming the defensive nature of trench warfare. H.G. Wells, who had predicted the horrors of a modern technical war, wrote to *The Times* in June 1915 complaining that 'available resources were not being used to the fullest extent' and that the British were falling behind Germany in technical achievements.[9] His outburst prompted members of the public to send in an avalanche of ideas.

The Admiralty responded in a typically British way by forming a committee. Known as the Board of Invention and Research, its task was to assess the flood of proposals and to suggest new areas of investigation. Many of the great and the good in the world of science – including Sir J.J. Thomson, Professor of Physics at Cambridge and discoverer of the electron, and Sir Charles Parsons, inventor of the steam turbine and director of his own engineering company – were appointed to the board's various sub-committees. Other luminaries also advised the Board, including William Bragg of Leeds

University (father of young William Lawrence Bragg) and Sir Oliver Lodge, along with Sir Ernest Rutherford of Manchester University. Rutherford, originally from New Zealand, was another Nobel Prize winner who had just carried out his pioneering work on atomic physics and had discovered the atomic nucleus. But he put this research aside in order to work with Bragg senior on the use of underwater sound waves to locate enemy submarines. However, the most senior naval men at the Admiralty were constantly suspicious of the scientists, who despite their eminence were regarded as outsiders unfamiliar with the requirements of the navy. Thomson in his frustration later complained that top naval staff saw the board as an 'excrescence rather than a vital part of Admiralty organisation'.[10] The scientists were denied any contact with the young naval officers who were actually fighting the war at sea and, despite Thomson's requests, the Admiralty never created a central laboratory or proving ground where experiments could be carried out. As a consequence, although the Admiralty was distinguished by its intelligence operation, the Royal Navy fell well behind in more practical developments such as gun control mechanisms and the advanced use of on-board radio.

The War Office, too, was struggling to keep up with developments in the technical war. Colonel Louis Jackson, with the unlikely title of Assistant Director of Fortifications and Works, was in charge of new weapons for the army, while new forms of shells and explosives were the responsibility of the Ordnance Board and their experimental teams based at the Woolwich Arsenal. But in June 1915, as the Admiralty created its new board, so Lord Kitchener, the Secretary for War, set up a Trench Warfare Department that was to be responsible for both the research and supply of new weapons. The man in charge was Alexander Roger, an unusual choice as he was an accountant and the manager of an investment trust. It was clear that Kitchener wanted the new organisation to be run on financially sound principles and to be careful as to what new ideas it should invest in. Roger struggled at first to come to grips with the technical issues he faced but proved to be a good organiser and before long acquired enough authority to be successful in the post. The Trench Warfare Department also set up a Commercial and Scientific Advisory Committee, inviting many members of the Royal Society's War Committee to join.

At the end of May 1915, after *The Times'* revelations about Neuve Chapelle and the subsequent collapse of Asquith's Liberal government, a major change took place in the supply organisation for both the army and the navy. At the centre of the new coalition was a Ministry of Munitions that set out to reorganise the supply of war material on a more efficient footing. Former Chancellor of the Exchequer David Lloyd George, a flamboyant and popular Welshman, was the first minister appointed to sort out the shell shortage. He would set about transforming the supply of materials to the armed services. He also created yet another group in his new ministry, a Munitions Inventions Department, putting in charge Ernest Moir, a civil engineer who had constructed the defences at Dover harbour. Consisting of a panel of twenty engineers and scientists, the department started work in August 1915, sifting through ideas that had been submitted, calling for new areas of research and, in the wake of the first Zeppelin bombing raids, starting to investigate various forms of anti-aircraft weapons. Many of the members of this third group, including Professor J.J. Thomson, Horace Darwin, Richard Glazebrook and Frederick Lanchester, were also members of the Royal Society or were already serving on the other committees, while new faces included Sir Alexander Kennedy, the Professor of Mechanical Engineering at Imperial College, London. The new Ministry of Munitions group were allocated proving grounds at Wembley along with a small staff to carry out experiments.

The problem, however, was that with at least three different groups of scientists working in advisory roles for the Admiralty, the War Office and the Ministry of Munitions by the end of 1915, there was the constant risk of duplication of effort – although this could be reduced to a degree by the fact that the same men sat on several of the committees and mixed socially. And, though some of these groups were listened to and had the ability to carry out effective research, others were not and did not. The military classes were proving a tough nut for the scientific community, no matter how eminent and willing, to crack.

Frank Heath, Secretary to the Board of Education, came up with another approach to the problem. Before the war Heath had done much to reform the organisation of London University and establish what eventually became the University Grants Committee, the body with responsibility for the public funding of the entire university sector in Britain. Aware, as were the members of the Royal Society, of Britain's reliance on German industry for the production of certain manufactured items crucial for the war effort, including drugs, antiseptics, optical lenses and even tungsten for the steel industry, in early 1915 he proposed the establishment of a small group to advise the government on the co-ordination of scientific and industrial research. The following year this became the Department of Scientific and Industrial Research. Prominent amongst its members was once again Sir Ernest Rutherford. The small group of scientists who made up its membership were charged with identifying areas where scientific research was needed and parcelling the work out to university laboratories that were known to possess the necessary skills. The Department handed out grants and made postgraduate research awards to encourage work in areas deemed necessary for the war effort. In 1917, the Department established a Fuel Research Board and the following year a Food Investigation Board to examine options for healthy eating within a limited wartime diet.

Meanwhile, in addition to his administrative duties as the first secretary of this new department, Heath also had a hand in improving the safety of every British soldier serving in the army. In 1914, British troops had gone to war wearing nothing more protective than their field service caps. German soldiers, on the other hand, were soon issued with steel helmets. Heath and officials at the

National Physical Laboratory in Teddington realised that a metal helmet was needed to protect soldiers from the high-velocity fragments of shrapnel and earth thrown up by artillery bombardments. Copying the style of helmet adopted by samurai warriors in the Middle Ages, Heath quickly came up with a shape for a British tin helmet that resembled an upside-down soup bowl, pressed out of a single piece of steel with a rim to protect the ears and shoulders and a leather strap under the chin to secure it in place. Although the design changed and improved during the war, it would become standard issue to British soldiers for the next fifty years. By the summer of 1916, one million steel helmets had been produced and by the end of the war seven and a half million, including one and a half million for use by American troops.

The next deficiency to be addressed was the lack of a suitable grenade. Armies had used grenades since the seventeenth century, when the French had called their elite troops 'grenadiers', but as the range of muskets improved, so the distance between combatants had increased and the use of grenade throwers declined. However, the advent of trench warfare revived the need for a small bomb that could be thrown at the enemy's defences in skirmishes or raids, or as a prelude to an attempt to advance into his trench. The German army had anticipated the need for this and the infantry were well equipped with grenades, the most common being the stick grenade, known by British troops as the 'potato masher' as it looked vaguely like a kitchen implement. The thrower held on to a wooden handle that was attached to a cylinder containing the explosives. Just before he threw the grenade, the soldier pulled a cord, setting off a fuse that detonated the explosives within a few seconds. It was a simple but efficient device. The British Army had nothing comparable.

In the first stage of trench warfare, from late 1914, British soldiers improvised a variety of devices to throw at the enemy. Most of them simply consisted of a high explosive with a fuse attached packed into an empty jam jar or tin can, often with a few bits of old metal or sharp stones included. The fuse burnt at a standard speed of about two feet per minute, so any length of a few inches could be attached depending upon the timing the user wanted to set. The user would light the fuse with a cigarette or match and throw the jar. Known as 'jam pots' or 'hair brushes' according to their shape, the bombs were decidedly risky to use when carried out and thrown from craters in the middle of no man's land. Something better was clearly needed.

A Belgian company had come up with a form of grenade with its own lever that, when released, began the process of detonating the explosive. It was safer and more reliable to use. The Belgian inventor was taken prisoner early in the war before his grenade had been fully developed, but in January 1915 an engineering friend of his met with English inventor William Mills. The son of a shipbuilder from Durham, Mills had gone to sea as a marine engineer repairing undersea telegraph cables. Having become fascinated by metallurgy, in the 1880s he had opened Britain's first aluminium foundry in Sunderland. Keen to find new opportunities for utilising the metal, he put it to a variety of uses from golf club heads to castings for motor car and aircraft engines. But he had no experience of working with explosives when the Belgian engineer brought him the plans for the grenade.

With his engineering background, Mills quickly redesigned the grenade, submitting it to the Inventions Branch of the Royal Artillery. They saw potential, and Mills worked at the Royal Laboratory in Woolwich for several months improving and refining his design. The first version proved inadequate; the lever had a tendency to spring off the grenade, which then exploded prematurely, often injuring the thrower. After further work, however, he arrived at its final form – a small pineapple-shaped device made of cast iron, easy to grip and weighing about one and a half pounds. When it was charged with a detonator the user held on to the lever, which was secured by a safety pin. The thrower would hold the grenade in one hand and remove the pin with his other hand. In throwing the grenade, the lever fell off igniting the detonator, which went off after four or five seconds scattering its charge of iron fragments in all directions. Simple, practical and safe to use, the Mills grenade was to prove an effective and popular weapon in the close environment of the trenches.

Siegfried Sassoon provided a vivid illustration of the grenade's use, describing an incident in the early days of the Battle of the Somme in which, infuriated by the death of a friend from a sniper, he devised an ingenious new way of throwing two grenades at the same time. Sassoon had a reputation among his men for daredevil deeds, and was known as 'Mad Jack'. On this occasion, carrying a bag of Mills bombs across his shoulder, he charged single-handed into a German trench at Mametz Wood, from where the sniper fire had come. With a grenade in each hand, he extracted the safety pin from the one in his right hand and then pulled out the pin from that in his left with his teeth, before throwing the two grenades into the German lines and screaming out a hunting cry. When he reached the enemy position he saw the last German soldiers retreating along the trench. He is supposed to have single-handedly frightened off a troop of between fifty and sixty, who must have thought a substantial force was charging at them.[11] Sadly, but probably sensibly, after wandering around alone in the enemy trench for a while, he decided it was wiser to run back to the British line; once there he leapt into the trench and collapsed, laughing hysterically in nervous exhaustion at what he had done.[12] In this brave action Sassoon had briefly captured the enemy front line, alone but for his Mills grenades.

The Mills bomb became the staple grenade of the British Army for the rest of the war. In trench fighting, as Sassoon had discovered, it was often more useful than a rifle for infantrymen in flushing out German troops from trenches and dugouts. Mills set up his own factory in Birmingham to mass produce the grenades, of which seventy-five million had been made by the end of 1918. After the war, the Royal Commission on Awards for Inventors awarded Mills the sum of £27,750, equivalent to a little less than £3 million today. The Mills grenade remained in use in the British Army right up to the early part of the Second World War.

Another battlefield innovation, born out of the changing nature of the Great War, enjoyed a rockier path to acceptance. The Germans had observed how, during the Russo-Japanese War of 1904–5, a Japanese attack on the fortress of Port Arthur with conventional artillery had made little impact on the heavy concrete and steel fortifications. It was only when the Japanese brought up huge 11in siege howitzers capable of throwing giant 500-pound shells that they succeeded in capturing the port. Now, aware they needed to swiftly capture the forts in their advance through Belgium at the start of their war in the west, the German General Staff decided in 1914 to employ high-angled, short-barrelled heavy mortars in their assault upon the Belgian fortifications of Liège and Namur. Guns with high-angled barrels sent their shells high in a tight arc and, coming down at a near-vertical angle, they had extra destructive power in penetrating armoured fortifications.

So Skoda in Austria produced a 12in portable mortar, while Krupp produced a massive 16.5in mortar known as 'Big Bertha'. The latter fired a mighty 1800-pound shell, and was bigger than any British naval gun in existence at the time. The only drawback of this heavy, monster mortar was that it was very difficult to move and had to be transported in sections by rail. Both guns however contributed to crushing the vast Belgian forts, which instead of holding up the Germans for several weeks capitulated within a few days, barely delaying the German advance.

With the advent of trench warfare, however, there was a need for smaller mortars that could fire just a few hundred yards on to the enemy lines. Both the German and Austrian armies were equipped with light mortars for this purpose, but neither the French army nor the British had anything suitable. During a visit to the front, the editor of *The Morning Post* learned of the lack of such a weapon and discussed it back in Britain with a friend, Wilfred Stokes, the manager of engineering company Ransome & Rapier. Stokes immediately got to work and designed a simple device consisting of a 3in tube mounted on a base plate and supported by an adjustable bi-pod. A 20-pound shell could be dropped down the barrel and detonated through a cap that ignited the charge at the bottom of the barrel. Stokes estimated it could fire up to thirty shells a minute over a distance of about 350 yards – certainly enough to persuade the enemy to keep their heads down if an assault was planned.

Initial trials were disappointing and the army rejected the mortar. So Stokes went back to the drawing board and redesigned the shell to be more aerodynamic, and therefore more accurate in hitting its target. By April 1915 the mortar and its new shell had proved a success in trials – but the army still rejected the gun, claiming bizarrely that there was no demand for it. When a young captain who had been wounded at Gallipoli heard of the weapon, however, he was much impressed by it and managed to force Lloyd George and Churchill to watch a trial. They too were enthusiastic and Lloyd George agreed to recommend an order for one thousand mortars from the Ministry. The army were still hesitant but, after improvements had been made in the fuse that ignited the propellant used to fire the shell, the Ordnance Board finally gave grudging approval in September 1915.

One of the advantages of the weapon was that it was easy to produce and could be manufactured by companies that were not already fully committed to armaments production. By this time Stokes had already approached several companies who were willing to manufacture the mortar, but the Trench Warfare Department now took over production of new weapons and insisted on arranging its own orders. Consequently there was a further delay. When the Stokes mortar finally reached the front line in March 1916 it was only used at first to lay down smoke screens. However, during the course of the year it became increasingly common in the British lines and acquired a reputation as one of the best weapons of the war. It added considerably to the firepower that could be thrown at enemy trenches.

There were without doubt serious problems with the early versions of the Stokes mortar. But the slowness with which the army bureaucracy had accepted a weapon for which the soldiers fighting at the front had already expressed an urgent need, showed a marked reluctance on behalf of the military world to accept an idea that had come from outside.

One area in which technology, rather surprisingly, failed to improve the battlefield techniques of the Great War was army communications. Wireless radio had developed rapidly in the decade before the war but radios were too large and heavy to be portable. The Royal Navy used them extensively to keep ships at sea in touch with the Admiralty in London and to intercept the enemy's radio signals. However, the army did not use radio to anything like the same extent. There was no demand on the front line for a lightweight, portable radio transmitter and receiver (this would have to wait until the development of the 'walkie-talkie' in the next war). As a consequence, commanders behind the lines were limited to traditional methods of receiving information about what was happening at the front.

Rockets, flares and flags could be used to send basic signals but their effectiveness was dependent on the weather or relied upon line of sight. Sometimes pigeons could carry a message to headquarters if wind and weather were right, while runners were given written notes and sent back from front trenches to battalion headquarters. But runners took time to get back through devastated trench positions and often the situation had changed by the time they arrived. Otherwise, front-line positions had to rely on telephone communication, and telephone lines were extremely vulnerable to shellfire and frequently were destroyed. On one single day during the Battle of Verdun, French engineers had to repair ninety miles of telephone cable destroyed during a German barrage. Sometimes it was the observers in the air who could provide the most reliable information as to what was happening on the ground, but here too the forwarding of detailed information to the field commander who needed it most could be a slow process.

As a consequence, the generals in command rarely had much idea of what was unfolding after the launch of an offensive. From the moment the troops went over the top and advanced into the smoke and shellfire of no man's land they lost contact with their battalion headquarters, who rarely knew either their own troops' location or that of the enemy. As Hew Strachan has observed, 'As soon as the attackers left their own trenches they lost direct and real-time communication with their own command chain.'[13] This affected almost every single offensive action on the Western Front. Generals were notoriously slow to send up reinforcements after the first waves of troops had broken through the enemy line. Likewise, they were equally slow to stop the futile repetition of a failed assault with the dreadful loss of life this often brought. Perhaps, in a war of new technologies, it was a result of their age and their lack of expectation of new technology. Maybe it was an acceptance that the 'fog of war' would always descend once an attack began and nothing could be done to clear it. And the army generals were a conservative bunch. By contrast with officials at the Admiralty, who tried to micromanage the deployment of ships at sea hundreds of miles away, generals in both the Allied and enemy armies sat behind the lines viewing the strategic progress of a battle by following the action on maps, but were reluctant to get involved with tactical decision-making, which was delegated down the system.

For whatever reason, tens of thousands of lives were lost because of the failures of the command-and-control system, while the generals themselves acquired a reputation for aloofness, remoteness and shocking incompetence that has endured for a hundred years. Alan Clark in the early 1960s penned the phrase 'lions led by donkeys' to sum up the contrast between the obstinacy and weakness of the generals and the heroism of the men.[14] This view has been much challenged by military historians in recent re-evaluations, but it is accurate to the extent that the 'donkeys' often gave up any attempt to control an assault from the moment it started. Communications technology failed the military command structure badly and once again the army was reluctant to look outside to deliver solutions.

The First World War is often described as an artillery war. By the end of it there was an artillery gun for every thirty yards of the Western Front from the Channel to the Swiss border, while a single gun might fire as many as 200 rounds in a twenty-four-hour period in a heavy engagement. During the Battle of the Somme each field artillery brigade required 24,000 rounds every week. It is estimated that as many as an incredible 1400 million shells were fired during the First World War by all nations and that shellfire caused approximately two out of three of all battle casualties.[15] Entire battlefields were transformed into lifeless, eerie moonscapes. The range of explosives needed for all this destruction was therefore immense.

All the shells and bullets fired during the First World War required a combination of explosives to function. Most of these explosives had been developed in the hundred years prior to 1914, some even earlier; the war saw no significant new inventions or advances in the chemistry of explosives. But it did call for the production of explosive chemicals on a previously inconceivable scale. Moreover, each round fired during the war usually relied upon a set of explosions. First there was the propellant, often inside a cartridge, used to propel a bullet or shell down the barrel of the weapon. This would itself need a detonator to set it off. When a bullet hit its target it would usually cause damage simply by passing through it, or in the case of a solid object by penetrating it, but when an artillery shell found its target a set of further explosions took place. A detonator would explode, setting off the main explosive that would blow the forged steel of the shell casing into fragments, spreading its charge in the form of energy over the surrounding area. Michael Freemantle has noted that a single artillery round could involve up to eight separate explosions: a percussion cap explosive would set off the primer that would act as an ignition charge to explode the propellant, which would send the shell down the barrel and off towards its target; when the shell reached its target a detonator would set off a booster, which would ignite the fuse that set off the main explosive charge in the shell.[16] Not all shells were so complex, but a wide variety were fired off in the war for many different purposes, including high explosive shells, shrapnel shells, incendiary shells, armour piercing shells, smoke shells, and many others. All of them relied upon different types of explosive.

Gunpowder had been around for at least a thousand years and by the twentieth century was known as a 'low explosive'. The combustion of gunpowder, like most other explosives, led to the release of energy and the production of a variety of gases. Gunpowder in the First World War was used both as a propellant and, early in the war, in mining operations to blow up the enemy's trenches. Its problem as a propellant was that it produced a lot of smoke, which would give away the position of the gunners. So the armies and navies of all sides preferred a smokeless propellant called cordite, which had been developed in the 1880s by British scientists.

Most explosives used during the war released far more energy than gunpowder and were therefore known as 'high explosives'. The core element of many high explosives, nitroglycerine, was a highly unstable compound that detonated when jolted or heated. It was the Swedish chemist Alfred Nobel who discovered how nitroglycerine could be turned into a useful explosive, patenting the result as dynamite in 1867. Dynamite was safe and easy to handle and could be set off with the use of a basic detonator. Eight years later he invented another compound of nitroglycerine, calling it gelignite, and in 1887 he developed a smokeless propellant, Ballisite. Nobel made a huge amount of money from his inventions and used his fortune to establish the Nobel Foundation in Stockholm that from 1901 has worked with the Swedish Academy of Sciences to award Nobel Prizes.

Another high explosive, trinitrotoluene – known as TNT – was invented in Germany in the 1860s. It became a staple of many artillery shells during the war, often in combination with a variety of other chemicals. TNT was made as a liquid; stable and relatively safe to handle, it could be poured into shells, although it needed a powerful detonator to explode. Ammonia or ammonium nitrate was yet another type of high explosive that formed the basis for many shells and was often added to TNT to create a new compound called amatol. As the supply of TNT could rarely keep up with the demand, amatol became increasingly common as the core content of high explosive shells. Ammonal was another combination of ammonium nitrate, this time with aluminium, and was about three times more powerful than gunpowder. It became the explosive most commonly used in mines under enemy lines, notably, before the Battle of the Somme when the Allies dug a series of huge mines under the German lines, filling each one with up to twenty-four tons of ammonal. The explosions of these mines just before 7.30 a.m. on 1 July 1916 marked the opening of the battle. The huge explosions could be heard in southern England and the crater from one of them, 300 feet in diameter and 90 feet deep, is still there today.[17]

However, with the rapid development of its chemical industry during the late nineteenth and early twentieth centuries, Germany was able to increase massively its production of explosives. By 1906 there were about 500 chemists working in industry in Britain by comparison to 4500 in Germany.[18] German chemical giants like BASF, Bayer and Hoechst were forging ahead producing entirely new products. Germany's technological superiority was symbolised by the Haber-Bosch process; introduced in 1913, this was enormously advanced for its day, working at pressures equivalent to about 200 atmospheres and operating at temperatures of 600 degrees Centigrade. The process produced ammonia from nitrogen and hydrogen, providing from the beginning of the war an alternative way for German industry to manufacture explosives. From ammonia and its product nitric acid it was possible to manufacture a large range of high explosives, such as TNT. In 1913, Germany produced 8700 tons of ammonia through the Haber-Bosch process. By 1916, this had risen to almost 100,000 tons.[19] As the British naval blockade on Germany tightened, denying the country access to a range of raw materials, the process of fixing nitric acid became the single most important source for the production of high explosives.

Britain and France, like many other countries, had imported most of their synthetic chemicals from Germany before the war. From 1914, Britain had to make its own dyes and chemicals, a task that presented a major challenge. Soon after the outbreak of war, a committee of industrial chemists was formed under the chairmanship of Professor John Haldane, the brother of the reforming Secretary of War, and in November, the War Office appointed Lord Moulton, a leading Fellow of the Royal Society, to be responsible for the supply of explosives. Moulton worked closely with the Research Department at the Woolwich Arsenal, where the small staff of eleven at the start of the war grew to more than 100 chemists and physicists; they included a group of female chemists taken on as 'probationary assistant analysts'.[20]

The lack of chemists available to work in British industry became an issue of real concern during the war years. So many young scientists had enthusiastically joined up when war was declared to do their bit that by the end of 1915, the Royal Society suggested it might be necessary 'to consider the methods of exempting persons from active service who are essential to the work of research in the laboratories of the country'.[21] A month later the eminent men of science proposed drawing up a 'census' of every scientist in the country. They sent out a letter to universities and major research institutions asking for the name of everyone who had passed through 'an Honours course in one of the scientific subjects (Maths, Physics, Chemistry, Engineering, Botany, Zoology, Geology, Physiology) or obtained a diploma in one of the technical subjects'. This would be the first national listing in British history of all the country's key scientists. During the early months of 1916, when plans for national conscription of men between the ages of eighteen and forty-one were being drawn up, the Royal Society lobbied the government hard, arguing that chemists – like coal miners and munitions workers – should be regarded as an exempted profession and not drafted into the armed services. Some retired colonels who did not understand that Britain was fighting an industrial war believed that the war would be won at the front and not in the factories, and so thought anyone not in uniform was a shirker who should be sent off to fight in France. But the gentlemen of the Royal Society persisted. Moreover, realising that, unlike William Lawrence Bragg, many scientists were already in the army and navy wasting their time doing work that did not draw upon their scientific skills, they suggested sending to the War Office a list of all scientists in the army in order 'to facilitate the more effective employment of such men'.[22] The War Office's response is not recorded.

In developing a native chemical industry, chemists in Britain had to look for new solutions to the challenge of manufacturing explosives in sufficient quantities to meet the demands of war. One of the solvents required in large quantities to manufacture cordite was acetone. Before the war, only small amounts of acetone had been needed, and it was manufactured by heating wood to produce what was known as wood vinegar. But it took about 100 tons of wood to produce a single ton of acetone and, like so many other essentials, supplies of wood were simply no longer available on the scale that was now needed. A chemist at Manchester University by the name of Chaim Weizmann came up with an alternative process of producing acetone by making use of bacteria he had discovered during the fermentation of maize and other starches.

A Russian Jew who had settled in Britain in 1904, Weizmann wrote to the War Office soon after the outbreak of war offering details of his new process, but he never received a reply. In 1915 he had more luck with the Admiralty when Winston Churchill, then still the First Lord, showed an interest. The Royal Navy relied on the use of cordite to fire the heavy guns on its ships; being smokeless, it did not give away their position to the enemy. Churchill set Weizmann a challenge to produce one ton of acetone through his new process. He achieved this at a trial in a gin factory in East London. Weizmann went on to take a sabbatical from Manchester in order to work for the Admiralty at a laboratory in London. He spent a year on the experimental production of a variety of new solvents and his invention was used at a factory in Poole that produced acetone by the fermentation of rice.

Weizmann was one of the many men of science who came forward, in his adopted country, to make his contribution to the war effort. But he had another interest; he was a leading Zionist who lobbied hard behind the scenes for political support for the establishment of a Jewish homeland in Palestine. In 1917 the Foreign Secretary, Arthur Balfour, issued the Declaration in his name that pledged Britain's support for a Jewish homeland, and in the following year Weizmann was appointed as the leading figure on a government commission to Palestine. Weizmann later became the leader of the World Zionist Organisation and the first president of the state of Israel in 1948.

In addition to the gradual mobilisation of the scientific community, vast new munitions factories were built under the auspice of the Ministry of Munitions. The largest at Gretna in southern Scotland on the border with England, employed more than 16,000 workers and at its peak produced one thousand tons of cordite each week. Chemists were trained in industrial practices and sent out to run the new factories. Martin Lowry, a young chemist from Guy's Hospital in London, took charge of a new Gun Ammunition Filling Department at the Ministry of Munitions; his task was to find new ways of producing TNT, as there was little experience of its manufacture in Britain. A government-financed company called British Dyes was set up to fill the gap left by the unavailability of German synthetic products. By the end of the war, thirty factories were producing nearly one thousand tons of TNT each week. Other factories produced the amatol which, combined with TNT, became the principal agent in millions of British shells.

The expansion of these new munitions factories was rapid and required a small army of workers to operate them. With so many men needed at the front, many of the recruits were young women who flocked in to take up the new jobs. Photographs show vast factory floors with rows of women pouring chemicals into shell casings. By the end of the war 947,000 women were working in munitions factories across Britain, becoming popularly known as 'Munitionettes'. Many of them came from textile mills where women had traditionally been employed in factory work, but, to quote one report at the time, 'they came also from Scottish fishing villages, from Irish bogs, and the workrooms and villas of English provincial towns.'[23] There was a crisis for the middle classes with the sudden lack of servants as so many women left domestic service to take up better-paid work in the factories.

The manufacture of munitions could bring a plethora of health risks. Women working with TNT often suffered from nausea, vomiting, headaches, tightness of the throat, rashes and blisters. Often their skin would turn yellow with poisoning from the chemicals they were handling; such women were referred to, one hopes sympathetically, as 'canaries'. Lilian Miles witnessed her black hair turn green and later remembered, 'you'd wash and wash and it didn't make no difference ... Your whole body was yellow.'[24]

Although the government laid down strict regulations for operating the new munitions factories, accidents inevitably occurred. On the afternoon of 2 April 1916, about 15 tons of TNT and 150 tons of ammonal blew up at the Explosive Loading Company factory at Faversham in Kent, killing over 100 people. The blast, which became known as 'the Great Explosion', could be heard as far away as Norwich. In January 1917 about fifty tons of TNT exploded at a factory in Silvertown on the Thames to the east of London. Seventy-three people were killed, hundreds injured and most of the factory was destroyed.[25] However, despite the risks, women overall very much liked working in the factories; it brought them a level of pay, friendship and new opportunities that they would never otherwise have enjoyed. Looking at photographs of these 'women at war' recently, the American feminist Sandra Gilbert has observed that, 'liberated from parlours and petticoats alike, trousered "war girls" beam as they shovel coal, shoe horses, light fires, drive buses, chop down trees, make shells and dig graves.'[26] The daughters of these

Great War pioneers would find a similar opportunity for freedom when classed as 'mobile women' a generation later in the next war.

By the end of the war, the Ministry of Munitions was a huge organisation. There was a staff of 65,000 running two hundred factories and employing more than three million workers. The factories not only manufactured and filled shells but also produced rifles, guns, aircraft and hundreds of other components needed for the war effort. When it came to machine guns, for instance, Haig had said at the beginning of the war that two machine guns per battalion 'were more than sufficient'. Kitchener, on the other hand, recommended four. Lloyd George as Minister of Munitions said, 'Take Kitchener's figure. Square it. Multiply by two. Then double again for good luck.'[27] The army had begun the war with 1330 machine guns. During the war, under the management of the Ministry of Munitions, 240,506 were produced. In August 1917, during the battle of the Menin Road Bridge, the British Army was able to employ 240 machine guns over a 4000-yard front – equivalent to one machine gun every seventeen yards.

One of the winners as the demand for explosives grew between 1914 and 1918 was the United States. As Britain and France were denied access to German output they not only began to increase their own production but also turned to new suppliers. Exports of explosives from the US in 1914 were valued at $6 million; by 1918 they had grown more than sixty-fold to a remarkable $379 million. In the 1920s and 1930s, America would become a world leader in the chemical industries and Germany would never recover its pre-war ascendancy.

The chemical industry had done much to increase the impact and effectiveness of explosives and, as shortages of raw materials became apparent, to provide new processes for the manufacture of explosives. But it was for something far more ghastly and terrible that the industry would become notorious during the Great War.

8

The Yellow-Green Cloud

On the late afternoon of 22 April 1915, a beautiful clear spring day, Captain Louis Strange was flying his RFC reconnaissance aircraft over the front line just to the north of Ypres, near Langemarck. Looking down, he spotted to his surprise a yellow-green mist floating from the German front line over the French trenches. Although he did not know it, he was witnessing a ghastly escalation in the science of war – the first use of poison gas. The soldiers in the front line facing the gas cloud were a French Territorial and an Algerian division, who struggled to breathe as the hideous cloud enveloped them and filled their lungs. They began to panic and then fled. Those that died were mostly left lying on their backs with fists clenched, their faces and lips slowly turning blue. A vast gap appeared in the French line, but the aerial observers reported that the Germans were slow to take advantage and British and Canadian troops soon rushed in to fill the breach. In the chaos of the asphyxiating chlorine gas attack, it was only aerial observers who were able to keep the high command informed of what was happening on the ground.

Although this incident in the spring of 1915 was the first significant use of poison gas in a modern war, the development of chemical warfare went back a long way. For hundreds of years, assailants had catapulted diseased animals into a town or castle during a siege in order to spread infection among the inhabitants. Greek fire, an early form of incendiary, had been thrown at enemy troops since ancient times. But during the late nineteenth century the growth of the modern chemical industry exponentially increased the scale of the killing that was possible by the use of poisonous gases. In 1899, in an attempt to lay down the international rules of war, the First Hague Convention had agreed that all parties should 'abstain from the use of projectiles, the object of which is the diffusion of asphyxiating or deleterious gases'.[1] Great Britain and the United States of America did not sign the Convention, but Britain did sign the declaration at the second Hague Convention in 1907. Nevertheless, before the war, Sir William Ramsay, a prominent British chemist and Nobel Prize winner, carried out experiments with a form of tear gas and similar work took place in France, where the police expressed an interest in using the gas. As this gas was felt to fall outside the terms of the Hague Convention, a series of experiments to find a more powerful type of tear gas took place at Imperial College in London's South Kensington in early 1915. The new gas was given the name SK, after the location. But no use of it was made on the battlefield at the time.

With the unexpected stalemate that followed the beginnings of trench warfare in late 1914, however, the German high command began to look for alternatives to conventional artillery shells for dislodging troops from well-entrenched positions.[2] Colonel Max Bauer, the man responsible for the supply of munitions to the German army, turned to Professor Fritz Haber, the distinguished industrial chemist who had invented the Haber-Bosch process before the war and who later received a Nobel Prize in Chemistry for his pioneering work in synthesising ammonia. Director at the time of the Kaiser Wilhelm Institute for Physical Chemistry, Haber willingly offered his services to the German army and agreed to carry out some tests. His initial attempt, shells filled with a combination of high explosives and tear gas, was first tried on the Eastern Front in February 1915, but the gas was diffused by the wind and the cold temperatures prevented it from vaporising, dramatically lessening its effect. The Russians did not even notice or record the use of the gas.

Haber then suggested using cylinders as a far more practical way of releasing a concentrated form of gas. And he proposed the use of chlorine. Bayer was already producing chlorine industrially and it could be made in liquid form to fill the cylinders that would then be transported to the front line and discharged. As it was released the liquid chlorine would turn into a vapour and would blow across no man's land towards the French and British trenches. Being much heavier than air, the chlorine gas would cling to the ground and sink into trenches, accumulating at the bottom. But in time it would disperse in the air and allow attacking troops to move forward without fear of being gassed.

The German General Staff were initially troubled by the morality of carrying out a gas attack like this, but were assured that it was technically not in breach of the Hague Convention as the poisonous gas would be released not from 'projectiles' or shells but from cylinders. There was also talk of a French attack using tear gas shells in March 1915 providing a justification, although if such an assault ever took place only very small quantities were used.

The German chief of staff, General Erich von Falkenhayn, was less bothered by the morality of the issue than by the possibility that the Allies would quickly retaliate and use poison gas against the German lines. However, Haber assured him that because of the superiority of the German chemical industries, this would be impossible. With this reassurance, Falkenhayn decided to use chlorine gas released from cylinders in the offensive being planned at the Ypres salient in the spring. The attack would form the opening of what became known as the Second Battle of Ypres.

Bauer formed a special Pioneer Regiment and started to train men in the use of gas. Several prominent chemists were recruited into this regiment, the first example of German 'scientists in uniform' (one of them, Otto Hahn, twenty-five years later, played a significant role in the attempt by the Nazis to produce an atom bomb). At a trial run on 2 April chlorine was released and proved so effective that both Haber and Bauer, supervising the experiment, were mildly gassed. Over the next two weeks, 1600 large cylinders weighing 187 lb each and standing more than four feet high, along with more than 4000 smaller cylinders, were filled with liquid chlorine and carried up to the front line near Ypres. But for the cylinders to be effective it was necessary for the wind to carry the vaporising poison in the direction of the Anglo-French front line. The Pioneer Regiment waited for the right conditions, but realised from the prevailing winds that they had selected the wrong place and so moved the thousands of cylinders to a new site further north near Langemarck. Having waited several more days for the wind to blow in the right direction, at 5 p.m. on 22 April they launched the first ever cloud of poison gas, ushering in a new era in the horrors of warfare.

At first, the French and Algerian troops in the front line thought that the yellow-green cloud approaching them was some form of smoke screen covering the advance of German troops. However, as the foul, bleach-smelling gas reached the trenches the men began to complain of pains in the chest, nausea and a horrible burning sensation in their throats. Chlorine kills by entering a person's lungs and causing such irritation that the lungs flood with fluid and the person literally drowns. Victims' lips and faces go blue as they are starved of oxygen. It was a terrible death, coming slowly and unstoppably to its first victims who had no idea what was happening to them. There were 4000 casualties, of which more than 1200 were fatal. The French line completely broke as thousands of men fled in panic, opening up a gap four miles wide. Even the gunners behind the lines abandoned their weapons and ran. But the Germans advanced slowly, with only minimal protection against the gas, and as they came upon the dead or those screaming with the agony of asphyxiation they were not encouraged to penetrate much further. Besides, Falkenhayn had minimal expectations for the attack and had provided no reserves. As a consequence the Germans failed to exploit the opportunity before them. British and Canadian troops on either flank fired at the advancing enemy and then retook the abandoned trenches. Within a few days the German line was almost back where it had started.

The Pioneer Regiment made two further small-scale gas attacks over the next two weeks. The first was against the Canadian troops who had been on the right flank of the French at Langemarck; the second, rather bigger, was against the British trenches positioned at what was known as Hill 60 on 1 May. Although British chemists had successfully identified the gas the Germans had been using as chlorine, there had not been enough time to prepare and issue proper gas masks, so the men at the front were told to soak field dressings in bicarbonate of soda and hold them over their mouths. This would neutralise the effect of the poison. A whole variety of improvised techniques followed in which dipping any available cloth, towel or handkerchief in an alkaline solution was found to be effective. Where nothing else was available the men were told to dip a cloth in urine and hold it over their mouths. Even this had some effect.

When, however, British soldiers came up against a poison gas attack for the first time in the attack on Hill 60, there were substantial casualties. A group from the 1st Battalion of the Dorset Regiment bore the brunt of the ghastly assault. Out of 600 men, 90 died of gas poisoning in the trenches and 207 were admitted to dressing stations, of whom 46 died quickly and another 12 after horrible suffering. Seventeen of the men poisoned in the first few hours of the attack were taken to a Casualty Clearing Station near Bailleul, but the next day only three survived. Sergeant Major Ernest Shepherd of the Dorsets wrote in his diary that they had been told the purpose of the gas was only to stupefy the defenders, but 'We soon found out at a terrible price that these gases were deadly poison ... The scene that followed was heart-breaking ... Had we lost as heavily while actually fighting we would not have cared as much, but our dear boys died like rats in a trap.'[3]

The need for some sort of immediate protection for front-line troops was urgent. The French found that a chemical designed for developing photos and known as 'hypo' was effective in dampening a mask. A group of chemists came up with a design for a tight-fitting respirator and the French army ordered one million to be produced. But this would take time. So before they could be manufactured and distributed the army called for all the mine rescue breathing masks in the whole of France to be gathered together and gave them out to officers and machine gunners. Several French scientists were drawn in, including Dr André Kling, the director of the Paris municipal laboratory. Over the next few months Kling was allowed excellent access to the French front line; quickly identifying the gas as chlorine, he started to explore more effective forms of protection against it.

In London, Lord Kitchener summoned Professor John Haldane to the War Office and asked for urgent advice. An expert on gas poisoning in mines, within days Haldane had designed a primitive respirator consisting of a pad of three layers of cotton gauze that was to be dipped in a sodium solution, placed over the mouth and tied around the back of the head with tape. The problem was that it was not easy to tape the gauze around the head in a hurry and it only provided about five minutes of protection before the cotton needed to be dipped in the solution again. So before long this was replaced by a rough flannel hood with celluloid window patches for the eyes, soaked in a solution of 'hypo'. A soldier could quickly pull this over his head and it provided greater protection. When Haldane tried it on he thought it was too uncomfortable to use, but a group of younger chemists who were in the army thought it worked well. The 'Hypo Helmet' looked like some scary medieval hood, but was more effective than the cotton respirator and so was adopted as its replacement. By July almost every British soldier in France had been issued with one of these anti-gas hoods. Once more the scientific experts had rallied around in a crisis, and had contributed their knowledge to reducing the effects of this new form of warfare. But there was still a strong sense that the army high command was not paying sufficient attention to what the men of science were telling them.

On 24 May both types of protection were put to the test when the

Germans launched another big gas attack against British lines along the Ypres salient. This time the men did not panic, and although the casualties were severe the line held. The gas soon passed behind the trenches, and determined machine gun and artillery fire from the defenders prevented any major German advance. On the following day the Germans called off their offensive and the Second Battle of Ypres came to a close. British losses in the five-week period of the battle amounted to 59,000 men, although only a small number of these had succumbed to gas.

There was a widespread sense of outrage at the German use of poison gas in the spring battle of Ypres. *The Times* denounced it as an 'atrocious method of warfare' and a 'diabolical contrivance'.[4] It was seen as yet another example of German beastliness and as contravening the rules of war. The soldiers, more practically, felt that when they could not even breathe the air around them they became helpless victims who had to endure slow, ghastly, asphyxiating deaths, or watch their comrades suffer a tortuous end. The possibility of breathing poisonous air arouses a primitive fear, and throughout the war that fear was to undermine the fighting ability of soldiers on both sides. 'The mere mention of gas,' a battalion commander in the Black Watch remembered, 'could put the "wind up" the Battalion.'[5]

However diabolical poison gas might be, there was no doubt in the minds of the Allied general staff that they must prepare to employ it as soon as possible. At GHQ in France, Haig's head of intelligence, General Charteris, wrote quite simply in his diary, 'We shall of course now have to use gas ourselves, as soon as we can get it going.'[6] Colonel Jackson, in charge of the use of new weapons in the British Army, took advice from the War Committee of the Royal Society. And despite their indignation at the German crime, the eminent men of science seem to have been perfectly willing to help. Professors Baker and Thorpe, prominent chemists based at Imperial College where the earlier testing on tear gas had been carried out, were quite happy to recommend different types of gas for use at the front and to advise on the best ways of directing it at the enemy.

There was only one firm in Britain, Castner-Kellner at Wallsend on the Tyne, that was capable of producing chlorine and George Beilby, a member of the Royal Society War Committee who was also a director of the company, put the War Office into immediate contact with them. Colonel Charles Foulkes of the Royal Engineers was charged at GHQ with forming special companies that were to prepare to carry out gas attacks. Chemists or chemical students from universities who were already in the army were recruited into these companies, just as German scientists had joined their Pioneer Regiment.

The British government reviewed the situation and despite outward declarations of outrage and horror at the German use of gas, approved the use of gases 'which were as harmful, but not much more so, than those used by the enemy, though preparations and experiments might proceed for the employment of more deadly things'.[7] Testing began during May, but there were problems with the gas cylinders, which began to leak at the welded joints where the valves were supposed to discharge the chlorine. After trying to resolve this, further tests in June were witnessed by Foulkes and a group of officers. Surprisingly, no scientists attended, the British Army being keen, unlike the French, to keep the men of science at arm's length. Foulkes was impressed, and a large order for liquid chlorine was placed with Castner-Kellner on behalf of both the British and French armies, as France did not have a single company capable of producing the poison in bulk. Over the next few months, thousands of heavy gas cylinders were filled and dispatched to the front, where the special companies of chemists manhandled them up to the front-line trenches.

Meanwhile, in early 1915, the Germans had carried out the first bombing raids along the English coast from giant Zeppelin airships. In April, London was bombed for the first time. There was understandable concern that instead of high explosives, the Germans might start dropping poison gas bombs. Sir Edward Henry, the Commissioner of Police, wrote from Scotland Yard to the Royal

Society asking for advice on precautions. The members of the Physiology Committee, a sub-group of the main War Committee that included Professor John Haldane, wrote back to him on 5 July assuring him that if the Germans dropped gas bombs on the streets of London, 'the enormous ventilating power of the atmosphere will prevent any dangerous concentration of gas'. The men of science advised that if gas bombs penetrated a building then the building should be evacuated, but there would be no need for panic. They also advised against issuing gas masks to the general public as they would not be necessary. Masks, they claimed, 'are necessary for the use of our troops because, if gas reaches the trenches, the men must stay on duty there in spite of it'. However, as long as people were directed away from the scene of an attack, the low concentrations of gas likely to be present would cause no general harm. The price at which gas masks were being offered to the public, said the scientists, was 'indefensibly high' and labour spent on their manufacture could be more usefully spent elsewhere, for example in making masks for the troops.[8] The Metropolitan Police followed this guidance and, again on scientific advice, recommended against the use of chemical liquid fire extinguishers as well because standards could not be guaranteed, preferring the use of water and sand. Fortunately gas bombs were never dropped on civilian populations during the war.

During the summer of 1915, Lord Kitchener instructed the British commander-in-chief in France to plan a new autumn offensive around the coal mining town of Loos to support a major French offensive on their right. General Sir Douglas Haig, the man in command of First Army, was to lead the attack, but he felt that Loos was the wrong place to launch an assault and was not keen. When, however, in late August, Foulkes laid on a demonstration of the use of chlorine gas, Haig was enormously impressed. He put his initial qualms about the use of gas to one side and, having apparently convinced himself that it might be the device to bring a breakthrough on the Western Front, decided to use it during the attack.

In the weeks leading up to the offensive Haig became more and more enthusiastic and wrote to the Chief of Staff that 'decisive results are almost certain to be obtained' by the 'very extensive' use of gas. He wrote to his wife even more enthusiastically predicting that within a month the British Army might be 'a good distance on the road to Brussels'.[9] Of course, the successful use of gas depended on the wind. But Foulkes continued to impress Haig with his confidence, telling him that the men of the special companies would always make the final decision as to whether or not wind conditions were right to use the gas. A mood of excited optimism prevailed at GHQ in the period before the attack.

When dawn came up on the day selected to launch the offensive, 25 September, the wind was blustery and conditions varied along the front. Haig was uncertain whether to launch the gas attack but knew the British had to keep in step with the French offensive. At about 5 a.m. he stood outside his headquarters and his ADC lit a cigarette. The two men watched as the cigarette smoke drifted in little puffs towards the north-east, in the direction of the German lines. This simple test to check the direction of the wind is often said to exemplify Haig's naivety and folly, but in reality it seems to have done little to convince him. Foulkes' guarantee that the men launching the gas would be able to use their initiative seems to have given him more reassurance that it would only be discharged where conditions were favourable.

However, almost everything went wrong that morning. Some of the cylinders still leaked and in places gas poured out into the British trenches. So many different types of gas canister had been called into use that frequently the chemists from the special companies found they had been issued with the wrong-sized spanners to turn the release valves. Moreover, although the specialists were supposed to decide if the wind was right, staff officers repeatedly overruled them and insisted that they launch the gas regardless. In one case the operator was told he would be shot if he did not release his gas.[10] Consequently, while in some places the gas blew straight into the enemy trenches as intended, in other places it hovered in no man's land, then blew back into the British trench as the wind changed direction. The men ended up being poisoned by their own gas. As if all this was not bad enough, troops had been issued with a new form of helmet, known as the P Helmet, supposed to give greater protection and allow them to advance close behind a cloud of poison gas. It involved breathing in through the nose and out through a valve held in the mouth, but it proved difficult to use and men wearing it quickly became exhausted. Moreover, thanks to the rain that fell that morning, chemicals in the fabric leached out on to the wearers' skin. Many men experienced throat problems and thought mistakenly they had been gassed; when others lifted the helmets to get some fresh air, they breathed in the gas that had blown back into their trenches and so were poisoned.

If there was chaos in the British trenches, however, there was panic in the enemy lines. It was the first time the Germans had found themselves on the receiving end of a gas attack. In many places the line broke, and the British made considerable advances during this first day, capturing the town of Loos during the morning. But it was the same old story of trying to advance on the Western Front. Haig did not have sufficient reserves to turn the breakthrough into a rout. When the reserves were finally drawn up and deployed, the Germans had recovered and fought back fiercely, and the balance by then had shifted back in their favour. The French attack on the right also failed. Within three days most of the fighting was over and any chance of a decisive victory had passed. Haig had worn down the enemy but his soldiers had suffered further considerable loss of life. And the Allies had lost any moral superiority they could have claimed against the Germans for their use of chemical weapons. Both sides would now be in a race to develop ever more ghastly and harmful forms of this new kind of killing.

The reality was that from 1916 onwards, poison gas simply became another weapon in the arsenal of war. Its use became a familiar tactic in the process of attrition, intended to maximise casualties and lower the morale of the enemy. Sometimes the Germans released gas simply to cause casualties and made no attempt to advance in its wake. Often gas was used at night when it was more difficult to detect. Sometimes, at dawn, smoke was released first. The troops in the opposing trenches would race to put their masks on. When they realised it was only smoke they would take their masks off, and *then* poison gas would be released. But neither side had a monopoly of success or deviousness in the use of gas. On one occasion in April 1916, the wind changed direction when the Germans were releasing cylinders of gas and the cloud was blown so quickly back into their lines that even the members of the Pioneer Regiment discharging the gas were poisoned. Increasingly there was competition for scientists on both sides to come up with more lethal forms of gas than chlorine and to find better ways of delivering it into the enemy's lines, a sort of chemists' arms race.

Fritz Haber, the man who had pioneered the use of poison gas in the war, was put in charge of development and supply for the chemical section of the German War Department. He built up a staff of some two thousand, many of whom were trained chemists, and advised the General Staff on all matters relating to the policy and supply of gases. The Kaiser Wilhelm Institute became the principal centre for experimentation and Haber's close links with the German chemical industry enabled him to draw upon the expertise of industrial chemists when it came to placing orders for a new poison. Haber appears to have been proud of his wartime work and later became justly known as the 'father of modern chemical warfare'.

In Britain, no single chemist had an equivalent central role in the development of poison gas. A testing ground was established at Porton Down on Salisbury Plain, where the terrain was thought to be closest to that of Flanders, but many scientists grew frustrated that there was no central body to coordinate and advise on the development of chemicals for war. As Professor Starling of the Royal Society complained, there would always be delays as long as 'we have no head of the gas services [for military supplies] in England'.[11] It was clear that Germany would lead the next stage of the use of gases in the war and that the French and the British would largely be playing catch-up.

After a particularly virulent attack on their lines near Verdun in November 1915, French chemists detected the use of phosgene. Elements of it were found on shell fragments and during post-mortems performed on some of the dead. Phosgene, the chemical carbonyl chloride, is over ten times more toxic than chlorine. Far more difficult to detect as it had only a mild smell of fresh hay, what was even worse was that phosgene did not always have an imme-diate effect and victims were often unaware they had inhaled the gas. Sometimes it would take several hours for symptoms to appear. Its use marked a deadly new escalation in chemical warfare.

One month after detecting its first use by the Germans, the French government itself authorised the manufacture of phosgene. This time, instead of releasing it from cylinders, it was incorporated into shells in liquid form that vaporised when the shell landed. The first phosgene shells were fired by French artillery in the Battle of Verdun in February 1916, while the British used phosgene shells during the Battle of the Somme in the summer of that year. The Germans also began to use phosgene in shells rather than releasing it from cylin-ders as it meant they were no longer dependent on wind conditions.

Even more lethal was the compound known as diphosgene, a chemical containing twice as many chlorine atoms in each molecule as phosgene. Both sides used this in artillery shells from 1917 onwards. Diphosgene could also be made in liquid form and simply poured into shells. When the shell exploded it released a heavy vapour that would stay on the ground in the vicinity of the explo-sion for up to thirty minutes without dispersing. German chemical giants Bayer and Hoechst produced more than four hundred tons of the gas every month, and diphosgene became the most common form of shell gas to be used in the war.

It was clear that a more effective form of mask or respirator than the hoods distributed in 1915 was needed for protection from these more lethal gases. Once again the Royal Society was called upon to advise the Royal Army Medical College at Millbank in London on defences against 'poisonous and irritant gases and corrosive fluids'.[12] Bertram Lambert, a chemistry lecturer at Oxford, and Edward Harrison, a pharmaceutical research chemist, came up with a form of respirator that offered increased protection. This entailed wearing a satchel containing in a box a filter of charcoal and lime permanganate granules, connected by a corrugated rubber tube to a tight-fitting mask. Boots the chemists in Nottingham started to manufacture the new respirators for artillerymen and machine gunners, but they were too large and cumbersome to wear in the trenches. So, in 1916, Harrison designed a smaller version which went into mass production and eventually came into use throughout the British Army, where every soldier was personally fitted with his own respirator. Harrison became famous for designing the respirator, which saved many lives. And the familiar shape of the gas mask that everyone would carry in the Second World War had been born.

In 1917 there was a further escalation in the development of chemical warfare with the use of dichlorethyl sulphide, a combination of chlorine and sulphur better known as mustard gas. The Germans first used mustard gas that summer in the Third Battle of Ypres, firing fifty thousand shells of the gas into the British lines on the single night of 12 July. Mustard gas is in fact an oily liquid that forms droplets in the air when the shell carrying it explodes. It was toxic not only when breathed in, but also when the droplets soaked into uniforms or even boots, attacking the skin and causing horrible burning blisters. None of the gas masks available by the end of the war offered any protection against mustard gas; it often blinded the soldiers it affected, burnt the skin, created a choking sensation and caused internal bleeding.

Adding to the horror of an attack was the fact that it could be many hours before symptoms appeared, and the accounts of those suffering from mustard gas are dreadful. After a few hours men would start to feel a terrible pain in their eyes. Often blindfold, they would be evacuated, each man's hands on the shoulders of the next man in a line led slowly by an orderly. Then the victims would start to suffer from the appearance of a series of horrible, yellow, oozing blisters. Their lungs would fill with a frothy and bloody liquid which, after a few days, they might start to choke up. If the contamination was severe enough, the gas would kill very slowly and painfully; it might take four or five weeks to die, and to see a friend die such a death must have been truly shocking. Mustard gas was the most horrific of the First World War gases.

Vera Brittain was a VAD, a voluntary nurse, serving in the Base Hospital at Etaples during the Third Battle of Ypres. At any one time she had up to ten tragic victims of mustard gas to care for, and she described 'the poor things' as 'burnt and blistered all over with great mustard-coloured suppurating blisters, with blind eyes – sometimes temporally [sic], sometimes permanently – all sticky and stuck together, and always fighting for breath, with voices a mere whisper, saying that their throats are closing and they know they will choke.' Brittain wished that those who thought that 'God made the war' could see the victims of 'such inventions of the Devil'.[13]

The Germans fired around 5000 tons of mustard gas in the last sixteen months of the war and the British suffered around 125,000 casualties from it, although the vast majority of these survived. Mustard gas was slow to clear and would remain in the soil and the landscape for days after being used. Fritz Haber considered it be 'a fantastic success' and the 'king of all the battle gases'.[14]

Although by 1917, the Germans much preferred the use of shells to spread gas than cylinders as they were not dependent upon wind conditions and could reach targets several miles behind the front lines, a British Royal Engineers officer named Captain William Livens, an engineering graduate from Cambridge, invented a new method of firing gas, flammable materials and high explosives into the German lines. He devised a heavy mortar with a 4ft-long steel barrel and an 8in diameter, buried in the ground at an angle of 45 degrees. A huge drum of 40 lb of liquid gas and high explosives, fitted with handles, was dropped down the barrel and was fired by electrical ignition with a dramatic flash. The bomb could only be projected a few hundred yards and so the device, simple to build and able to be manufactured in large numbers – was called the Livens Projector. A special brigade was trained to use them and the Livens Projector was first fired in anger at the Battle of Arras in April 1917. Hundreds of Projectors lined up along the front line were capable of simultaneously propelling thousands of tons of toxic gases over the enemy lines. They could not hit precise targets but could saturate an area of the enemy front. In March 1918, 3730 Livens gas bombs were to be fired into the German lines around the town of Lens in a single day.[15]

In October 1917, Winston Churchill, now Minister of Munitions, at last created a Chemical Warfare Committee drawing together the expertise of scientists from the Royal Society, the universities and industry. But it was too late to fundamentally change the direction of the chemical war and create an Allied advantage over the Germans in chemical weapons. There were never enough gas shells available to meet the demands of the British generals and so a favourite remained SK, the tear gas developed in 1914. Although less toxic, it was slow to disperse and forced enemy troops to keep their respirators on for long periods, reducing both their efficiency and their morale. Meanwhile, British chemists disagreed about the best way to manufacture mustard gas and the delay meant that the British Army only started to use it in the last six weeks of the war.

The Germans, in contrast, used vast numbers of gas shells in their offensive on the Western Front in 1918. They coded them according to colour: a green cross indicated lung irritants, a blue cross sensory irritants and a yellow cross mustard gas. Green and blue shells were fired against defending infantry, while yellow shells were fired against artillery as mustard gas remained in the vicinity for longer. At one point, four out of every five shells fired against the British artillery were mustard gas shells, and it was reported that after the German bombardment of Armentières, the gutters of the town were running with the liquid.

Chlorine, phosgene, diphosgene and mustard gas were the principal poison gases used during the First World War. But many more were tested, tried out or occasionally thrown at their enemy by all sides. They included chemical variants of cyanide that got into the bloodstream and created heart failure; a form of arsenic gas developed in the United States and thought to be the most powerful killing agent of the war; and chloropicrin, a chemical first developed by the Russians but later used by the Germans. British soldiers called this the 'vomiting gas', as it induced nausea and vomiting and injured the stomach and intestines.

In fact the various participants used forty-six gases, thirteen smoke agents and nine different chemical incendiaries during the course of the hostilities, in an ever more hideous escalation of the chemists' war. And by the end of the war, it has been estimated, between them the participants had manufactured an extraordinary 176,200 tons of chemical gases, while about 65 million chemical shells had been fired – though remarkably this huge number equates to only about 5 per cent of the artillery shells fired on all fronts during the course of the war. Historians have argued about the effectiveness of the poison gases used during the war and there is a problem with analysing the records. Not only is it difficult to obtain accurate casualty figures from the Russian armies that fought on the Eastern Front, it is also thought that the Germans might have reduced the numbers of their own casualties. In addition, it is impossible to know how many of those wounded during the course of the war died years later of respiratory or other diseases as a consequence of being gassed. As a very rough approximation, however, there were between 1.2 and 1.5 million gas casualties in all the armies of the First World War. Of these, about 93 per cent survived, although roughly 10 per cent of survivors had some permanent disability caused by the gas. So the total number of terrible deaths due to gas in the war was somewhere between 80,000 and 100,000.[16]

Of the hundreds of thousands of casualties caused by the use of these shells, one has a particular resonance. Four weeks before the end of the war, a barrage of mustard gas shells fired by British gunners temporarily blinded a lance corporal who was a runner in a Bavarian infantry regiment serving on the German front line south of Ypres. Many of his associates were killed in the bombardment. By the time the war came to an end he had been evacuated to a convalescent hospital in Pasewalk, Germany, where he recovered his eyesight. The lance corporal's name was Adolf Hitler. In the war that he unleashed, two decades later, as Chancellor of Germany, despite the appalling atrocities that otherwise took place, Hitler never ordered the use of chemical weapons on the battlefield.[17]

9

Breaking the Stalemate

Hiram Percy Maxim, the American inventor, took out patents on a large number of inventions from the 1860s onwards. These ranged from an elaborate mousetrap to the electric light bulb – the latter causing a dispute with Thomas Edison, who invented the incandescent light bulb. Maxim became chief engineer for one of the American new technology companies, the US Electric Lighting Company, and came to Britain to represent that company in 1881. The following year, during a visit to Vienna, he is supposed to have met an American who, dismissing the new technologies of the late nineteenth century, said to him, 'Hang your chemistry and electricity! If you want to make a pile of money, invent something that will enable these Europeans to cut each other's throats with greater facility.'

Heeding this advice, in 1884 Maxim patented a new form of machine gun. The key advance in his weapon was that it used the energy of the gun's recoil to eject the used cartridge and power a spring that loaded the next round. Earlier weapons that were capable of multiple firing, like the Gatling gun that had caused havoc in the American Civil War, required the operator to turn a crank handle to feed the bullets into the firing mechanism. Maxim's new weapon was a genuinely automatic machine gun. Early versions were supposed to be able to fire 600 rounds a minute.

Like all inventors, having come up with his idea, Maxim then had to try to sell it. Initially, the reception must have disappointed him. The British Army, as so often with new devices, were unimpressed, commenting, 'Why use fifty bullets when one will do?' But the German army were more interested and placed an order, as did the Imperial Russian Army. A few years later even the British came round and ordered 120 Maxims which were used in various imperial wars. There was sufficient demand eventually for Maxim to merge his own business with the large armaments producer, Vickers, so that in 1897 the company became Vickers, Son and Maxim. The Vickers machine gun became the standard British Army machine gun for decades to come, while several versions were produced in Germany, France and elsewhere in the first decade of the twentieth century.

Maxim went on to design a steam-engine powered flying machine that never flew and a far more successful amusement ride called a 'Flying Machine', an example of which is still in operation at Blackpool's Pleasure Beach. He became a British subject in 1900 and was knighted the following year. Meanwhile, the impact of his new, easy-to-use, automatic machine gun began slowly to sink in. Theoretically, one relatively unskilled man equipped with a machine gun could halt the advance of hundreds of well-trained and well-drilled infantrymen. But it was not until 1914 that most military commands really began to learn this lesson. Leading military historian Basil Liddell Hart wrote in 1930 that Maxim's name 'is more deeply engraved on the real history of the World War than that of any other man. Emperors, statesmen and generals ... found themselves helpless puppets in the grip of Hiram Maxim who, by his machine gun, had paralysed the power of attack.'[1]

The mobile war unleashed by the German army in August 1914, as its troops marched in vast numbers through Belgium and across France, had been intended to achieve a decisive victory over the French army and the tiny British Expeditionary Force on their left. More than 120 divisions and three million men confronted each other in the battles of Belgium and France that month. After the

German reversal at the Battle of the Marne in September, both sides tried to outflank each other, moving north towards the sea. As they spread out, the armies rapidly dug trenches to hold their positions. Soon the trenches were supported by barbed wire (an American invention to keep cattle inside vast grazing lands), and by machine guns. The Germans desperately tried to turn the Allied line by capturing the town of Ypres in Flanders and throwing back the British and Belgian armies, separating them from the French. But even quickly dug defensive trenches proved too powerful as defensive lines for the mighty German army to overwhelm. On 8 November, General von Falkenhayn, the Chief of the German War Staff, informed the Kaiser that it was no longer possible to maintain the offensive in Flanders because 'the barbed wire cannot be crossed.'[2]

Within a few weeks the British had come to the same conclusion. Lord Kitchener wrote to Sir John French in January 1915 giving his view that 'the German lines in France may be looked on as a fortress that cannot be carried by assault and also cannot be completely invested.'[3] The Edwardian British Army that had gone to war in 1914 had never envisaged a static, defensive battlefield. It had planned for a war of manoeuvre and of dramatic flanking movements. But Kitchener now saw an enemy front line akin to a fortress running for 450 miles from the sea to the Alps. For the first time in history, there was no flank to turn. The Allies could lay siege to the German lines, but they would find it nearly impossible to successfully assault them.

And so, in this context, a few imaginative individuals started to think about designing something that would overcome the stalemate caused by the machine gun, trenches, barbed wire and the firepower of the enemy's artillery. The eventual result was one of the few completely new and original machines to come out of the war. By 1918, it would finally deliver its promise on the Western Front, giving birth to an entirely new form of armoured, mobile warfare. Several people would share its parentage and the process of its gestation would be slow. And many of them would have to struggle against military authorities that were shockingly slow to pick up a new idea.

In order to solve a problem, the first task is to define what exactly that problem consists of. This was where Colonel Ernest Dunlop Swinton came in. Swinton was the son of a judge in the Indian civil service who decided to become a professional soldier. He joined the Royal Engineers and served in the Boer War, where he won the DSO. But he was already known as something of an intellectual, and so he was marked out for an unusual career path in the decidedly unintellectual world of the Edwardian army. His role as author of the official history of the Russo-Japanese War of 1904–5 had a profound influence on him as he studied the use of entrenched positions, the power of machine guns and artillery, the deployment of heavy siege weapons and the terrible impact all of this had on the casualty figures. He also wrote a novel, *The Defence of Duffer's Drift*, in which a Lieutenant 'Backsight Forethought', straight out of military college, has to learn how to defend a river crossing; the book became a classic on military tactics. In 1913 Swinton was appointed assistant secretary to the Committee of Imperial Defence, the organisation at the centre of British thinking about the nation's military role. His boss was Maurice Hankey, the all-powerful secretary of the committee and perhaps the most influential civil servant of the era. Hankey seems to have admired Swinton and the two men got on well together.

A few weeks after the declaration of war, Swinton was asked to go to France as an official war correspondent and write communiqués on behalf of GHQ. These he composed in clear, unemotional terms, and they were released through the official Press Bureau as coming from an anonymous figure named 'Eye-Witness'. The time he spent observing and reporting in France enabled Swinton to reflect on the nature of the fighting that, after the early battles, quickly settled into the trench warfare of the Western Front, a form of warfare he was already familiar with from his study of the Russo-Japanese conflict. Swinton discussed his thoughts with his boss, Hankey, who over Christmas 1914 wrote a short paper on the war and the problems of military stalemate. Hankey pointed out that in previous wars, whenever new problems had arisen a means had been found to overcome them. He argued that new ideas were needed to solve the problem of machine guns, barbed wire and trenches. His own suggestion, which probably came from Swinton, was of a giant bullet-proof machine, pushed from behind, with a roller in front that would crush the wire and advance to the enemy front line, where it would disgorge its soldiers straight into the enemy's trenches.

Swinton and Hankey's idea was not entirely new. H.G. Wells had in 1903 written for the *Strand Magazine* – famous as the journal in which Arthur Conan Doyle's detective Sherlock Holmes first made his appearance – a short story in which, with remarkable prescience, he had predicted the deadlock of trench warfare and had envisioned how to break through it. In his story, called 'The Land Ironclads', Wells described an imaginary war in which a correspondent watches as the enemy suddenly appears using a device

> the size of an ironclad cruiser, crawling obliquely to the first line of the trenches and firing shots out of port holes to its side ... It might have been from eighty to a hundred feet long ... its vertical side was ten feet high or so ... The thing had come into such a position to enfilade the trench, which was empty now, as far as he could see, except for two or three crouching knots of men, and the tumbled dead. Behind it, across the plain, it had scored the grass with a train of linked impressions like the dotted tracings sea-things leave in sand.

This extraordinary mechanical monster advances forward and crosses the trench moving on 'thick stumpy feet, between knobs and buttons in shape – flat broad things reminding one of the feet of elephants or the legs of caterpillars'. Wells called these feet 'pedrails'. Aghast at the spectacle he is witnessing, Wells's correspondent reflects that it is a case of '"Manhood" versus "Machinery"'.[4]

A machine like this was in 1903 purely an invention of its author's fertile mind. However, machinery of this sort, particularly using some of the devices Wells describes, was already in development in the early years of the twentieth century. Robey & Co. of Lincoln, agricultural machine manufacturers, produced a heavy steam tractor fitted with a set of feet on its back wheels that enabled it to cross rough ground; the feet were designed by Bramah Diplock and were called 'pedrails'. Wells picked up on this invention in a military journal. Another engineer, David Roberts, took this a step further and designed a track system to replace the wheels on a heavy haulage agricultural tractor.

The development of tracks would be crucial to the story of the new machine, but in Edwardian Britain it attracted no interest at the time. British farms already had all the mechanical devices that farmers wanted. In America, however, where farms were much larger and huge tracts of virgin land were still available to be turned over to agricultural use, the new form of propulsion attracted far more attention. An American group, the Holt Company of California, bought the patent in 1910, registering the trade name 'Caterpillar' for the new tracks, and several American companies began to develop machines moved by such tracks.

In a separate technological development, armour plating had been used to protect naval vessels at sea for decades. It had been developed by the big armament and metallurgical companies like Vickers and Cammells in Britain and Krupp in Germany. The steel plating they manufactured for warships was up to ten inches thick, far too heavy for any land machine to carry, but they also made armour plating a quarter of an inch thick, sufficient to provide a shield from bullets fired from a distance of one hundred yards. In purely technical terms, therefore, many of the elements needed to produce an armoured, tracked fighting vehicle already existed when war came in 1914.[5] What was needed was the inspiration to draw these elements together.

It was Winston Churchill at the Admiralty who took the project to

the next stage. Churchill wanted to prevent the Germans from occupying the Channel ports in Belgium. Since the army was too
stretched to provide extra men, Churchill committed the Royal
Marines, who came under Admiralty command, to the continent to
reinforce the port of Antwerp. He also instructed the Royal Naval
Air Service to send a squadron to Ostend to attack German targets.
As the RNAS developed differently to the Royal Flying Corps, part
of its role became to defend Britain, and Churchill wanted it to fulfil
this purpose by using it aggressively to bomb Zeppelin bases in
occupied Belgium. The mechanically astute minds of the naval men
now came up with the idea of fitting armour plating to some of their
vehicles in order to send raiding parties against German troops.

Antwerp fell to German forces in October but the development
of armoured naval vehicles continued. A site was acquired at
Wormwood Scrubs in west London where testing and further
development continued. A real specialism developed here, and by
the end of the year armoured cars with revolving turrets equipped
with machine guns were in existence. In the static positional war of
the trenches there was no real opportunity to use mobile armoured
cars; however, the several enthusiasts continued to work on
improved forms of armour plating, vehicle design and means of
propulsion. Among their number was Albert Gerald Stern, a successful and wealthy banker who had taken a commission as a
lieutenant in the Royal Naval Reserve. He donated his own Rolls-
Royce to the venture and was recruited to the workshops at
Wormwood Scrubs.

The War Office now rather belatedly began to get in on the act.
Two Holt tractors with caterpillar tracks, bought in the United
States, were given a trial in February 1915 at Shoeburyness in the
Thames estuary. The tracks proved effective at crossing rough,
waterlogged ground but no one present could imagine how to turn
a tractor into a fighting vehicle. It was agreed that the machine
could prove useful as a gun carrier and a small order was placed,
although most artillery officers thought that horses were far more

flexible and manageable in hauling guns over almost any terrain. As the War Office pondered the future, the initiative still lay with the Admiralty.

From Winston Churchill's point of view, the war at the beginning of 1915 was mired in stalemate. What torpedoes and mines were to his sailors at sea, machine guns and barbed wire were to the soldiers on the Western Front. Churchill was fascinated by Hankey's idea of using a giant roller machine to get men into enemy trenches. Realising that something entirely new to the arsenal of war was needed to penetrate the enemy lines, which grew in sophistication and defensive power as every month passed, he wrote to Prime Minister Asquith on 5 January suggesting that the army should take forward the idea of bullet-proof steam tractors with caterpillar tracks to cross enemy trenches. He made the further point that it would not cost much to manufacture such a machine and if it failed only a small amount of money would have been wasted.

Asquith forwarded Churchill's letter to the War Office and asked them to respond. Churchill was amazed when, after waiting seven weeks, he received a reply from the Master General of the Ordnance rejecting the proposal outright because 'the project is not likely to lead to success'.[6] So he decided to go ahead himself, and on 20 February he formed a Landships Committee at the Admiralty under the chairmanship of naval engineer Eustace Tennyson d'Eyncourt, the Director of Naval Construction, who at first could not see what any of this had to with him. But d'Eyncourt would become one of the key drivers of the project over the coming months, while Albert Stern moved from Wormwood Scrubs to became the energetic and forceful secretary of the new committee.

There were two alternative routes of development open to the Landships Committee. One was a vehicle powered by huge rolling wheels like a giant steamroller. The design the committee explored for this option had wheels 40ft in diameter, two at the front and one at the back. This colossus, known familiarly as the Big Wheel, would be able to fire 4in guns, the sort of weapons usually found on a naval destroyer. At 100ft long and 46ft high, it was closest to the monster imagined by H.G. Wells. The second was a tracked vehicle, smaller and possibly a little faster than the Big Wheel steamroller. But most of the experts were doubtful that caterpillar tracks were the way for-ward; they thought the tracks would break up or collapse under the pressure of crossing barbed wire or rough ground.

While the members of the Landships Committee were consider-ing the two alternatives, they received a third design: a machine that looked like a land submarine with a conning tower in the centre, steered with a ship's wheel and using pedrail feet for propulsion. Like the other two designs, its purpose was to carry up to fifty men in bullet-proof safety across no man's land to storm the enemy trenches. The committee members asked for models to be made and in March Churchill, off his own bat, authorised the use of £70,000 of Admiralty funds (approximately £7 million in 2014 money) to make prototypes. Churchill was taking a chance, committing naval fund-ing for land vehicles that were so speculative. But, stung by the War Office's rejection, and feeling that someone needed to get the wheels rolling, he decided to go ahead. It was fortunate that he did.

A small agricultural machine maker in Lincoln, William Foster & Co., started work on the Big Wheel option based on giant steam trac-tors they had built before the war, while the Metropolitan Carriage, Wagon and Finance Company in Birmingham was eventually asked to develop the tracked option. The land submarine idea became mired in disagreement and soon fell by the wayside. The committee asked another veteran engineer, Colonel Crompton, to pursue the different options, and with immense energy and vigour he pushed and cajoled each manufacturer to come up with the goods. When Crompton visited France in the spring of 1915, however, to see for himself what conditions at the front really looked like, GHQ refused him permission to travel there; he was, after all, working for the Admiralty. The army's obstructiveness to the development of a machine to cross no man's land and break the deadlock on the Western Front was becoming scandalous.

When, at the end of May, Churchill was forced out of the Admiralty, one of the principal supporters of the first stage of the machine's development no longer had a position with real influence. D'Eyncourt, however, kept the project alive. In July the new Ministry of Munitions took over research and development, which was now to be directed by Colonel Louis Jackson and his Trench Warfare Department, although the members of the Landships Committee continued to advise and Stern remained secretary of the new committee that was formed. Swinton once more entered the scene and succeeded in generating some interest from within the War Office. With more army involvement, it was resolved that the principal object of the new machine should be combat rather than the carrying of troops. This required a smaller construction than some of the more fanciful grand designs that had been bandied about. The Big Wheel version was abandoned and the committee decided that the future lay with tracked vehicles.

In August, construction of a prototype at last began at Fosters in Lincoln under the guiding hands of William Ashbee Tritton, the managing director, and engineer Major Walter Wilson, who was appointed overseer. More examples of tracked vehicles were brought over from America, and Tritton and Wilson made an inspired decision that the tracks should pass all the way around the body of the vehicle. In repeated trials the tracks kept breaking off and failed to work; if they could not be made to function, the whole project would fail. But the team working on the design were determined to succeed. Eventually, years of experience in solving mechanical problems enabled Tritton and Wilson to invent tracks that did finally work. They came up with pressed steel tracks riveted to heavy-duty castings that linked the track in a continuous loop.[7] It was at last the crucial technical breakthrough.

Finally, in December 1915, the first version of the device, nicknamed 'Mother', was demonstrated in Lincoln. Swinton, Stern, d'Eyncourt and many of those who had been working on the project for nearly a year were delighted with what they saw. The machine consisted of a large steel girder frame with armour plating riveted to it, thicker at the front than at the sides. The idea of a gunned turret had been abandoned and 6lb naval guns were fixed in sponsons on the side of the vehicle. Two sets of tracks, nearly two feet wide, went around the whole frame. 'Mother' was powered by a 13.5 litre, 105 hp Daimler-Knight petrol engine. The exhaust was directed up through the roof, but the small cabin area was both extremely noisy and dreadfully hot as every member of the crew was only a few feet away from the massive engine. Moreover, the petrol was stored in two tanks inside the vehicle, making it a highly dangerous place to be when under enemy fire. 'Mother' was 31ft long, nearly 14ft wide and 8ft high. The driver steered it via two tail wheels attached to the rear. It needed a crew of eight. There was some discussion as to what to call the machine. As its existence was top secret it had to be given a name that would not attract attention when it was transported around the country. Observers felt it looked a bit like a water carrier but Stern did not want to be known as secretary to the WC Committee, so someone came up with a more generic name, the 'tank'. It stuck.

On 2 February 1916 the new 'tank' was shown off to a group of VIPs including the First Lord of the Admiralty, Arthur Balfour, the Minister of Munitions, Lloyd George, and the Secretary of War, Lord Kitchener. It passed through a variety of obstacles, traversed shell holes and crossed an imagined trench. Bouncing about slowly but heavily, it met the two main conditions of being able to climb a vertical face five feet high and cross a ditch eight feet wide. Everyone seemed impressed and an immediate order for forty of the machines was placed. Only Kitchener was unconvinced. As he left he was heard to say that the war would never be won by such 'pretty mechanical toys'.[8]

Despite Kitchener's scepticism, the next stage of the development of the tank progressed relatively smoothly. A range of companies around Britain started to manufacture tanks, and the nickname slipped into more general use. Before long the order was increased to 150, while Swinton was put in charge of managing supply and organisation. There was now a need to train commanders and crews for the new fighting vehicles. A unit was created called the Heavy Section, Machine Gun Corps and men were drafted into it from across the army. Training took place in conditions of great secrecy on a private landed estate at Thetford in Norfolk. Armed guards were placed around the estate to keep observers out. There was a shortage of naval guns, so only half of the tanks were armed with the 6lb cannons; the other half were fitted with a Vickers machine gun on either side. Starting to think about how the new machines should be used, Swinton wrote a paper calling for the use of massed tank formations on largely dry ground. He came up with detailed instructions for the tank captains on how to operate their vehicles. But everything was beginning to move at such high speed that most commanders barely had time to read the new guide. In mid-August the first tanks were sent to France covered with massive tarpaulins so as not to reveal what they were. The crews that went with them had only spent eight weeks in training for this entirely new form of warfare.

In July 1916, Sir Douglas Haig, promoted to commander-in-chief of the British Army on the Western Front, launched his 'Big Push' on the Somme to help relieve the French at Verdun. After five days of intense artillery bombardment, tens of thousands of men went 'over the top' on the morning of 1 July. Most were from what was called the New Army, formed out of the volunteers who had responded to Kitchener's appeal to do their bit for king and country at the beginning of the war. Army commanders thought the recruits would not be up to much and so ordered them to advance at walking pace in long lines across no man's land. They thus proved easy targets for the German machine gunners, most of whom had survived the artillery bombardment by sheltering in bunkers dug deep in the chalk landscape. The opening of the attack turned into nothing less than a massacre. On the first day, the army suffered 57,000 casualties, including 19,000 deaths. Some battalions in the first wave were almost completely wiped out. Organised as they often were into groups of men from a single locality – the so-called 'Pals' Battalions' – streets and sometimes whole towns across Britain and Ireland went into mourning. It was by far the worst day in the history of the British Army.

Haig wanted to use his new weapon as soon as he could. He grew as excited about the potential of the tank to achieve a breakthrough, as he had been about the use of poison gas at the Battle of Loos: 'I hope and think they will add very greatly to the prospects of success and to the extent of it,' he wrote to the Chief of the Imperial General Staff on 22 August.[9] He put great pressure on Swinton to get the tanks to the Somme as soon as possible, but there were very few available, and even fewer trained crews. Those that assembled in France became the centre of immense attention as literally hundreds of officers took time out to visit them and quiz their crews about how they worked.

The tanks were rushed into action in one of the later phases of the Battle of the Somme. At 6 a.m. on 15 September, they went into combat for the first time near the village of Flers. Only forty-nine tanks were ready for action that day, not enough to make much difference even if they had all performed magnificently. As it was, seventeen broke down or failed to reach their starting positions, while the remaining thirty-two rumbled forward as best they could.

They still had a 'shattering effect' on the front-line German troops. According to one German regimental history, the men 'felt quite powerless against these monsters which crawled along the top of the trench enfilading it with continuous machine gun fire'.[10] However, if there had been little time to train the tank crews, there had been even less to co-ordinate their use with the infantry. Everyone seemed to expect miracles. But they were disappointed. Not only did the crews struggle with the noise and clatter of their own machines, but for the first time they had to navigate across a shell-pitted battlefield and stand up to hostile enemy gunfire.

The tanks headed off twenty minutes ahead of the infantry. Nine did creditably well and dealt with the obstacles they encountered. Nine more made some progress, if so slowly that the infantry soon overtook them. The other fourteen broke down during the advance or became stranded in shell holes or trenches. Many tanks were hit by enemy fire, although most of them survived and managed to limp home. Ten were abandoned on the battlefield, remaining there as rusting hulks for the rest of the war. It was hardly the revolution that the enthusiastic proponents of this new arm of warfare had expected.

Haig and the army high command have frequently been criticised for throwing away the advantage of surprise by using this new military technology for the first time in such a limited way.[11] There were definitely too few vehicles available, tools for maintenance were inadequate, the crews were not well enough trained and no one had worked out effective tactics. Swinton was deeply opposed to their use in September 1916, regarding it as premature. However, the fact was that Haig could equally well have been criticised if he had failed to use this new machine of war when it became available to him, during one of the bloodiest campaigns of the war. And once they arrived in France, it would have been very difficult to keep their existence a secret for long. There would always have to be a first use, and the Battle of the Somme was not a bad opportunity to try for a breakthrough. The real criticism is that the military authorities had failed to perceive the need for a new machine capable of advancing across terrain dominated by the concentrated firepower of the enemy's artillery and machine guns. It was not the tank's initial failure to shift the balance of the battlefield from defender to attacker that was the problem. It was the delay in getting such a machine into use at all.

Haig at least was not downhearted with the performance of the tanks. He placed an order for a further 1000 machines. Others were less impressed, a memo from GHQ in October concluding that 'in the present stage of their development' the tank was to be regarded as no more than 'an accessory to the ordinary method of attack'.[12]

Lessons were however learned from battlefield experience and improvements were made. The next generation of tanks were manufactured with thicker armour and more powerful engines. The pair of tail wheels were abandoned and the tank was steered by varying the speed of the tracks. The Vickers machine guns were replaced with lighter Lewis guns. A new training ground was established on sandy heathland at Bovington in Dorset to train the thousands of men needed to crew and maintain the tanks. In July 1917, a new formation, the Tank Corps, was created and tank crews began to develop their own esprit de corps. However, the weeks that followed marked the lowest point in the use of tanks.

During the Third Battle of Ypres, better known as Passchendaele, which opened on 31 July, immense artillery barrages turned the fields of Flanders into seas of mud. The water table in this region of reclaimed swampland was always high and men and machines faced the horror of drowning in the mud. Many tanks, weighing 28 tons, simply sank in the quagmire. Everywhere movement slowed to a snail's pace; at one point it took a group of tanks nine hours to cover a single mile. Passchendaele proved to be a tragic disaster. Nearly half a million British soldiers were killed, wounded or posted as missing. But the battle strengthened the case of those opposed to the use of the tank. If tanks could not operate in the real conditions of war, it was argued, then they were simply of no value.

The opportunity for the tank to show what was really possible came in the month following the end of the battle of Passchendaele. An original proposal by a staff officer in the Tank Corps, Colonel J. Fuller, for a series of tank raids was developed into a plan for a full offensive which was launched along a six-mile front at Cambrai on 20 November 1917. This time, the choice of well-drained, dry chalklands meant the terrain was at last right for armoured vehicles. Three hundred and seventy-eight tanks were assembled and readied for combat. In addition there were grapnel tanks for destroying swathes of barbed wire, supply tanks, bridging tanks and gun carriers, all manufactured in factories across England.

At last the number of tanks and trained crews available had reached the level at which something resembling a massed attack could take place.

Cambrai was not just a tank battle, it was a full scale artillery-infantry offensive as well.[13] However, this time the artillery adopted new tactics. With the extensive use of sound ranging and aerial reconnaissance to identify the major German gun positions, a brief artillery bombardment of only about ten minutes took place before the assault began and the gunners then adopted a 'rolling' or a 'creeping' barrage, moving forward at an agreed rate just in front of the advancing troops. The first day saw remarkable progress against German trenches that formed part of the Hindenburg Line. At times the advance was so rapid that the tanks had to halt to avoid driving into the rolling barrage. By the end of the first day, tanks and infantry had penetrated up to four miles and had passed through the first, second and third German defensive lines. The casualty rate was less than half that of a day's average losses at Passchendaele, and the distance advanced was beyond that gained in three months in Flanders. Church bells rang in Britain to celebrate the victory.

As so often on the Western Front, however, insufficient reserves were available to exploit an early advantage and the enemy recovered quickly. Cavalry troops tentatively moved into one of the gaps in the German lines but soon withdrew when they came up against machine guns. The Germans, who had had more than a year to refine their anti-tank tactics, used artillery very effectively against the British behemoths. One lone gunner is supposed to have taken out nine tanks.

Within a few days the British offensive had stalled. More than 250 tanks were out of action, roughly one-third due to enemy action and the rest as a result of mechanical breakdowns or because they had just got stuck. After more than a week of intense fighting, the tank crews were exhausted. On 30 November, the Germans counter-attacked in force, using small groups of stormtroopers to lead the assault after a brief, concentrated artillery bombardment – a tactic developed on the Eastern Front. By 7 December they had retaken most of the British-occupied ground, and when the fighting stopped the lines ended up pretty much where they had started. But Cambrai had shown that new tactics by both sides could work – with the artillery's creeping barrage, the use of stormtroopers and of course the use of tanks to crush heavily defended positions. Many of these tactics would be used again effectively in the following twelve months. The Battle of Cambrai had hinted at what was possible, but the real heyday of the tank came in the latter stages of the conflict, if only after the most serious Allied reversal of the war.

After the 1917 Revolution in Russia, a peace treaty was negotiated with the Bolsheviks. In that same year America entered the war. The Germans rushed troops from the East to the Western Front, sending west forty-four divisions, three-quarters of a million soldiers, during the course of the winter. On 21 March, General Erich Ludendorff, the joint commander-in-chief of German forces since Falkenhayn's resignation in the summer of 1916, launched a huge offensive on the Somme. The Germans soon recaptured all the ground the Allies had gained so painfully in 1916. On 28 March, when the Germans attacked at Arras, the British line held. And when Ludendorff attacked in Flanders on 9 April, British generals were forced to abandon Passchendaele, taken at such cost the previous summer. The British Army lost 100,000 men killed or captured within a few weeks, and Field Marshal Haig, usually so calm and unflappable, came near to panic. On 12 April he issued his famous Order of the Day: 'With our backs to the wall and believing in the justice of our cause each one of us must fight to the end.'[14] Even as he wrote the words, he began to prepare for a British withdrawal to the Channel ports.

In the midst of the crisis, Marshal Foch was appointed overall Allied commander-in-chief, with the French, British and American armies reporting to him. On 27 May, Ludendorff attacked in the south along the Aisne river. His troops advanced ten miles in a single day. Soon the Germans were at the river Marne, as far as they had reached in 1914. Their long-range Big Bertha howitzers shelled

Paris, barely more than forty miles away. On 15 July, they attacked near Rheims in the Champagne district. At each point of the Germans' assault the Allied lines had fallen back. But nowhere had they completely broken. Just as Allied generals had found in 1915, 1916 and 1917, Ludendorff now discovered it was possible to make initial advances but nearly impossible to exploit the advantage. By the mid-summer of 1918 there were a series of salients where the German lines protruded into the Allied defences.

Then, on 18 July, as the German advance slowly ran out of steam, the French army launched a counterattack. Massive Allied reserves that Foch had held back were thrown into battle. On 8 August Haig attacked in the north at Amiens. The lessons of Cambrai had been learned. There was only a brief bombardment and 456 tanks were deployed. Haig's troops advanced six miles in a day. It was the turning point. Ludendorff later described 8 August as 'the black day of the German army'. The French army attacked in the centre on 10 August and again on 17 August, while the British launched a further attack at the end of the month. And on 12 September the American army attacked for the first time in the war as an independent force, overwhelming a German salient at St Mihiel.

For its last months the war became, once again, one of movement. But this time it was thanks to a combination of science, technology and new tactics. Tanks made giant mechanised thrusts to punch their way through enemy lines. Once they had broken through in one place, they attacked in another. Rolling artillery barrages kept moving forwards just in front of the advancing troops. Aircraft flew thousands of sorties in support of the troops on the ground. The Ministry of Munitions oversaw the production of vast numbers of shells, machine guns, artillery pieces and tanks, more than enough to replace the losses suffered earlier in the year. By force of arms and the triumph of science, the huge British Army, the biggest ever sent into battle, along with the French, Canadians, Australians and Americans, forced the German army in the field into a full-scale retreat. In early November the Germans sought an armistice, and on

11 November the guns fell silent. Ludendorff blamed his defeat not on the Allied generals but on 'General Tank'.

The British had not been alone in developing tanks during the First World War. The French had simultaneously developed their own armoured vehicles but, having had little luck with their 'heavies' in the offensives of 1917, they changed policy and built for 1918 much smaller and lighter two-man Renault tanks, fitted with a rotating turret and a 37mm gun. The Germans had tried to design and construct their own tanks from the beginning of 1917 but progress was desperately slow. The fifty British tanks they captured in the counterattack at Cambrai amounted to more than their own manufacturers had built by that time. And of course, in describing the tank, it must be remembered that the principal objective of these slow-moving, heavy monsters was to cross no man's land, break through barbed wire entanglements, destroy machine guns and penetrate the enemy's fortified defences. They were very different from the fast-moving tanks of the Second World War with their increased firepower and tremendous manoeuvrability. The concept of *Blitzkrieg* or lightning war led by mobile panzer groups was still twenty years in the future.

When it was all over, the story of the tank came under objective scrutiny. It had not been the machine that had won the war, but it had certainly helped. The question was widely asked, who had *invented* the tank? The public wanted to know and there was much speculation. For several days in October 1919 the Royal Commission on Awards to Inventors, which dealt with many different wartime inventions, considered the question of the tank. The hearings were conducted in a formal and judicial manner at Lincoln's Inn before Mr Justice Sargant. The Treasury was concerned that so many claimants would come forward that it would be forced to pay out a small fortune in award money. Consequently, both the Attorney General and the Solicitor General, along with two leading counsels who represented the Crown, clearly tried to limit the damage to government funds.

Winston Churchill was the first to appear. He was not claiming an award himself, but was an expert witness who gave his own testimony on the path that led to the invention. Thirteen separate claimants came before the tribunal and presented their case. Mr Justice Sargant took a strong line, dismissing the cases of those public officials who in his view had merely been doing their job in supporting the work of the various committees that had advised on the different stages of development. In all, hundreds of thousands of words were recorded in evidence.[15] Finally, and probably fairly, awards of £1000 were given to both Ernest Swinton and Tennyson d'Eyncourt; Tritton and Wilson shared £15,000 between them.

Once again, in the laboratory of war, individuals with good, fresh ideas and a new approach had won through against an often stubborn, obdurate military leadership. At Tank Corps headquarters the real significance of the Battle of Cambrai had been clear. It was written that Cambrai marked 'a new epoch in warfare, the epoch of the mechanical engineer'.[16] It was a victory for science, of sorts. The Inventions Commission did not take it upon itself to condemn the military authorities for their lack of enthusiasm in finding a machine to cross no man's land. But their tardiness, at times amounting to obstruction, had cost tens of thousands of lives. If large numbers of tanks had been available earlier, the final victory might have come sooner. Basil Liddell Hart wrote that the real opponent of the tank was not the Germans but 'the British General Staff'. 'The numbers manufactured,' he wrote, 'sufficed to bring victory; but they could not bring back the dead.'[17]

Part Four

Doctors and Surgeons

10

The Body

With all the high explosives, shells, gas and chemicals being thrown at frail human bodies along the Western Front, it is not surprising that the First World War would witness some of the highest casualty rates and most horrific wounds ever experienced in war. But with the help of outside consultants, the mobilisation of the nascent pharmaceutical industry and the rallying of the medical profession when some 11,000 doctors left their civilian practices and joined the army, the medical services responded with remarkable effect to the challenge of repairing bodies damaged or mutilated by the slaughter of an industrial war.

'This is our Butcher's Shop,' said one medical officer to a visiting journalist in a Casualty Clearing Station at Corbie near Amiens. 'Come and have a look at my cases. They're the worst possible, stomach wounds, compound fractures and all that. We lop off limbs here all day long, and all night. You've no idea.'[1] Amputation was certainly still one of the traditional options to which military surgeons resorted in a desperate attempt to prevent infection from spreading. That had been the case for centuries. But in the fifty preceding years, medical science had advanced in leaps and bounds and now offered a variety of cures and surgical responses to the huge volume of casualties generated by the war. Antiseptics were available, there was a greater awareness of the need for sanitation and a new understanding of the science of bacteriology.

Every army since the beginning of time has had to be able to deal with its wounded in battle. Soldiers' morale is without doubt related to how they feel they will be treated if wounded. And by the early twentieth century, the morale of the citizens' army that went to fight the Great War was closely tied up with the performance of its medical division. The availability of good medical services became a vital part of military efficiency. It was not only that good medicine would enable more men to return to the front quickly, but the provision of even the simplest bandage and dressing clearly gave men a huge psychological as well as physiological boost. It made them feel they had not been abandoned. As one RAMC corporal put it, 'A clean white dressing ... seems to reassure a wounded man strangely. It makes him feel that he is being taken care of.'[2]

R.H. Tawney, later one of the leading economic historians of his generation, wrote that the worst aspect of being wounded was being 'cut off' from others; he felt an overwhelming sense of reassurance when he finally received medical attention. 'I knew he was one of the best men I had ever met,' Tawney later wrote about the doctor who found him in no man's land. 'He listened like an angel when I told him a confused, nonsensical yarn about being hit in the back by the nose-cap of a shell. Then he said I'd been shot by a rifle bullet through the chest and abdomen, put a stiff bandage around me and gave me morphia ... [and] in a grovelling kind of way, I worshipped him.'[3]

Most armies throughout history have suffered more casualties and deaths through disease and fever, like dysentery and typhus, than from battle itself. In the American Civil War there had been two deaths from disease for every one from battle wounds. Even in the Boer War this had only been reduced to 1.8 deaths from disease to one from wounds. However, in the First World War, the ratio was to be dramatically transformed. That war saw just one death from disease to every ten from battlefield injuries. It was the first conflict in which fewer men died from related diseases than from battle wounds. This massive reduction in deaths from disease, achieved by improved sanitation and health, was a great triumph in itself. Mortality levels from wounds dropped equally dramatically, from more than four out of every ten men wounded in the American Civil War to one out of ten in the Great War.[4] The way in which the medical services approached the hideous wounds arising from a modern industrial war was to show how effective science could be at saving lives in war.

In the Crimean War of the 1850s, the care of those wounded in war had become an issue of great public concern. Mary Seacole and Florence Nightingale had shown up the inadequacy of medical support for the wounded and Nightingale had become something of a deity in Victorian Britain. The public began to demand that soldiers who risked their lives for the nation in battle should be cared for properly if injured. Traditionally, army officers had looked down on medics, who as non-combatants were not given the same status as the rest of the officer corps, but this began to change in 1898 with the formation of the Royal Army Medical Corps (RAMC).

In the Boer War at the end of the century, the army medical services scraped by, but the challenges of a European war would be on a different scale. From 1906, Haldane's army reforms created a new system to establish a network of clearing hospitals behind the lines, fully equipped with the sort of operating theatres and other facilities found in civilian hospitals of the day. Haldane's medical reforms were another aspect of his belief that science could help improve military efficiency. In 1907 a new RAMC college was established at Millbank and became part of London University's Faculty of Medicine. The Director General of the Army Medical Services was the energetic and progressive Sir Alfred Keogh, who saw eye to eye with Haldane. He recruited many top London consultants into the Territorial reserve and helped to bring the military medical services much closer to those of the civilian population. Until his retirement from the RAMC in 1910 to become Rector of Imperial College,

London, Keogh was to carry out substantial reforms in military medicine,

In line with Haldane's other reforms, Keogh tried to supplement the small regular army medical service with support from the Territorial Force. Volunteers in the St John Ambulance Brigade and from the Officer Training Corps at universities provided the extra manpower. And a new emphasis was laid on preventative health for soldiers, a manual on *Military Hygiene and Sanitation for Soldiers* in 1908 concluding that 'disease prevention is synonymous with military efficiency.'[5] Sir William Leishman, Professor of Pathology at the Royal Army Medical College, developed new forms of inoculation against typhoid fever. If typhoid had been as endemic as in previous wars it would have been a major scourge in the vast British Army, but Leishman's innovations succeeded in keeping levels of the disease right down. It was later calculated that his inoculations had saved more than 130,000 lives and prevented about 900,000 soldiers from being invalided out of the army. As his obituary for the Royal Society noted, 'For this achievement, Leishman must be accounted to have been one of our most successful generals in the Great War.'[6]

When it came to medical care, the emphasis fell on evacuating wounded men from the front to hospitals where proper care could be provided. Motorised ambulances were to be used to speed up the evacuation process. Much was done, meanwhile, to establish a system that could return wounded soldiers to the battlefield as soon as possible after they had been treated. This was ultimately the purpose of all army medical officers. And the War Office was acutely aware that the public would not stand for failures in the care and treatment of the wounded.

Despite all the pre-war preparations, in its typically bungling way the British Expeditionary Force that travelled to France in August 1914 did not take with it a single motorised ambulance. Although the army possessed such vehicles and there had been a debate about the need for them, the Director of Military Operations, Sir Henry Wilson, ruled that horse-drawn ambulances would be sufficient as

The aircraft in which John Moore-Brabazon made the first powered flight by a Briton in Britain on 30 April 1909; it looks like a giant box kite.

Winston Churchill, First Lord of the Admiralty (1911–15); he was so keen to encourage naval flying that he even tried to learn to fly himself.

Charles Rolls at the wheel of one of his luxury vehicles; but he was also a pioneer aviator who tried to get the army interested in powered flight.

An A-type glass plate camera being handed to an observer in a Vickers FB5; despite the primitive look, photo intelligence developed as a sophisticated science during the war.

Alliott Verdon Roe beside one of his early flying machines; his company, Avro, went on to make iconic aircraft in the First and Second World Wars.

William Lawrence Bragg, a scientist who went into uniform and developed sound ranging for the army; the youngest person ever to win a Nobel Prize, at twenty-five.

Baron Rayleigh, one of the top scientists of the day, who worked with government committees on explosives and aviation, and for the Royal Society War Committee.

Colonel Ernest Swinton who reported from the front in 1914–15 and realised the need to break the stalemate of entrenched defences and barbed wire.

Captain Philip Joubert de la Ferté who carried out the first aerial observation flight of the war in August 1914 but twice had to land and ask where he was.

Brigadier-General Sir David Henderson, the first commander of the Royal Flying Corps; he and his moustache inspired a generation of early flyers.

A group of early aviators study a map in front of a BE2 biplane on the Western Front; they directed the artillery at the enemy's guns.

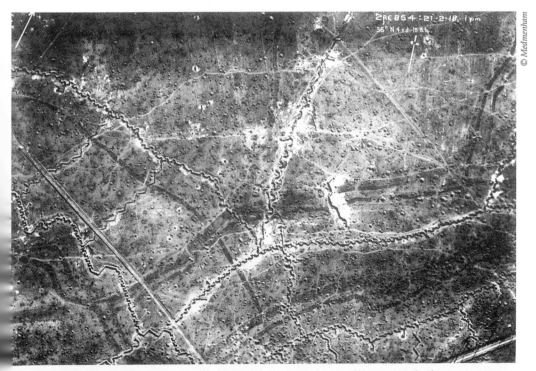

An aerial photo of German lines, February 1918; ten million aerial photos were distributed during 1918.

Admiral Reginald 'Blinker' Hall (centre), the key figure behind the code-breaking and intelligence gathering at Room 40 of the Admiralty; taken while still captain of HMS *Queen Mary*.

Naval actors; two naval officers disguised as the skipper and the owner of the *Sayonara*, the ship Hall sent to spy along the Irish coast.

Vickers machine gun team on the Somme, July 1916; they wear Hypo helmets to protect them from gas attack.

Wilfred Stokes, engineer, shows off the mortar he invented and some of the projectiles it could fire.

Eustace Tennyson d'Eyncourt, naval engineer; to his surprise Churchill put him in charge of developing the tank.

One of the early tanks prepares for action at Cambrai, November 1917; the tank was a completely new machine that emerged out of existing technologies during the war.

Scottish soldiers using improvised cotton-pad respirators, May 1915; when the Germans first used gas it took some time to develop an effective mask.

Soldiers demonstrating the use of a Hypo helmet, 1916; it provided more efficient protection from a gas attack.

Wounded men on stretchers laid out around a Regimental Aid Post, the first tier of medical care on the Western Front, not far from the front trench, 1916.

Victims of mustard gas that caused horrible yellow blisters and blindness, being led away, 1918.

Charles Masterman, politician and literary figure who started Britain's official propaganda campaign.

John Buchan, one of the most successful novelists of the era who wrote profusely to promote the British cause and ran the propaganda department in 1917.

Alfred Harmsworth, Lord Northcliffe, before the war, the press baron who turned official government propagandist in 1918.

Max Aitken, Lord Beaverbrook, the Canadian magnate and newspaper owner who became Minister of Information in 1918.

Recruiting Posters, 1914; left, the famous Alfred Leete poster of Kitchener. Right, women are encouraged to send men off to their deaths.

Anti-German hysteria, 1914; *Punch* cartoon shows Kaiser Wilhelm standing over dead Belgian women and children.

Harold Gillies, plastic surgeon, who did remarkable work reconstructing the faces of some of the most horribly mutilated victims of the war.

rawings of one of Gillies' patients, before and after surgery, made by Henry nks, the celebrated artist.

Women workers in a munitions factory; filling shells with explosives was a dangerous job but nearly one million women came forward for this work.

J.B. McDowell, official cameraman, with his hand-cranked film camera; cumbersome and heavy, this type of camera was used to film the Battle of the Somme.

© IWM Q70165

Frames from the famous 'over the top' scene in the *Battle of the Somme* film; staged at a trench mortar school behind the lines, these images appeared to show men falling as they went forward and shocked cinema audiences around the world.

© IWM

© IWM Q70168

Staff and patients at Craiglockhart War Hospital, Edinburgh, a Victorian Hydro-spa converted to treat officers suffering from shell shock; Captain William Rivers (front row: sixth from the left) treated Siegfried Sassoon here and Arthur Brock treated Wilfred Owen who then began to write some of the most famous poetry of the war.

it would take time to clear the bodies immediately after a battle. It had been estimated that the BEF might have to endure up to about 3000 casualties in its first engagement and Wilson believed horse-drawn ambulances would be adequate to evacuate this number of wounded. As a consequence, when the number of casualties in the BEF's first battles at Mons and Le Cateau at the end of August was much higher than expected, it proved impossible to evacuate many of the wounded, who had to be left to die or to be captured in the retreat that followed. J.P. Lynch, a private in the RAMC, wrote in his diary of one small incident, 'I was sorry to have to leave 14 men behind in the barn but there was nothing to do as we had no room for them.'7 The public in 1914 were unwilling to accept such poor treatment and when the story was reported there was an outcry. Kitchener appealed to the public for funds.

Many people in Britain and around the Empire responded by making generous donations. The citizens of Calcutta alone gave enough to equip an entire fleet of ambulances, while within a couple of months readers of *The Times* had donated the huge sum of £281,000 (about £28 million in 2014). By November this fund alone had paid for 93 motorised ambulances. By the end of the year there were 250 motorised ambulances with the British Army in France and Belgium and more were arriving daily, along with mountains of blankets, stoves and other supplies that people in Britain thought would be helpful. Accompanying the supplies were volunteers from the St John Ambulance Brigade and many privately funded groups of medical and nursing staff, all of whom tried to get as close to the battlefield as possible to provide their care. Their assistance was valuable, but the fact that so much medical care was provided by private charities made for tensions between the professional army medical men and the titled ladies who tended to run the charities. They were usually formidable women who were used to being listened to and treated with great respect, and who usually got their way.

Sir Arthur Sloggett, a career army medical officer, was the

Director General of Army Medical Services at the outbreak of war. A classic army type, with a jaunty walk, a cheerful, outgoing manner and a fund of witty stories for every occasion, he had been wounded in a cavalry charge at the battle of Omdurman, had served in the Boer War and been director of medical services in India for three years. After the early failures and the complications that arose with the privately funded charity workers, Sloggett decided to locate himself permanently in France, while Sir Alfred Keogh was recalled from retirement to take over the director general's role at the War Office. Despite their contrasting characters, Keogh and Sloggett got on well and between them now began to reorganise medical care for the British Army in a way better suited to a European war. Keogh ran the overall strategy from the War Office in London. This he was well equipped to do, with his established links to the academic world and his experience of the earlier army reforms. Sloggett managed the situation in France and Belgium and built up good relations with commanders in the field. He had a particularly good relationship with Haig, and at one New Year's Eve dinner was 'the life and soul of the party with his yarns, some of which were libellous and few of which would have passed muster in a drawing room'.[8]

The army turned for outside help to a group of consulting surgeons, asking them to provide advice on technical and specialist matters and to tap army practice into the latest developments in surgery. Sir Anthony Bowlby was the most eminent of the group. Having been a volunteer surgeon in South Africa in the Boer War, Bowlby knew the army well. Returning to London after the war he had become a senior surgeon at St Bartholomew's Hospital, and in the army reforms had joined the local branch of the Territorial Force's medical services. Bowlby came up with several ideas for improving the army's medical care and argued forcibly for sensible modernisation of techniques and practices. He was particularly concerned that wounded men should be treated as quickly as possible, which in practice meant near to the front, writing that 'public and professional opinion was united in expecting casualties to be treated as thoroughly as possible at the front."[9] With Keogh in the War Office, with Sloggett in France and with Bowlby's expertise to draw upon, from 1915 the British Army began to plan for its medical services the system that would last the rest of the war.

The first general problem to appear was a major shortage of anaesthetics, and of other drugs such as antiseptics. Like other branches of the chemical industry, the British pharmaceutical industry before the war had been dependent upon imports, especially from Germany. John Anderson from the National Health Insurance Commission, the body responsible for the supply of drugs, turned to the Royal Society and asked its War Committee for advice. The Royal Society approached several universities asking for help in making drugs for the army and received thirty-two responses.[10] The brightest young researchers in many universities had already joined up and gone to fight, but St Andrew's University managed, under the supervision of Professor James Irvine, to carry out important work over the next year and found ways of synthesising a range of anaesthetics that were in desperately short supply. The same department later in the war, ironically, was to carry out important work on the manufacture of mustard gas.

There were of course many different types of wounds to be treated. Bullet wounds could be quite clean; a bullet entered a limb and exited on the other side without bringing any bacterial infection, unless it had first ricocheted off something. However, by far the largest number of wounds in the First World War were from shell fragments or shrapnel. These could bring far greater risks, not only from the damage caused when the shrapnel penetrated the body but also from infection. Because the soils of northern France were intensively cultivated and had been heavily manured for centuries, the bacteria picked up by an exploding shell could cause as much damage as the explosion itself. A typical fragment of a few square inches would be propelled from a shell that had exploded in the soil at high speed; when it hit a man it would first penetrate his outer clothing, which would often be covered with mud and trench filth.

Carrying with it the bacteria it picked up, the shell fragment would then pass through his underwear, which might well have been unwashed for several days. It would penetrate his skin, severing veins, and might well cut a muscle or go through a main artery before hitting the bone, which would shatter sending smaller fragments in all directions. The fragment might then exit from the body, making an even larger hole on the way out and causing more tissue damage. Obviously the nature of the injury depended upon where the shell fragment hit the body, but the problems of treating a single wound like this were often complex and varied.[11]

Tetanus was a major source of infection in wounds on the Western Front. Tetanus is a disease caught from bacteria found in the intestines of animals and which are present in heavily cultivated soil. Once the bacteria infects a wound it poisons the muscles through their nerve endings and causes contractions, usually of the head and neck, a condition often known as 'lockjaw'. Before the era of antibiotics, the only way to treat tetanus was through the injection of a serum, and this was widely carried out on wounded men. Similar to tetanus but more severe was gas gangrene (which has nothing to do with gas warfare), an infection which could be combated by no known serum. It too came from bacteria in the soil, and when it became established in a wound the only option was to amputate. However, as understanding grew about the need to keep the wound clean, cases of gas gangrene infection dropped markedly, from 10 per cent of all wounds in 1914 to about 1 per cent by 1918.

In addition to these infections a wounded man might also be suffering from shock, perhaps delayed, and from blood loss. Blood transfusions were not generally performed in the early twentieth century as it was not known how to keep blood supplies for any length of time; any transfusion of blood took place directly from the artery of a comrade into the wounded man's vein. As the war progressed, however, Australian and Canadian units began to use transfusion more frequently, and slowly methods improved for storing blood in refrigerated conditions, if only for a few days. The

spread of blood transfusion techniques was one of the most important surgical advances in the war and helped greatly to improve the
prospects of recovery for wounded men. But more than anything, a
man's recovery depended upon how quickly he could receive
proper medical attention. The speed with which a wounded man
could be given the right treatment meant the difference in most cases
between life and death.

To deal with the scale and complexity of the injuries incurred on
the Western Front, the RAMC needed a system that it could manage
strictly. Sir Alfred Keogh was an excellent administrator and
Haldane described him to Kitchener as 'one of the best organisers I
ever knew'.[12] From his senior position in the War Office and with his
wide experience of setting up the Territorial Force hospitals, he was
able to construct a system that adapted to the needs of the war and
was capable of dealing with the terrible scale of the wounds inflicted
over the next few years.

Every soldier in the British Army was now to carry in a pocket in
his uniform jacket a basic field dressing consisting of two gauze
dressings, cotton wool and a bandage, packed inside a waterproof
parcel. The idea was that if he was hit by a rifle or machine gun
bullet, a soldier nearby would pull out the dressing and wrap it
around the entry and exit points of the wound. But the dressings
were often not large enough to cover shrapnel wounds, so a man
wounded by a shell fragment would usually have to wait until a
stretcher-bearer appeared. There were sixteen stretcher-bearers to
every battalion, every one of whom was trained in basic first aid and
carried in his haversack a medium first aid kit containing larger
gauze dressings to cover shrapnel wounds. The intention was to
keep a wound as clean as possible until the injured man could
receive proper medical attention; in reality, however, men often lay
wounded, sometimes for hours, in the fetid conditions of no man's
land, with mud, water and filth everywhere, and even if wounds
had not already been infected by the bacteria carried on the shell
fragment they were likely soon to become infected.

By the end of 1914, the war in the west had ossified into the static war along the trench lines of the Western Front. This enabled armies on both sides to develop a system of medical care that relied upon a method of getting wounded men back from the front line to an established hierarchy of medical aid positions. In the British Army, stretcher-bearers would first take a wounded man to the Regimental Aid Post or Advanced Dressing Station. Each battalion had its own medical officer, a qualified doctor who would work with a small group of orderlies. The medical officer was a familiar figure, living with the men in the trenches, and in addition to dealing with daily issues relating to the soldiers' health, like regular colds, aches and pains, he was the first port of call for the wounded. His duty was to run the Regimental Aid Post, which was usually only two or three hundred yards behind the front line, situated where possible in an old farm building or in dugouts. Being so close to the front, the posts were frequently exposed to enemy fire and were only marginally safer than the front line itself; as a consequence more than a thousand medical officers were killed during the war.[13]

Often, during a major advance, the situation in the posts became chaotic as large numbers of wounded were brought in. There were often only minimal supplies of clean water and rarely enough space. But the intention was to examine and clean up every wound if possible, sometimes by coating a wound with iodine or setting a fracture in a splint. There would not be time for much in the way of medical intervention, but as soon as possible a field ambulance would remove the wounded man further from the front. The Field Ambulance units themselves included dressing stations, but their principal purpose was to convey wounded men to the next tier of care, known as the Casualty Clearing Station (CCS).

The CCS was effectively a small hospital a few miles behind the front, out of range of the enemy's artillery. By 1916 the British Army had set up about fifty of them in France and Belgium, and each could accommodate between 500 and 1200 wounded men. There would be about ten medical officers in a CCS, supported by many orderlies and even a few female nurses. This was as far forward as the army allowed women during the war. Despite the opposition of some doctors, who thought that having women this near to the front brought unnecessary risks and burdens, the presence of female nursing staff was usually a great encouragement to the men and good for morale. The CCSs included wards for different types of patient, modern operating theatres, a laboratory, X-ray machines and a full medical supply depot. Under pressure from Sir Anthony Bowlby, as the war progressed the CCSs would increasingly come to specialise in particular treatments, such as abdominal wounds or head injuries. They were usually located at central points near road junctions or railway sidings. Some were made up of large buildings requisitioned for the purpose; others consisted of rows of marquees and temporary Nissen huts laid out across green fields.

As soon as a wounded man arrived at a CCS he was classified in one of three categories that determined the type of treatment he received. This was not a new system in the Great War, the principles went back as far as the treatment of the wounded in Napoleon's army. Soldiers with slight injuries were treated quickly and returned to the battle zone as fast as possible to clear space for others. The second group, those needing more serious attention, were treated to the highest level available at the CCS. This might entail the amputation of infected limbs or the cutting out of wounded flesh to prevent infections from spreading or getting hold. The third group, those regarded as being beyond hope, were left on one side. There was a cruel arithmetic to military medicine. It was thought that if a man was too far gone there was no point trying to treat him when the time could be spent on saving perhaps three or four others. Every CCS was therefore surrounded by slowly dying men, often in terrible agony, calling out for help, for water, for attention. And each CCS would have attached to it a cemetery that during the course of the war grew inexorably in size. It was horrible, but this was the reality of war.

In most cases, the wounded never got further than the CCS. If they were not returned to their units immediately after treatment, they would be sent back after being given a short time to recuperate or recover. The seriously wounded were, however, evacuated from the CCS and sent back 'down the line', usually in a specially equipped ambulance train or barge. Early in the war, trains were like almost everything else in desperately short supply. But through voluntary contributions and sponsorship from the private railway companies in Britain, their numbers grew considerably. By 1916, there were more than thirty such trains, fitted out to transport badly wounded men. Some carriages contained wards with beds, others had seating, and each train had a complement of at least three doctors and forty orderlies.

The next tier in the medical hierarchy to which the trains and barges took the wounded men was the base camp General Hospital. There were three main general hospitals located along the French coast: one at the port of Boulogne, another just north of Etaples along the coastal dunes and the third a few miles south of Rouen. These were huge establishments with up to 14,000 beds and full facilities to treat the victims of gas attacks and wound infections, and to carry out surgery. Most of their nursing work was carried out by women. The presence of women continued to cause controversy, although in the main they were far more popular among the wounded than the male orderlies. Having a sympathetic young female figure to talk to was usually very important to the men when they felt at their weakest and most vulnerable. Many of the nurses, as volunteers, were from the middle classes, and so in the main they regarded themselves as on a par with officers, but in their caring role they were able to gain the respect and admiration of many of the 'other ranks', normally from the working classes. There were occasional stories of liaisons between nurses and wounded officers, although such relationships seem to have been extremely rare and were certainly frowned upon by the authorities.

Most of the young women had been thrown into situations completely outside their previous experience, and for many this was as significant in their lives as fighting in the trenches proved to be for their brothers. Having felt that student life at Oxford was 'like Nero fiddling while Rome was burning', Vera Brittain left her studies at Somerville College, Oxford after only a year to work as a volunteer nurse. She explained her reasons for taking up nursing as 'not being a man and able to go to the front, I wanted to do the next best thing.' And the experience of nursing horribly wounded men in military hospitals had an immense impact on her; she wrote in a letter that 'after seeing some of the dreadful things I have to see here, I feel I shall never be the same person again.'[14]

Army policy was to return even the more seriously wounded to their units as soon as possible. But for men who needed further treatment or convalescence, the next stop on the journey was back to 'Blighty'. Six hospital ships, each with the capacity to carry 2500 men, constantly ferried the wounded back across the Channel. From the Channel ports they were taken by train, often to Charing Cross station in London where huge crowds regularly gathered to cheer them as they were carried across the station to waiting ambulances. In December 1914, there had been 40,000 beds set aside in hospitals in Britain for war wounded. By the end of the war this number had grown to more than 360,000.

Back in Britain, non-commissioned soldiers and 'other ranks' were still kept under strict military discipline, doing useful tasks where possible and ordered to carry out drill when fit enough. Officers, on the other hand, were allowed to plan their own sick leave and convalescence, presumably on the basis that as a matter of honour they would return to their regiments as soon as possible. But as the shortage of men became acute, so in 1917 this loose system was tightened and all sick and wounded up to the rank of colonel had their convalescence scheduled for them. However, officers were still put in different wards and sent to separate hospitals from the men they commanded.

As the war progressed the system continued to provide more specialist medical care at each tier. But at every stage, the evacuation of the wounded brought its own transport problems. The stretcher-bearers were usually terribly overworked at times of offensive action. The mass of mud in the Flanders salient meant that it might take eight men rather than four to carry a stretcher and by the time they arrived at an aid post the stretcher-bearers were often utterly exhausted. There were never enough stretcher-bearers to cope at times of heavy action. Nor were there enough field ambulances to carry men to the CCSs, while those available were often without suspension, so that despite being motorised they provided a bumpy and horribly uncomfortable ride. At the next stage of evacuation, the ambulance trains had to run largely on single-track French railways where priority was given to munitions trains carrying supplies up to the front and troop trains bringing up reinforcements, so journeys could take many hours longer than expected as trains with the wounded waited in sidings. But however rudimentary, the lives of hundreds of thousands of men were saved by a system that in the main worked, and which improved as the war progressed.

Many aspects of the medical care of the wounded during the war were debated intensely at the time. The professional army surgeons did not always see eye to eye with their civilian advisers, while the civilians regularly argued among themselves as to the best way to solve the new problems thrown up by the war. One prominent member of the Medical Research Committee, Sir Almroth Wright from St Mary's Hospital, Paddington, argued that the army was taking the wrong approach to the treatment of the wounded. He set out to investigate the problem of wound infections at the Base Hospital in Boulogne, establishing a small laboratory and taking with him to study the use of antiseptics many of his researchers from St Mary's, among them the young Alexander Fleming, who fifteen years later would discover penicillin. Wright argued that the rapid evacuation of the wounded increased the likelihood of war wounds becoming infected. By 1916 most military surgeons in France were trying to prevent infection by cutting out damaged tissue and irrigating the wound with a new antiseptic solution pioneered by the

American Dr Alexander Carrell in France and the Scottish chemist Henry Dakin. The use of the Carrell-Dakin solution to sterilise wounds rapidly became widespread. But Wright believed this was ineffective and suggested the use of a form of hypertonic saline solution, which entailed packing septic wounds with tablets of salt.

Wright also believed that treatment should be standardised in line with the latest scientific research. Suggesting that the Medical Research Committee should coordinate this treatment by studying the nature of injuries across the entire Western Front, he wrote in 1915 that every doctor should work 'not as he individually thinks best but as part and parcel of a great machine'.[15] This was one aspect of the rationalisation of care that the Medical Research Committee strongly believed in. At the end of 1916 Wright publicly criticised the army, offering a direct challenge to Keogh and the members of the army medical corps. Seeing his insistence on standardisation as nothing more than a way of introducing his own ideas on treating wounds, they united in opposition to Wright's attack. Sloggett insisted that one of Wright's pamphlets should be banned on the grounds that the ideas it put forward were dangerous, and after meeting with Wright, the army's chief consulting surgeon, Sir Anthony Bowlby, wrote in his diary, 'Wright, who knows nothing whatever about wounds at all, and also nothing about Carrell's treatment, talked the usual rot about his pet salt treatment which everyone else has given up.'[16]

The controversy went as far as the Secretary of State for War at the time, Lord Derby. When he was persuaded not to take Wright's criticisms seriously, the matter was dropped, at least at the official level, but it was a sign that in the rapidly expanding world of medical research, there were usually at least two answers to most questions. And it showed that the relationship between the Royal Army Medical Corps and the civilians of the new Medical Research Committee was not always an easy one.

The Battle of Loos in September 1915 proved a turning point in the establishment of the military evacuation system. British losses exceeded 50,000 and the pressure on the medical units was severe. General Haig, commander of the assault, paid particular attention to the treatment of the wounded and was shocked when he visited a CCS and found that doctors had been working continuously without rest for seventy-two hours. Haig respected both Sloggett and Bowlby and under the influence of the surgeons agreed that more surgery should be carried out in the CCSs, shifting the emphasis to treating wounded men nearer to the front. When Haig became commander-in-chief towards the end of the year, he maintained close links with the medical care services, and his confidence in them might have encouraged him in sustaining the war of attrition on the Western Front.[17]

Before the launch of Haig's Somme offensive in the summer of 1916, the army made meticulous preparations for medical care of the anticipated casualties. With estimates that there might be up to 10,000 wounded men each day to treat and evacuate, fourteen CCSs were prepared behind the lines where the assault would take place. For each one an ambulance train was equipped and readied and six ambulance convoys were on standby. Three days before the assault all the CCSs were cleared of non-emergency work, and dental and optical care was postponed. On 1 July, the bloodiest day in the history of the British Army, all these preparations were overwhelmed. The actual number of wounded that day amounted to nearly 40,000, half of whom had suffered serious injuries. But only 22,436 were treated during the day by field ambulances or in CCSs, all of which struggled desperately to keep up.[18] The remainder never got beyond the Regimental Aid Posts or, more probably, were still out in no man's land waiting to be brought in and fed into the system. At many places up and down the front, truces agreed with the Germans on the night of 1 July allowed stretcher-bearers to go out from both sides and bring in the wounded.[19] Nevertheless, many wounded men had to wait for hours, some for days. One, Private Matthews of the 56th (London) Division, lay in no man's land stuck fast in mud for fourteen days. Miraculously his wounds did not turn septic and he survived.

Howard Somervell was twenty-six and had only qualified as a doctor at University College Hospital, London, in 1915. He volunteered on qualifying and went straight into the RAMC. On 1 July he was a surgeon at a CCS at Vecquemont, a large tented hospital that had prepared for 1000 patients. Like the other CCSs along the Somme front, Vecquemont was overwhelmed by the numbers of wounded, as a line of ambulances one mile long waited patiently to unload their human cargoes. Somervell wrote that 'the whole area of the camp, a field of five or six acres, was completely covered with stretchers placed side to side, each with its suffering or dying man upon it.' Working in one of the four operating theatres, he had the chance occasionally to make 'a brief look around to select from the thousands of patients those few fortunate ones whose life or limbs we had time to save. It was a terrible business.' He learned quickly the principles of military surgery: 'we rapidly surveyed them to see who was most worthwhile saving. Abdominal cases and others requiring long operations simply had to be left to die. Saving life by amputation, which can be done in a few minutes, or saving of limbs by the wide opening of wounds, had to be thought of first.' Writing nearly twenty years later, Somervell recalled,

> Even now I am haunted by the touching look of the young, bright, anxious eyes, as we passed along the rows of sufferers. Hardly ever did any of them say a word, except to ask for water or relief from pain. I don't remember any single man in all those thousands who even suggested that we should save him and not the fellow next to him ... There, all around us, lying maimed and battered and dying, was the flower of Britain's youth – a terrible sight if ever there was one, yet full of courage and unselfishness and beauty.[20]

After the seismic shock of the first day on the Somme, the system gradually recovered. On the second day, more than 33,000 wounded men were treated. Sloggett and Bowlby met with General

Rawlinson, commander of Fourth Army, who was in direct command of the offensive. Medical officers were rushed in from quieter parts of the front with the intention of carrying out more urgent surgery in the CCSs. The evacuation system began to catch up with the massive congestion caused on the previous day. By the third day, the numbers being moved on from the CCSs actually exceeded the numbers arriving. And by the end of July the CCSs had treated about 96,000 men, of whom approximately 10 per cent had been operated on. It was a remarkable number to have coped with.

The battle rumbled on for several months, and when Haig visited a CCS in September, he was astonished to meet a surgeon who had carried out eighty-two operations in the previous ten hours.[21] But he was encouraged by the cheerful mood of the wounded men he met; their morale seemed to be high, despite the horrors they had endured. Possibly it was for this reason that he felt able to continue with the offensive for another two months until the winter rains of November brought the fighting season to an end.

Haig lost 420,000 men killed, wounded, missing and captured during the Somme. But despite these appalling losses the British Army had survived intact. The hard-pressed medical services had played a large part in enabling the army to fight on and preventing a catastrophic collapse in morale.

The next major offensive, the Battle of Arras in April–May 1917, had the dubious distinction of causing the highest daily average death toll of any battle fought by the British during the war. Twelve CCSs and twenty-eight ambulance trains were prepared for the battle, proportionately far more than had been available at the Somme. Again, the medical systems struggled but finally coped with the vast scale of casualties. And again the local army commander, this time General Allenby, paid frequent visits to the CCSs to gauge conditions and assess morale. Once more, what he saw made him feel that the offensive could continue. And it did, although this time it lasted for only a month.

One consequence of the battles of the Somme and Arras was that the CCSs began to specialise. By the Third Battle of Ypres in the summer of 1917, many CCSs had in addition to dealing with all common injuries their own specialisms in the treatment of chest or abdominal injuries, fractures or gas casualties. This enabled the provision of better care nearer to the front and resulted in an increase to a rate of about 25 per cent of all casualties being operated on at CCSs. The numbers wounded in the battle were not quite so high as had been anticipated but the conditions at the front, especially the thick, exhausting, omnipresent mud, were the worst that had been known in the war. Sickness rates shot up, as did cases of infection, especially gas gangrene, for which the only remedy was amputation.

The process of treatment and evacuation had lasted through the great battles of 1916 and 1917 but was hard pressed in the aftermath of the German offensive of March 1918. With the return of a more mobile form of warfare, the established tiers of care were difficult to sustain. The Germans advanced up to twenty miles in places and some CCSs were overrun completely, many of the seriously wounded being left behind in the care of small teams of volunteers who faced inevitable capture. New systems had to be quickly improvised, and field ambulances and trains were used with increasing regularity to move the wounded further from the front line.

After the German offensive had petered out, and especially from August 1918, when the Allied armies began their own advance, mobility was key. New hospitals and CCSs were established as the armies moved forward. The medical corps had to show great flexibility in this final phase of the war, as inevitably they received very little advance notice that would allow them to plan for receiving the casualties of new assaults. Nevertheless, by the autumn of 1918, many new mobile CCSs had been set up, each one supported with forty motorised lorries. It was reckoned that a CCS, with 500 or so beds, operating theatres, X-ray machines and all its supplies, could be packed up, moved off and reopened in less than forty-eight hours. This was a creditable achievement, especially as the daily casualty rate was still counted in the thousands. Inevitably gaps developed between the advanced dressing stations and the CCSs as the armies advanced. Often the railway system could not cope, and more sidings and rail links had to be laid, all of which took time even if only a matter of days. 'The difficulty is railways,' wrote one exasperated senior medical officer in October.[22] One answer was to use more ambulances to bridge the growing distances involved.

In addition, during the last two months of the war large numbers of German wounded were brought in and treated alongside Allied soldiers. A huge wing of the Etaples base hospital was given over to the exclusive care of German prisoners. Overall, the last few months of fighting in the summer and autumn of 1918 showed how well the medical services had been integrated into the operations of the British Army, and how flexible and capable those services had become in treating the large numbers of terribly wounded and mutilated men still being brought in.

For many of those with severe injuries, the problems were permanent. About 41,000 British servicemen with amputated limbs were given prosthetic replacement body parts. There were no major new developments in prosthetics during the war and the techniques remained essentially Victorian, with the use of wooden and leather limbs and attachments. But the scale of the demand for replacement legs, arms and hands provided a generation of doctors with a mass of experience that it would have usually taken a lifetime to pick up. Queen Mary's Hospital in Roehampton was the centre of the use of prosthetic limbs and nearly two-thirds of all amputees were sent there; it became the centre of excellence for the use of prosthetics and for the care and rehabilitation of patients. The work carried out became important in the maintenance of morale, by showing soldiers and their families that everything possible would be done for those severely wounded to make them feel and act normally once again. A story filmed at Roehampton by British Pathé newsreel cameras recorded men showing off their prosthetic legs, walking, marching, taking part in tug-of-war and even playing football with some nurses. It was a propaganda exercise, of course. The fragments of film that survive show the men cheerful with a strong sense of camaraderie and proud of their artificial limbs.[23] Although it was no doubt staged for the cameras, however, there are many comments about how cheerful groups of amputees could be. They were no doubt pleased still to be alive. The problems often came later as they struggled to assimilate back into civilian life. The concept of 'broken warriors' and 'help for wounded heroes' was well established by 1918, and the sight of the blind and limbless victims of the Great War on crutches or struggling with artificial limbs and attachments became commonplace in the decades following the end of the conflict.[24]

A recurring issue throughout the war was finding the right balance between general care for the bulk of the wounded and specialist care for those with particular wounds. Military medicine traditionally favoured general care that could be applied quickly across the board. But the ghastly nature of the wounds suffered during the war years provided an opportunity for specialists to try out new ideas and develop new expertise. One of the most remarkable specialisations was that of Harold Gillies and the face reconstruction work that he carried out, first in Aldershot and then at Queen's Hospital in Sidcup, south-east London.

The First World War seems to have generated an excess of head wounds – possibly because soldiers were exposed at times above the parapet of a trench, or perhaps simply because more patients with head injuries survived than in previous wars thanks to improved general medical treatment. Some of these wounds were monstrous: parts of the face had been blown away, jaws shattered, noses removed and cheeks destroyed. In the cases of explosions at sea, men suffered from terrible facial burns, as sailors did not then wear face masks. The only treatment at the start of the war for anyone who had suffered such a trauma and survived was to pull the edges of the wound together and sew up the face. This closed the wound but did not replace the lost tissue, and it was left for nature to do what it could. It was into this world that a young New Zealand surgeon entered in 1915.

Harold Delf Gillies had been a brilliant youth, a great sportsman, a fine artist in watercolours and a talented violinist. After attending school in both New Zealand and England, he went up to Gonville and Caius College, Cambridge in 1901 to study medicine. Although he was only slight, he rowed in the Oxford and Cambridge Boat Race three years later and helped Cambridge to a big win. He went on to play golf for the university, and the sport became a particular passion. He was an argumentative student who wanted to do things his own way and did not hesitate to speak his mind and disagree with his tutors, a trait which did not endear him to the academic community. He finished his medical training at St Bartholomew's Hospital in London and qualified in 1906, becoming house surgeon to the senior lecturer on surgery and a specialist in ear, nose and throat surgery. Clearly a talented surgeon, he soon became assistant to Sir Milsom Rees, one of the leading London surgeons of the day.

In 1914, then aged thirty-two, Gillies volunteered for the Red Cross and worked in Belgium and France for a few months. While in France he visited Hippolyte Morestin, the most famous plastic surgeon in Europe, and was allowed to watch an operation in which Morestin removed a cancerous growth from the face of a patient and rolled up a flap of tissue from under the jaw to repair the cheek wound. For Gillies, this was a road-to-Damascus type moment, of which he later said that 'it was the most thrilling thing I had ever seen'.[25] He left Paris feeling 'a tremendous urge to do something other than the surgery of destruction'. Instead he became passionate about the surgery of what he called 'reconstruction'.[26] He put the case to Sir Anthony Bowlby that the many face and jaw casualties of the war posed entirely novel challenges to the military surgeon and that a new establishment devoted to this work was needed. Bowlby in turn persuaded Sir Alfred Keogh of the necessity for such a unit and in January 1916, Captain Gillies was told by the War Office to report to the Cambridge Military Hospital in Aldershot and open up shop.

Before long a stream of men with terrible wounds began to appear at his unit. A nurse vividly remembered men arriving 'with half their faces literally blown to pieces, with the skin left hanging in shreds and the jawbones crushed to a pulp'. None of the patients could eat solids and they were mostly fed on a mixture of whipped eggs, milk and sugar, known as 'egg flip'. As the nurse recalled, 'Hardest of all was the task of trying to rekindle the desire to live in men condemned to lie week after week smothered in bandages, unable to talk, unable to taste, unable even to sleep and all the while knowing themselves to be appallingly disfigured.'[27] For Gillies, the reconstructive surgery he now began to apply was not just about learning the technical processes of how to replace bone, cartilage and skin, but was about restoring a man's features and his confidence in himself. He called it 'a strange new art'.[28] He soon became very good at it.

There were no textbooks to guide Gillies in this new work, and he and the talented team he gathered around him had several elements to master. Applying anaesthesia to men with blocked or wounded air pipes was a problem that took some time to solve. Chloroform was often used, but this was extremely unpleasant for the patient and sometimes escaped, with the risk that it would affect the surgeon as well. Sometimes it proved better to operate on a man in a sitting position than lying down. A team of dental surgeons were also needed and a small group began to work closely with Gillies.[29] And as a final sign of the artistic nature of how he saw his work, Gillies asked artist Henry Tonks to sketch men's faces before and after treatment. Tonks had taught at the Slade School of Fine Arts before the war, specialising in anatomical drawing, and had taught such luminaries as Augustus John and Stanley Spencer. When war came he joined up as a humble medical orderly, and when Gillies discovered him at Aldershot he asked him to join his team. Tonks's pastel drawings of faces before and after surgery provide a permanent tribute to the work carried out by Gillies.[30]

Gillies' key technique involved reopening original wounds that had been stitched together at the CCS and adding flaps of skin tissue removed from other parts of the body. It was a slow process that often required more than one operation, but when it worked it could be extremely effective in restoring the features of a man's face. At the start of an operation, Gillies would mark out, usually on the man's chest, the shape of the face he intended to cut out, including the skin for the eyes, nose, cheeks, or for the whole face if needed. He soon developed a variety of different techniques for rebuilding different parts of the face, and gave them exotic names like the Bishop's Mitre Flap, the Caterpillar Flap and the Transposition Flap. He also performed some bone grafting. With these techniques, noses were restored, jaws rebuilt and gaping holes filled, enabling men to face the world again – or, at least, in the case of some of the less disfigured, to return to the trenches.

The chief consultant at the hospital in Aldershot, Sir William Arbuthnot Lane, soon began to take an interest in Gillies' radical techniques. In the summer of 1916, anticipating the arrival of a large number of cases from the Battle of the Somme, Lane showed his confidence in the new form of surgery by allocating an extra 200 beds to Gillies' ward. But this was nothing like enough; within a ten-day period, two thousand men had arrived with desperate and grotesque face wounds that needed urgent attention. Some had had much of their face blown away. Many were encased in bandages, unable to speak or eat; some were unable to see. Gillies and his team worked around the clock, seven days a week, doing what they could to heal some of the worst wounds that any of them had witnessed. Gillies himself was constantly present, visiting the wards. He became a familiar figure, always encouraging and cheerful, known for his plain talking and for attending to all the wounded, officers and men alike. Sometimes a man had to endure up to twenty operations over several months before his face was put back together again. Aware of the need for patients to have faith in him in order to be able to confront the long process ahead of them, Gillies exuded an air of calm and confidence. 'Don't worry, sonny,' he would say to a horribly mutilated patient, 'you'll be all right and have as good a face as most of us before we've finished with you.'

The face is so much part of our identity, of our sense of who we are, that it is difficult to imagine the psychological effect of having one's face so badly burned or disfigured that one is unrecognisable. Worse still, as was sometimes reported with Gillies' patients, a man's wife or children could be so repulsed at the ghastly sight of their husband or father that they did not want to see him again. Mirrors were consequently banned in the ward. But with many of the patients having lost the will to live, Gillies and his team did what they could to restore not only a man's face but also his self-confidence and his ability to go out and interact with people once again. And Gillies had a remarkable success rate, although by no means did he solve every case. Slowly, the press began to pick up on the reconstruction work that he and his team were doing. When reports got back to the army about Gillies' extraordinary achievements, without doubt they helped to lift morale and instil confidence that a man would be cared for no matter how cruel his wounds.

By the summer of 1917, the ward in Aldershot had become insufficient to deal with the number of cases arriving. Keogh decided the RAMC should set up a completely new hospital and acquired land at Sidcup in south-east London, taking over an eighteenth-century country estate. The nursing and medical staff moved into the mansion house and a large oval of single-storey wards were built in the grounds, equipped with operating theatres, X-rays, psychotherapy rooms and a large admission block, as well as a small area for Gillies to play golf. Tonks was given a studio in which to draw and another studio was provided for a group of sculptors who made plaster casts of patients' faces in order for Gillies and his team to plan procedures in detail before operating. Gillies and the team arrived at Sidcup from Aldershot on 18 August 1917 and were almost immediately overwhelmed with patients arriving from the Battle of Passchendaele. 'We literally put down our suitcases and picked up our needle holders,' Gillies later wrote. 'Is there a better way to open a hospital?'[31]

Once again, the team worked at full stretch, struggling to cope with the huge volume of patients. Sir William Arbuthnot Lane had great ambitions for the new establishment, and wrote to Gillies saying, 'I want to make Sidcup the *biggest* and *most important* hospital for jaws and plastic work in the world and you consequently a leader in this form of surgery.'[32] And indeed, Queen's Hospital soon became famous for its reconstructive surgery. Surgeons came from Australia and Canada to practise alongside Gillies, and observers arrived from the United States to watch and learn, taking many ideas away with them.

On 3 October, Gillies was carrying out an operation on an able seaman who had been horribly burned in a cordite explosion at the Battle of Jutland, eighteen months before. It was one of the worst burn cases the team had ever had to deal with. The man's nose, lips and eyelids had been destroyed. 'Appalling' was the word Gillies used to describe it. They cut a large piece of skin from the man's chest and folded it over his face with the lower end still attached to ensure an adequate blood supply. When stitched into position it was given an additional blood supply from two thinner strips of skin raised from the shoulders. It was while raising these strips that Gillies had a flash of inspiration. 'If I stitched the edges of those flaps together, might I not create a tube of living tissue which would increase the blood supply to the grafts, close them to infection, and be far less liable to contract or degenerate as the older methods were?'[33] Two weeks later the growth of the new skin had progressed well and over a period of eighteen months the restoration of the seaman's face was completed. This new technique using what Gillies called 'tube pedicles' transformed skin grafting. It increased both the likelihood of the new skin taking to the face and the possibility of making shattered faces recognisable once again. The tubes of skin could be applied to any part of the face and became a common feature in the hospital's wards. The technique marked another significant breakthrough in the development of Gillies' work.

By late 1917, there were one thousand beds available at Queen's and a group of local satellite hospitals. Several distinguished visitors came down to see what was going on, most of them going away horrified but impressed. The Prince of Wales came to visit in 1918 and, against advice, insisted on visiting two wards that were known to the staff as 'the Chamber of Horrors'. He apparently came out 'looking white and shaken'.[34]

By the end of the war, the Queen's Hospital had become internationally famous. Gillies and his team had carried out more than 11,500 major face operations, pioneering along the way nothing less than a new form of plastic surgery. Compliments poured in. Some of the men treated returned to civilian life and reported that they had been able to carry on normal lives with barely any signs of the terrible scars from which they had suffered. Others still felt difficulties assimilating into society and lived the rest of their lives alone or out of sight of most people. Overall it was an extraordinary achievement for Gillies and the dedicated team of men and women working with him, carrying out what must have been the most unpleasant form of surgery imaginable. And it was an extreme example of how the laboratory of war could provide an opportunity for medicine and science to advance.

Like so many participants in the Great War, the military medical services that went to war in August 1914 made some dreadful errors at first. But with the help of outsiders and consultants, and by the development of new specialisms, great strides were made and the slaughter of industrial war was tackled more effectively than might be imagined. The RAMC showed considerable flexibility in adapting and developing its procedures to deal with circumstances that had not been anticipated when the war began. The medical services available to the British Army that won victory in the autumn of 1918 were dramatically improved by comparison with those of the army that had gone to war four years before. The strength of the RAMC had gone up nearly ninefold, from 18,000 officers and men in 1914 to 160,000 in 1918. Ironically, it was the Home Front that suffered. There were shortages of both general practitioners as well as consultants at home, and in many of the large inner-city hospitals work was undertaken by unqualified medical students.[35]

Of the two million British soldiers who received battlefield wounds in France and Flanders from 1914 to 1918 and were treated in military hospitals, about 7 per cent died of their injuries and 26 per cent were returned to duty.[36] The survival rates were much higher than those of previous wars and they suggest that the knowledge that good medical care existed must have helped keep up the morale of the British Tommy even during the worst days of killing on the Western Front. About three-quarters of a million British soldiers, sailors and airmen died during the Great War.[37] That number was terrible enough. But had medicine and science not rallied to the help of the armed services, it would have been far higher.

11

The Mind

In November and December 1914 an increasing number of men with strange and unusual problems began to arrive in the Casualty Clearing Stations behind the newly established front lines of the Western Front. The men showed no visible signs of physical injury. They had not been hit by machine gun bullets, nor had they been struck by shrapnel. Their limbs had not been damaged. They had no apparent wounds to the head. Some of them had minor cuts and bruises but nothing more severe. But they all seemed to display similar strange symptoms that puzzled their doctors. They were suffering from peculiar forms of paralysis. Many were described as having 'the shakes'. Some could not stand up or walk normally. A few appeared unable to speak coherently and were stuttering badly; others had been struck completely dumb and could not speak at all. Most appeared to be in a state of stupor and a few had completely lost their memory. Still others seemed to find it difficult to see clearly. Many seemed to have lost their sense of taste or smell. Some vomited repeatedly.

The doctors who tried to attend to them had never seen such symptoms before and were unsure how to respond. Many of the men were sent back to England with 'nervous and mental shock' but the War Office began to grow alarmed at the numbers of men being evacuated. After a few months, 7–10 per cent of officers and 3–4 per cent of other ranks had been sent home to recover. At this rate, the British Expeditionary Force, having already lost one-third of its strength in a single battle at Ypres in October, would cease to be a fighting force in a matter of months.[1] Something clearly had to be done.

The army turned first to an unlikely candidate. Charles Samuel Myers was a lively intellectual from a wealthy Jewish family in London. In 1914 he was in middle age, an academic who had spent most of his career helping to establish the new science of psychology. He had grown up enjoying a wide range of interests including mountain climbing and tennis and was a talented violinist, but when he went up to Gonville and Caius College, Cambridge in 1892 he chose to concentrate on science. After graduating he went on to St Bartholomew's Hospital in London but opted not to go into medical practice. Instead he joined a Cambridge anthropological expedition to the Torres Strait, where for the first time the techniques of modern science were to be applied to the study of so-called 'primitive' peoples of New Guinea. Myers spent a year studying the attitude of the natives to music while his Cambridge colleagues, William Rivers and William McDougall, recorded marriage customs and other aspects of their tribal society.

When Myers came back from New Guinea, he joined Rivers in teaching psychology at Cambridge, and did much to lay down the foundations of this new science in the decade before the war. In 1912 he raised funds to establish at the university an experimental laboratory in psychology, the first of its kind in Britain. Soon after this he became a Fellow of the Royal Society. When war was declared, he sought a role in the growing conflict but was turned down by the War Office because of his age. So he went to Paris to visit the internationally renowned neurologist Jules Dejerine, before securing himself a position at the Duchess of Westminster's War Hospital at Le Touquet, one of the private hospitals attached at this early phase of the war to the British Army. It was here that in the last weeks of

1914 a group of soldiers arrived suffering from various forms of mental neurosis. Other doctors avoided these patients, but for Myers they provided a fascinating insight into a new type of disease.

Myers treated one soldier who had been trapped for hours in barbed wire in no man's land. While out there several 8in shells had burst near him. The man, who had been cheerful and positive before this terrifying experience, was eventually brought back to the British lines in a pathetic state, crying and shivering in a cold sweat. His escape was described as 'a sheer miracle'. He appeared to be suffering from blurred vision and felt a burning sensation in his eyes, making him panic that he was going blind. Writing about this case several years later, Myers described it as a turning-point: 'It was clear to me that my previous psychological training and my present interests fitted me for the treatment of these cases.' He concluded that although the soldier was not wounded, he had suffered some form of physical concussion from the proximity of the shell explosions and that 'the high frequency vibrations' had caused 'an invisibly fine molecular commotion in the brain'.[2] Myers believed that the man was now displaying the symptoms of this physical disturbance. He wrote up the case, along with a few others, in the doctors' journal *The Lancet* in early 1915 and described it, using a term coined by the soldiers themselves, as 'shell shock'.[3]

The name caught on immediately, and rapidly became the generic term for a wide range of peculiar mental symptoms that doctors could not easily explain. It was simple, but had instant resonance with the strange cases that were coming in from the front. Doctors used the words shell shock to describe every sort of nervous breakdown. Although the stalemate of trench warfare was only a few months old, it was quickly realised that this form of immobile war was a new phenomenon. Men hunkering down in a trench with shells constantly landing all around them but unable to exercise the instinctive human response to get away, were suffering from extreme forms of anxiety or stress. The shells were far more lethal than in previous conflicts and now consisted of high explosives that could be fired from miles away in rapid succession. Sudden, horrific and seemingly random death became a feature of trench life that every soldier had to live with. Tom Pear, a young academic psychologist, wrote that conditions in the trenches were unique: 'Never in the history of mankind have the stresses and strains laid upon body and mind been so great or so numerous as in the present war.'[4]

The public soon became fascinated. Over the next few months, the press devoted hundreds of pages to the subject, analysing its causes and describing its effects. Broadly speaking, the newspaper debate created a sympathetic attitude among the public towards the victims of shell shock. Within the army, too, the term rapidly became commonplace and the public reaction to it forced the authorities to acknowledge its existence as a specific condition.

It was in this context that, in March 1915, the army turned to Myers to deal with what was happening. He was given a commission in the RAMC and quickly dispatched to the Base Hospital at Boulogne. As a psychologist, Myers believed not only that he understood some of the causes of shell shock but that he could help to treat them as well. As an army doctor he accepted that his responsibility was to cure wounded men and get them back to their units as soon as possible. Within days of arriving he was dealing with several extreme cases. One young soldier was convinced he was still in the trenches and spent all his time dodging shells while hiding under his bed. Other men had dreadful stoops and could not walk straight. Myers treated one man who had been struck mute by a dreadful experience in the trenches, and within days was able to restore his power of speech. Using hypnosis at times to try and cure patients, Myers brought to his new role a sort of evangelical energy. He was thoroughly committed to showing that if the causes could be understood, shell shock could be treated successfully.

The army found it difficult to cope with the growing number of shell shock cases. Soldiers, it was firmly believed among the military, needed regular, tough training and discipline, a harsh regime out of which the pride, cohesion and strength of a fighting unit would emerge. In the military view, soldiers were either fit and capable, or sick and wounded and so unable to fight. Forms of mental disorder were somewhere in between the two and, in this simplistic way of seeing things, complicated matters dreadfully. Soldiers were supposed to put up with difficult conditions and show a stiff upper lip in the face of adversity. In this context, mental breakdown – then called hysteria – was traditionally thought of as being un-masculine. Named from the Greek word for the womb, hysteria had until not long before the war been thought of as a woman's condition. Showing signs of hysteria was therefore seen as an indication of weakness. Moreover, a hysteric might let down the rest of the troop.

As with all such matters, the army soon began to distinguish between the mental disorders of officers and men. It was believed that the men in the trenches could do nothing but passively watch and wait while shells exploded all around them. Their symptoms, of being struck deaf or dumb, or being paralysed, were often diagnosed as hysterical. But this term was rarely applied to officers, who had duties to perform and responsibilities to attend to, and so were seen as being more active. For them the more scientific term 'neurasthenic' was used. Neurasthenia was thought to be a consequence of the long, gradual wearing down of an officer's emotional strength, a build-up of anxiety that often showed itself in symptoms of depression or suffering from bad nightmares.

The army was struggling to understand what has become quite obvious over the century since the Great War. Men under fire can suffer from a mental breakdown caused by the trauma they have endured. Today it is called post-traumatic stress disorder. Its symptoms are often physical but the cause is psychological. However, in the Great War many traditionally minded commanding officers thought that all forms of mental illness were a display of weakness. Their perception of how to treat a mental breakdown was governed by a moral judgement. But it was soon realised that the strange phenomenon could affect anyone. Even the best soldiers with the strongest 'nerves' and the most reliable officers, those who had a faultless record, could suddenly snap and break down. Many senior officers found themselves torn between the traditionalist view that all neuroses were some kind of weakness, and a realisation that this could happen to any soldier, even to their own sons serving dutifully at the front. But, for those officers, what lay behind everything was the fear that if the situation got worse and some sort of mass hysteria broke out, the army would collapse and cease to exist as a fighting organisation. This of course had to be prevented at all costs.

So the reaction of many senior figures – and indeed of several medical officers – was to suspect that many of those exhibiting symptoms of shell shock were in reality what in army parlance were called 'skrimshankers and malingerers', that is they were pretending to have shell shock as a means to escape the horror of the trenches. Medical officers complained to Myers, 'We have seen too many dirty sneaks go down the line under the term "shell shock" to feel any great sympathy for the condition.' '"Shell shock" should be abolished' was another response.[5] Myers himself accepted that there were several instances of soldiers swinging the lead and trying it on. Men would turn up at a CCS and when asked what was wrong with them would boast, 'Suffering from shell shock, sir.' The army clearly had to pick out the malingerers and return them to their unit as fast as possible. Senior officers expressed this as the need to prevent 'wastage'.

There was no single cause for the variety of symptoms that contributed to the condition known as shell shock. Many men felt a considerable sense of guilt and shame in being diagnosed, a sense that they had let their peers down. This intensified victims' anxiety. Some men would hold on as long as they could, but the accumulated stress would finally make them snap. Captain H. Kaye, a temporary doctor in the RAMC, observed one officer who had carried on despite witnessing the almost total destruction of his battalion three times in heavy fighting during the Second Battle of Ypres in the spring of 1915. He then found himself in a dugout that received a direct hit, killing three of his comrades but leaving him visibly unhurt apart from singed hair. He busied himself burying his comrades and still carried on. Then, after a shell landed nearby, he was buried by debris for several hours before being dug out. Still he continued in command. It was only when the quartermaster brought up a string of horses, not knowing that the officers for whom they were intended had been killed, that the man completely broke down and cried for days.[6] It was the cumulative effect of the horrors of trench life that did for many.

The cause of shell shock became the subject of much earnest debate among the medical community, both within the army and to a degree in the press. Was it caused by the physical consequences of being near an explosion, such as the sudden change of atmospheric pressure? By the mental strains of fighting in a modern industrial war? Or by the emotional reaction to witnessing some horrific scene, like seeing a friend blown to pieces or, in the case of Dr Kaye's patient, simply seeing a group of horses all of whose riders were dead? As more men sought treatment with an increasing range of physical problems that clearly had some mental roots, the doctors looked on in amazement. 'I wish you could be here in this orgie [sic] of neuroses and psychoses and gaits and paralyses,' wrote the Professor of Medicine at Oxford to a friend. 'I cannot imagine what has got into the central nervous system of the men ... Hysterical dumbness, deafness, blindness, anaesthesia [sic] galore.'[7]

In its typical way, the army decided to categorise cases to make the problem easier to deal with. Victims who had suffered from the explosive shock of a nearby shell were classed as 'Shell Shock W' (for Wounded). This was the original sense of the term shell shock – some sort of concussion caused by the proximity of an explosion. These men were acknowledged as being physically wounded and were given the care and dignity that they therefore deserved. Those thought to be suffering from some sort of hysterical response, a temporary breakdown of the nerves, were classed as 'Shell Shock S' (for Sickness). They were not classed as genuinely wounded and were to be returned to their units after rest, relaxation, regular meals and a period in a positive environment, during which doctors were to encourage them to feel better and to want to return to their duties. Sometimes this amounted to little more for a soldier than taking a break from the trenches for a few days and then being told by a medical officer to pull himself together and get back to his battalion. The final category, almost always reserved for officers, was neurasthenia, caused by prolonged mental strain and manifested by symptoms of chronic fatigue, headache and the loss of appetite.

This classification did little to improve the already confused situation. Obvious inequities soon became apparent. Men suffering from a genuine breakdown were not always given proper treatment, while others with similar or even less serious symptoms were evacuated to England. Myers noted the case of an artillery officer whose battery came under heavy bombardment during which he managed to keep going for as long as possible but eventually collapsed. He was diagnosed with a nervous complaint and categorised 'Shell Shock S'. Two of his men who gave way as soon as the bombardment began were categorised as 'Shell Shock W', having suffered according to regulations from the 'effects of an explosion due to enemy action'. According to Myers, the two soldiers 'by giving way immediately, became entitled to rank as wounded and wear a wound stripe'; the officer, by bravely carrying on, was sent down stigmatised as 'nervous'.[8]

One thing was clear to Myers: all attempts to treat shell shock victims should be carried out quickly and as near to the front as possible, if there was to be any hope of returning them to their units. Once they were removed from the theatre of war by being evacuated to convalescent homes in England, the time taken for recovery would increase considerably. Myers noted that a popular and strong regimental medical officer could often persuade a man to make a rapid return, whereas if once sent 'further down the Line, it may take many weeks or months before they are again fit for duty'.[9] He tried to persuade the army to set up specialist wards only a few miles from the front. This principle of 'proximity' has since become the basis for all modern military psychiatry, but there was hostility to it at the time, as the traditional army view was that it did not want to be 'encumbered with lunatics in Army areas'.[10] Myers at least had some success here, even if the approach to treating shell shock victims varied enormously.

During 1915 and the early months of 1916, as Myers toured the CCSs and Base Hospitals and examined some 2000 patients, debate continued to rage about the causes of the multitude of conditions. Frederick W. Mott, a neurologist at the Maudsley Hospital in south London, had not been to the front but had studied patients evacuated to London. He argued in a series of lectures that exposure to shell fire had a pathological effect on the body's nervous system, possibly caused by concussion to the brain from the impact of an explosion. But he also agreed that psychological factors played a part and that a man's state of mind before an explosion might well affect his reaction to it. Some people were more disposed to suffer from the horrors of trench life than others, Mott believed. He argued that fear was a biological instinct and the anticipation of death or mutilation was a major cause in what he called the neuroses of war.[11]

Another view came from Harold Wiltshire, an experienced physician who worked at a Base Hospital in France for a year before concluding that the symptoms of shell shock were entirely of a psychological rather than a physical nature. He observed that men who had lost limbs in shell explosions did not suffer from shell shock. In hospital they were often cheery and supportive of the medical and nursing staff. This contrasted with the morose gloom and lack of hope evident among patients in a shell shock ward. He believed that men suffering from shell shock had been worn down by the prolonged strain of trench warfare into a position where a further sudden psychological shock could tip them over the edge. He cited the example of a soldier who experienced mental shock having been ordered to clear away the remains of a number of comrades who had been blown to pieces by a shell. It was becoming clear to those working with patients that the term shell shock was, as Myers himself later put it, 'a singularly ill-chosen term; and in other respects . . . a singularly harmful one'.[12] In the vast majority of cases, Myers accepted, shell shock had psychological and not physical causes. But by this point, the term was too well established to be abandoned. It served its purpose in helping to delineate between traditional concepts of courage and cowardice under fire. This was the state of the debate when in the summer of 1916 everything went up a gear.

The Battle of the Somme opened on 1 July. Detailed preparations had been made, as we have seen, for the treatment of the wounded. But the medical authorities were overwhelmed by the number of shell shock cases. In one division the number of 'Shell Shock W' cases increased fivefold in a few weeks, amounting to one in six of all casualties. The number of 'Shell Shock S' cases were not recorded but Ben Shephard, a historian of military psychiatry, has concluded that the number of official casualties could probably be 'multiplied by at least three to give a real sense of the scale of the problem'.[13] Moreover, it was noticed that the incidence of shell shock in all its forms was much higher in some units than in others. Once again, the War Office became concerned that losses on this level would undermine the army's ability to continue waging war. It was clear that a tough stance had to be taken, even though Myers had spent a year trying to persuade the authorities to be sympathetic to the sad victims of shell shock. This approach now went into reverse.

One case clearly reveals the hardening of attitude. The 11th Battalion of the Border Regiment was a Pals Battalion recruited by the Earl of Lonsdale from the farm labourers, industrial workers and miners around his estate in Cumberland and Westmorland. On 1 July 1916 the battalion went over the top in the first-wave assault near Thiepval on the Somme, but was massacred by German machine gunners as the men tried to get through barbed wire that the artillery barrage had failed to cut. The unit lost 516 men (out of about 850) during the course of the morning, including its commanding officer, Lieutenant-Colonel Percy Machell, a popular figure with the men who at the age of fifty-four still joined in the assault and was hit as soon as he climbed out of the trench. All but three of the officers in the battalion were killed along with him.

Just over a week later, on the evening of 9 July, the remains of the battalion, about 250 men, were recovering behind the lines when the newly appointed officers were told to select 100 men to carry out a trench raid. The medical officer, Lieutenant Kirkwood, reported that many of the men claimed to be unfit for duty suffering from shell shock and were unwilling to go over the top again. He made out a certificate that explained in some detail why this was the case: the men were still suffering from the demoralising effect of the assault on 1 July, they had spent a week digging out the dead in an atmosphere of decomposed bodies, and they had been under continuous shellfire since going over the top. He concluded that 'few, if any, are not suffering from some small degree of shell shock'.[14] Ignoring this evidence, Brigadier J.B. Jardine ordered that the raid should go ahead anyway. It was a complete failure.

The army ordered a court of inquiry, which blamed the brigadier for not knowing the state of the men in one of his battalions. He, in turn, blamed the non-commissioned officers for failing to set a good example in the absence of officers. General Sir Hubert Gough partly blamed the men, saying it was 'inconceivable' that British soldiers could 'show an utter want of the manly spirit and courage which at least is expected of every soldier and every Britisher'. But the full weight of his wrath fell on the medical officer, Lieutenant Kirkwood. Gough said that 'so long as he is allowed to remain in the service he will be a source of danger to it ... [and] it is not for a MO to inform a CO that his men are not in a fit state to carry out a military operation.' Kirkwood's sympathy for the mental state of his men had hit a nerve. Although the 11th Battalion had suffered badly, it had by no means endured the worst losses on the first day of the Somme. Martin Middlebrook lists thirty-two battalions that had suffered casualties of more than 500 men on that day.[15] But the senior command realised it had to take a stand to prevent the collapse from shell shock becoming an epidemic and getting, in its eyes, out of hand. Kirkwood was dismissed from the army in disgrace.

Sir Arthur Sloggett, head of the RAMC in France, came to Kirkwood's defence, claiming he was a 'scapegoat'. He persuaded the authorities to reassign Kirkwood from his front line MO role to a humble position in a Base Hospital. But without doubt, the word would have got around. The army was going to take a hard line on anyone who was not tough on shell shock.[16]

In early August, GHQ issued new demands to reduce 'wastage' and cut down on the numbers of shell shock cases being evacuated back to England. This was a chance for Myers once again to argue that men with mental disorders should be treated as quickly as possible near the front. For the first time his own views coincided with those of the high command. Appointed to the senior position of Consulting Psychologist to the Army, he toured the front meeting generals and inspecting aid stations. For the first time a special treatment centre for shell shock cases was set up in a CCS and Dr William Brown, a former pupil of Myers, was put in charge of it. But Myers' star crashed as quickly as it had risen. The belief of senior commanders in his work had never been very genuine and with the Battle of the Somme over, in January 1917, he was summarily removed from his senior position. Myers returned home to England on two months' sick leave, frustrated and embittered by his experiences on the Western Front and his failure to change the views of the generals.

Myers' replacement was Gordon Holmes. The son of an Anglo-Irish landowner, Holmes had read medicine at Trinity College, Dublin and in 1901 started work as a house physician at the National Hospital for Nervous Diseases in Queen Square, London. Here he carried out detailed examinations of the brain and began to realise that its various parts, such as the cerebellum, the thalamus and the cerebral cortex, accounted for different sense perceptions. Holmes soon acquired a reputation as a leading neurologist. After the outbreak of war, he worked for some time at the Base Hospital in

Boulogne, where he came to the attention of Sir Arthur Sloggett. Holmes' views, it was noted, were far more in tune than Myers' with the attitudes of senior commanders in the army. He was a disciplinarian who believed that hysteria could easily spread from one soldier to another and had to be dealt with 'in no uncertain fashion', otherwise the army would soon be drained of its fighting spirit.[17] He believed the role of the military neurologist was merely to remove the physical symptoms of distressed soldiers rather than try to find their psychological causes.

Claire Tisdall, a young volunteer nurse, later recalled the shame and lack of compassion exhibited towards these tragic victims. She was collecting a group of patients from an ambulance train one day when another ambulance appeared unexpectedly. The second train was totally closed up, with all its windows barred and shut. 'What's this ambulance coming in?' she asked. 'Haven't we done the train?' 'No, sister,' came the reply. 'This is for the asylum; it's for the hopeless mental cases.' The young nurse later wrote, 'I didn't look. They'd gone off their heads. I didn't want to see them. There was nothing you could do and they were going to a special place. They were terrible.'[18]

Despite the hardened attitudes, official figures indicate that the incidence of shell shock in the Battle of Passchendaele in the summer of 1917, at about 1 per cent of the men who took part, was far lower than it had been during the Somme. Partly this might have been down to administrative changes. In order to reduce the numbers being classified as 'Shell Shock W', the army had invented a new category of 'NYDN', meaning 'Not Yet Diagnosed Nervous'. It is possible that thousands of men were put in this category and waited for long periods before the paperwork was completed to finalise their condition. Regimental medical officers were given much more authority to decide on cases.[19] And although there was still no consensus on what caused nervous breakdowns, more and more doctors diagnosed men as simply exhausted. They called for a week or two of rest and sleep, and a few firm words of reassurance.

William Johnson, who worked in No. 62 CCS in Flanders, claimed to be able to treat a large number of patients, even restoring speech to those who had become mute or restoring the memories of those with amnesia. He would summon a man into his office, ask him to lie down on a couch, and tell him to rest, close his eyes and give himself up to sleep. In a state of light hypnosis, Johnson would then instruct the patient to return to the trenches and relive his particular moment of shock. At first the soldier would twist and turn and sometimes cry out. But slowly, by describing his moment of terror, he would improve; it was as though describing the problem under hypnosis would lift it from his mind and slowly enable him to work out his neurosis. Sometimes, Johnson could cure a man in about twenty minutes. Getting men back to their unit was, after all, the objective of the military medical officer.

Of course, some cases could not be cured in this way, and men were still evacuated to England displaying the ghastly effects of shell shock. At a hospital in Devon, the freakish behaviour of certain shell shock victims was recorded on film. The short clips show tragic scenes of men who cannot walk and who roll about on the floor, who shake uncontrollably, who leap under the bed at the mere mention of the word 'bomb'. The filming took place at Seale Hayne Hospital on the edge of Dartmoor. Here, Dr Arthur Hurst practised. A great showman, Hurst made countless claims as to his ability to cure the symptoms of hysteria, which he was able to do using his strong powers of persuasion, sometimes backed with hypnosis, physical manipulation or even electric shock treatment. The expectation that a patient would receive a miracle cure was drummed into him from the moment he arrived at the hospital, while the build-up to meeting the doctor on the day of the treatment made it into almost a religious experience. The purpose of the films made at Seale Hayne was to show victims before and after their treatment with Hurst and certainly, the patients on film appear to be cured. After treatment they walk briskly and confidently by the camera, feed chickens or work happily in the hospital farm. Myers and other critics, however, were not convinced that removing the *symptoms* of hysteria was the same as curing the problem. There were stories that by the time some patients had arrived back in London on the train, they had developed new symptoms and were almost as bad as before.[20]

At a hospital just outside Liverpool, a different and far more radical approach was taken in an effort to solve the underlying psychological causes of shell shock. Maghull Hospital (later called Moss Side Hospital) had been built before the war as a large, airy and spacious establishment to treat epileptics. When the War Office requisitioned it, Ronald Rows was put in charge. A reformer of mental hospitals who had an interest in Freud, Jung and psychoanalysis, he encouraged a caring approach to the shell shock victims who soon filled the beds. Doctors were to talk to the patients and to use their dreams (or nightmares) to try to understand what was at the root of their fears. Frequently, finding the cause of the problem and talking it through would make the fear disappear, a form of psychotherapy.

Rows looked for the emotional origins of a soldier's problem. He soon attracted a rush of bright young academics from the universities who saw work at Maghull as a brilliant opportunity to look into men's minds and to apply the ideas of Dejerine and Freud. They included Tom Pear, who went on to become a professor of psychology at Manchester University, and Grafton Elliott Smith, Professor of Anatomy at the same university. Pear would tell his patients reassuringly, 'You are suffering from an illness. It's called mental illness. You are not mad and you are not a lunatic.'[21] Maghull was rare among British hospitals in showing real sympathy and understanding for its shell shock patients. It was a classic case whereby the war created a human laboratory for some of the brightest young psychologists of the day.

One of those who passed through Maghull, William Halse Rivers, has since become perhaps the most famous of the Great War psychologists. Rivers was in his early fifties at the time and had done a range of work before the war. He had studied medicine at St

Bartholomew's in London, where he became the youngest graduate in the hospital's history; he had worked as a physician in the National Hospital for Nervous Diseases in Queen Square; he had been a clinician at the Bethlem lunatic asylum; and in 1897 he had helped to establish the Psychology Department at Cambridge University. He was interested in the relationship between mind and body, and at Cambridge he carried out remarkable research with his colleague Henry Head by cutting the nerves in Head's left forearm and over a period of years mapping its recovery of sensory perceptions. In 1898 he had accompanied Myers on the anthropological expedition to the Torres Strait, where he had carried out detailed and painstaking work into the social and belief patterns of the local tribesmen. He went on from here to south-west India, where he continued his pioneering ethnographic studies.

When he arrived at Maghull in July 1915, Rivers must have appeared an unusual figure. A bachelor who rarely smoked or drank, he was something of a recluse and remained obsessed with his anthropological studies. But he slowly began to realise that Maghull offered him a way of opening a window into the minds not of distant tribesmen but of his own fellow Britons. The interpretation of dreams and the understanding of emotional problems provided new areas of study. In 1916, Rivers was commissioned as a captain in the RAMC, and later that year he was sent to Craiglockhart Hospital for Officers just outside Edinburgh, located in a dilapidated Victorian hydro spa that had formerly been used as a convalescent home. Here he was to play a memorable role in the development of techniques to heal shell shock.

Craiglockhart was a humane, friendly and supportive establishment for officers only. By day, the patients could play cricket, tennis, bowls, croquet or use the hydro-pool. But at night, many were still tormented by nightmares. Rivers went back to study Freud and spent more time analysing his own dreams as well as those of his patients. He increasingly felt sympathetic to the pacifists who saw the war as a great folly, and he grew uncomfortable about his role as a military doctor whose principal duty was to prepare men to be sent back to the front. Then, on 4 August 1917, a new patient arrived at Craiglockhart. His name was Siegfried Sassoon.

The relationship between Rivers and Sassoon became one of the most famous doctor–soldier relationships of the war. It was described extensively by Sassoon in his own writings and provides the core for Pat Barker's award-winning 1990s *Regeneration* trilogy of novels, the first of which was made into a film.[22] 'Mad Jack' Sassoon had been a brave soldier and had won an MC in 1916, but while recovering from wounds in London in 1917 he had issued a public declaration denouncing the conflict: 'I am making this statement as an act of wilful defiance of military authority, because I believe that the War is being deliberately prolonged by those who have the power to end it.'[23] However, instead of a court martial, influential friends intervened and arranged for him to be sent to Craiglockhart, supposedly suffering from shell shock. Whether or not Sassoon really was suffering from a nervous breakdown is a moot point, but he had certainly suffered from nightmares in which he saw wounded men from his battalion crawling towards his bed.

Rivers met Sassoon and the pair began a long series of conversations. Rivers had suppressed his homosexual tendencies, while Sassoon was becoming increasingly open about his own, and the two men soon became friends. Sassoon described Rivers as a 'father-confessor' figure – his own father had died when he was seven. After three months of 'treatment' either Rivers persuaded Sassoon to return to the front, or Sassoon himself decided that it was his 'mission' to rejoin his battalion. It was an extraordinary turnaround. He was sent at first to Palestine but eventually returned to France, where he was finally invalided out of the army after being wounded again in July 1918. After his experience with Sassoon, Rivers decided to leave Craiglockhart, and he became a psychologist at the Royal Flying Corps hospital at Hampstead in London. 'Aviators' Neurasthenia' was high in the RFC; in the later stages of the war it affected 50 per cent of all pilot officers, although roughly half of those were successfully treated and returned to full flying duties.[24] Rivers played his part here as well although it is for his treatment of Sassoon at Craiglockhart that he is best remembered.

Another patient treated at Craiglockhart was Wilfred Owen, who was more obviously suffering from shell shock. A sensitive young officer in the Manchester Regiment who had already started to draft some poetry, Owen had gone through some horrific experiences at the front. He had been sent with his platoon to a tiny flooded dugout in no man's land where he spent more than two days under continuous shelling. He had fallen into a cellar and been trapped there for three days, and on a later occasion he was blown out of a trench by a shell blast and covered with debris. Noticing how shaky he was, his fellow officers sent him to the regimental doctor, who diagnosed him with neurasthenia. He was first seen by William Brown at Myers' specialist centre in France. But he was bad enough to be sent back to Britain and to Craiglockhart, suffering from terrible nightmares in which he repeatedly saw the accusing faces of the men he had witnessed being blinded or gassed in front of his eyes, what he called his 'barrag'd nights'.

Depressed, anxious, with a strong sense of guilt and sleeping badly, Owen became the patient at Craiglockhart not of Rivers but of another doctor, Arthur Brock. Brock believed that every patient had it in him to cure himself, all he needed to do was to work at it. So he set Owen tasks like lecturing to the Craiglockhart Field Club, teaching at the Boys' Training Club, editing the house magazine, *The Hydra*, and writing poetry. Owen also introduced himself to Sassoon, whose work he greatly admired. Sassoon later wrote that he did very little for Owen apart from loan him a couple of books, but the combination of Brock's treatment and Sassoon's friendship changed Owen altogether, and his nightmares began to fade as he began to confront his experiences in poetry. He left Craiglockhart towards the end of 1917 a different man, more secure in himself and more confident in his writing. For several months he wrote some of the finest poetry of the war, sometimes revisiting previous experiences, as in

'The Sentry', at others confronting the utter waste and futility of the war, as in 'Dulce et Decorum est', 'Futility' and 'Strange Meeting'. Owen returned to the front in September 1918 and won an MC for bravery. Then, on the early morning of 4 November 1918, while crossing a canal in one of the final advances of the war, he was killed. He was twenty-five years old and the Armistice was signed a week later. Wilfred Owen later came to epitomise the innocent generation sacrificed in the Great War. But it was the treatment of shell shock at Craiglockhart that had helped to make him a great war poet.

Although the experiences of Sassoon and Owen at Craiglockhart are probably the most famous shell shock stories of the Great War, they were in no way typical. The pair were highly sensitive men, treated by remarkable psychologists in a very unusual hospital. As officers they received far more humane, sympathetic treatment than did the men, of whom there were probably six times as many victims. The attitude of wanting to conceal patients who were thought to be suffering from some form of madness was probably more common. Many of the men evacuated to England found themselves in wretched, run-down hospitals where bored assistants probably relied more on using electric shocks than on talking with them and trying to understand the basis of their neuroses.

The use of electricity to treat mentally disturbed patients had begun in the late nineteenth century and was politely known as the 'faradic battery treatment'. The extensive use of electric shock treatment by most of the armies of the Great War was not therefore strictly a wartime innovation. It was, however, understandably much feared by patients. But that was part of the intention. The doctor often delivered it with some theatricality. The nurses would talk to the patient beforehand and explain how wonderful the effects would be. The doctor would then appear and take charge of the patient, forcefully explaining that he must listen to his instructions while the electricity was being 'administered'.

At the National Hospital in Queen Square, Lewis Yealland from Toronto, who had worked in asylums in Canada, adopted the use of electric shock therapy with fervent zeal. He took on patients suffering from hysteria whose arms or legs were paralysed or who had lost the power of speech, and who had defied previous treatment. He used small amounts of electricity at first, but if a patient resisted he applied more powerful charges. Yealland saw it as one of the disciplinary measures necessary to jolt a man into a cure, and many of his treatments became a sort of battle of wills. Perhaps surprisingly, he had a great deal of success. Yealland described curing six cases of mutism one morning in less than half an hour by using 'electricity mixed with persuasion and encouragement'.[25]

One young soldier he treated had collapsed suddenly after two years' loyal and brave service and was unable to speak. Various treatments had been tried over a period of nine months, without success. As a last attempt at a cure the patient was taken to Yealland. He was brought into a darkened room, where Yealland announced, 'You will not leave this room until you are talking as well as you did.' He started the 'treatment' by applying through an electrode attached to the back of the man's throat a shock so severe that the jolt threw him backwards. After an hour the patient was able to say 'Ah', and another half hour later, he could pronounce some vowels. Stronger shocks applied to his larynx slowly enabled him to whisper the days of the week. At each stage, Yealland gave the patient bullying encouragement intended to push him further.

When he seemed almost able to speak again, his left arm began to shake, then his right arm, then both of his legs. Yealland told the poor man, 'It is the same resistive condition only manifested in another part.' Each tremor was stopped by the application of further pulses of electricity. After four hours of continuous treatment the man cried out, 'Doctor, doctor, I am champion,' to which Yealland replied, 'You are a hero.' At the end, the patient asked, 'Why did they not send me to you nine months ago?'[26] Today, it seems difficult to see the difference between treatment like this and torture, but at the time Yealland became famous for his successes and was sent patients from all over the country to 'cure'.

By 1918 the term 'shell shock' was no longer admissible in the army. Doctors, though, were still looking for new explanations for the nervous disorders that continued to appear. When the American army and its surgeons arrived in France they brought an enthusiasm for the new science of endocrinology, applying it to front-line neuroses. In situations of sudden and intense pressure, glands fire adrenalin into the blood, making the heart rate shoot up and helping muscles to perform. This was thought to be a primitive response of the body to 'flight or fight' at a sudden threat. In the trenches under shellfire, neither flight nor fight was possible and it was thought that maybe the products of the glands simply built up and the body was unable to properly process them. Doctors looked at physiology as well as psychology for explanations of neurotic behaviour but still no consensus was reached.

Of course, soldiers in the British Army were not the only ones to suffer from neuroses brought on by the horrors of the war. The French and German armies also experienced severe problems, as did the Americans when they joined the conflict. In the first year of the war it was estimated that 111,000 German soldiers passing through field and military hospitals exhibited some symptoms of psychological disturbances known as *Nervenschock* or *Granaterschütterung* (shell disorder).[27] There were several professors of psychiatry in German universities before the war, along with a network of free public clinics to cater for the mentally disturbed, while the French had been the world leaders in the study of hysteria and the development of the science of psychotherapy. However, the armies of both countries found it as difficult as the British to deal with this explosion of strange new cases. The French army, like the British, found the sooner and the closer to the front line a man could be treated, the more likely he was to be capable of returning speedily to his unit. According to French statistics, about one in three soldiers diagnosed with mental neuroses were returned to the front line within two to three weeks.[28]

The two major French thinkers on the subject profoundly

disagreed as to the causes of shell shock. Jules Dejerine believed that physical symptoms had emotional origins, and that until these were discovered and treated there could be no resolution to a man's mental disabilities. Joseph Babinski, a towering figure in the world of pre-war psychiatry, took an entirely different and far less sympathetic point of view, arguing that the physical symptoms were not caused by the horrors of war but by a victim's auto-suggestion, or by his wish to imitate others who had been taken out of the line. He believed that most patients were malingerers, and under his influence the French ushered in a tough regime known as *traitement brusque*, in which doctors confronted their patients in a battle of wills, shouting at them or administering electric shocks in an attempt to bully them back to work. Hydrotherapy, or the use of cold baths and showers, was also applied as a means of coercion. Dr Clovis Vincent meanwhile was one of the more enthusiastic practitioners in the school that believed in electric shock therapy. In a neurological centre at Tours, Vincent would apply painful levels of electricity while telling his patients to distinguish between the physical reality of pain and their traumatised state of mind. In one celebrated case, a poor soldier wounded in October 1914 who had passed through a succession of hospitals, when faced with Dr Vincent approaching him gleefully holding out two electrodes simply got up and punched the doctor in the face five times. When the soldier was charged with striking a senior officer, the ensuing military trial became something of a cause célèbre in France. While the professional medical world stood behind Dr Vincent, public opinion rallied to the soldier, and the man was finally given nothing worse than a suspended sentence of six months.

In Germany treatment was often equally harsh. At the start of the war, one German specialist, Max Nonne, had firmly believed that no German man could ever suffer from hysteria and that it was a condition unworthy of the German soldier. Confronted by hundreds of cases, Nonne had to rethink his outlook; trying hypnosis, he found that he was good at it and that it often had a beneficial effect. Most other doctors resorted to more familiar techniques, electrotherapy once again being commonly administered. Neurologist Dr Fritz Kaufman made his name thanks to the severity of the treatment he offered in a hospital near Mannheim. Insisting on a strict regime, he hoped to deter any soldier from malingering by the fact that word would spread about the harshness of the conditions. Kaufman believed that if a man had experienced a shock, the best way to treat him was to administer another, physical shock. He would give his patients electric shocks lasting between two and five minutes, while shouting at them that they should be ashamed of themselves and should snap out of their condition. Extraordinarily, in many cases this sadistic approach seemed once again to work. Not only did some of the men recover but they were enormously grateful to Herr Doktor for his treatment.

Sadly, in a few cases, men died undergoing this electric shock treatment. There were about twenty German victims during the war, while others committed suicide rather than face the treatment. By 1918 the German press and members of the provincial assemblies began to speak out about the harshness of the treatment, calling for a more sympathetic approach. The German army defended itself by arguing that if a few patients were lost, the vast majority, at least 50,000 men, had been successfully treated and returned to the front line. More humanely, the Germans were realistic in recognising that soldiers who had suffered a major nervous breakdown should never be sent back to the front, and as a result thousands of wounded soldiers were allocated to work as farm labourers or factory workers for the duration of the war.

For all the armies that fought in the First World War, the purpose of military discipline was to ensure that men obeyed orders when faced with the extreme pressures of the battlefield. And the ultimate way to enforce discipline was through a court martial. In order to discourage others, the military believed, it was necessary to make an example of men who had shown signs of cowardice or indiscipline. Strict military tribunals were held at which a man's commanding officer would be called to give evidence; sometimes a medical officer would be summoned as well, although MOs did not always receive a sympathetic hearing. During the war nearly 2700 men in the British Army were found guilty of cowardice or desertion. Some were classed as 'insane' and were sent to an asylum. For those who were regarded as responsible for their actions, the sentence was death by firing squad. One in nine of those found guilty were actually executed, a total of 306 men. Some of them had received only a basic hearing lasting just a few minutes.

These cases included Sergeant Walton, accused of desertion near Ypres in November 1914. He claimed this was a case of mental confusion and not a wilful act. But the court took no notice of his defence and he was executed on 23 March 1915. Private Harry Farr had been a professional soldier since 1910 and fought bravely until May 1915 when he was evacuated for shell shock. Five months later he returned to the front. Twice more he was kept in an aid station suffering from various forms of shell shock. When, on 16 September 1916, his battalion was sent into action during the Battle of the Somme, Farr pleaded he was sick but the doctors refused to treat him and sent him back up to the front. Farr went missing and was later arrested in the rear. At a court martial he was found guilty of cowardice and sentenced to execution. He was shot on 18 October 1916. His wife, Gertrude, was not at first told of his disgrace; only when it was explained that she would receive no widow's pension did the full story come out.

Private Earp was found guilty of desertion during the build-up to the Battle of the Somme, but his divisional and corps commanders recommended clemency because of his poor mental state. His case went up through the military hierarchy right to General Haig, who confirmed the death sentence. 'How can we ever win if this plea is allowed,' he wrote in the papers. According to Haig's biographer, Gary Sheffield, this indicates both the general's acceptance that Earp had shell shock, and his belief that to commute the sentence would be to legitimise the condition and open the floodgates to thousands of others who would see it as a way of escaping the trenches.[29] Haig clearly believed it was essential to maintain discipline at all costs. He wrote in his diary about the sentencing of Lieutenant E.S. Poole, another soldier with a clear case of shell shock, 'After careful consideration, I confirmed the proceedings ... Such a crime is more serious in the case of an officer than of a man, and also it is highly important that all ranks should realise that the law is the same for an officer as a private.'[30] Haig personally confirmed the death sentences of 253 men and three officers, all of whom were shot.[31]

Even during the war the executions aroused intense controversy. In December 1917 an MP raised the subject in Parliament, asking if any men who suffered from shell shock had been executed. The Under Secretary for War told the House that not a single soldier had been executed 'without being examined before trial and before sentence by a medical officer'.[32] However, thanks to widespread public sympathy for shell shock victims, the matter of military executions would not go away. In 1923 a War Office Committee of Enquiry Report failed to end suspicions that men who should have been regarded as victims of war had been shot rather than given medical treatment. Throughout the inter-war years, whenever the subject came up the authorities denied that there was a single case in which a man suffering from shell shock had been executed. As the records were kept secret it was impossible to verify this one way or the other.

After seventy-five years, however, the Ministry of Defence released the papers of the relevant courts martial, and in the 1990s the topic returned to the headlines. Once more questions were asked about the number of men 'shot at dawn', as the popular phrase had it. The evidence showed that some men's psychological condition had clearly not been taken properly into account. Harry Farr's case became notorious as his widow Gertrude, then aged ninety-nine, and his family led a campaign for justice. Following more questions in Parliament, Prime Minister John Major refused to issue any retrospective pardons, arguing that it was wrong to impose modern judgements retrospectively on historical cases. However, it was becoming clear that many soldiers had been executed despite suffering from various forms of shell shock, and public pressure continued to mount. At the National Memorial Arboretum in Staffordshire a special memorial to the men shot at dawn, showing a blindfolded soldier facing execution, was unveiled in 2001. Finally, in 2006, the Defence Secretary, Des Brown, who had strong personal views on the subject, announced a posthumous pardon for all 306 men who had been executed. Claiming that it would be invidious within the limitations of the surviving records to go through each case one by one, he decided to pardon them all in a single gesture. Their families were of course delighted.

Despite official denials issued over the last hundred years, the evidence is absolutely clear that some men who had psychologically broken down under the strains of the fiercest and most deadly war fought to date were subjected to appalling injustice. The army had never really known how to deal with shell shock cases and was always afraid that if it did not take a firm line, mass hysteria would break out and the fighting spirit of the men would collapse. But that does not condone the treatment of men who, instead of receiving sympathy and understanding for the terrible mental injuries they had suffered, were put up against a wall and shot. The German army executed 48 men and the French army 700 on similar grounds. It is a stain that hangs over not just the British Army, but over many of the armies that fought in the terrible, brutalising conditions of the Great War.

Part Five

Propagandists

12

The War of Words

When war was declared on 4 August 1914 Lord Kitchener, the newly appointed Minister of War, was the only prominent politician who believed it would not be over in a few months. In Cabinet he predicted a devastating war, adopting the full force of modern weapons that would last for at least three years. To fight this war, Britain would need massively to increase the size of its small, professional army, and accordingly, Kitchener made an immediate appeal for volunteers to join up. His appeal led to the production of one of the most famous posters of all time, in which a large picture of Kitchener staring out and pointing his index finger, appealing directly to everyone, was accompanied by a caption of just a few words: 'Your Country Needs YOU'.

The image was immensely powerful. A moustachioed and uniformed Kitchener not only represented the military leadership but epitomised Britain's imperial past; he was the victor of the Sudan, the hero of the Boer War, the commander-in-chief of India and the ruler of Egypt. His face embodied British history as he appealed to every man to join up and do his bit. Alfred Leete, the poster's designer, was a commercial graphic artist who had created adverts for brands like Rowntrees, Bovril and Guinness, as well as for the London Underground. Now he brought successful advertising techniques to the business of building new armies for war. Produced also in a version featuring the words 'Britons [Kitchener] Wants You', and below this, 'Join Your Country's Army', the poster first appeared in September 1914 and Kitchener was soon staring out from thousands of hoardings, shop windows, buses, trams, railway carriages and vans all over the country. It is an image that is still well known today.[1]

Posters aimed at recruiting young men into the army were the government's first attempt at public propaganda, and they were successful beyond even the wildest, most optimistic hopes of the War Office. Tens and then hundreds of thousands of men came forward in a fever of patriotic enthusiasm. Recruiting offices could not cope with the influx. Thousands had to be turned away and asked to come back a week or a month later. Kitchener had expected 100,000 volunteers in the first six months. In fact, by the end of September 500,000 men had already come forward, and about 100,000 volunteered in each of the following fifteen months, after which the numbers went into a marked decline. An all-party group, the Parliamentary Recruiting Committee, took responsibility for the recruitment campaign, producing dozens of different posters over the next year. The message was always simple and clear: 'Come Along Boys! Enlist Today', 'There's Still a Place in the Line for YOU, Will You Fill It?' Some of them appealed to women to encourage their sons, brothers or fiancés to join up, with messages like 'Women of Britain Say "Go!"'

Local figures printed their own posters; in recruiting a battalion of men on his estate near Penrith in Cumberland, the Earl of Lonsdale put up a poster in his racing colours of red, yellow and white that asked bluntly, 'Are you a Man or Are you a Mouse? Are you a man who will for ever be handed down to posterity as a Gallant Patriot ... [or] as a rotter and a coward?'[2] By spring 1916 it was estimated that 12.5 million copies of 164 different posters had been printed and distributed.[3] The result was that more than two million men had volunteered. But none of the later posters had the impact of the first, with Kitchener's bold in-your-face appeal. If the first attempt at domestic propaganda had proved a triumphant success, it was not the case with what came next.

In the first days of war, Kitchener at the War Office and Churchill at the Admiralty formed a press bureau to censor all news about the progress of the war and to control the flow of information from the military to the press. It was effective from the start: the press were not even informed that the British Expeditionary Force was preparing to go to France until three days after it had actually arrived. Journalists and editors were soon very unhappy about what they quickly nicknamed the 'suppress bureau'. But Kitchener had a decidedly old-fashioned view about the press. Seeing them as a nuisance, his main concern was to avoid releasing any information that could possibly be of use to the enemy. He wanted the newspapers to print nothing other than the brief reports of progress here or advances there that had been handed down to them. The wars Kitchener was used to were imperial campaigns fought in foreign lands and usually over quite quickly. He certainly had no conception of how a modern war would come to need a supportive press to keep up public morale at home.

In addition to setting up the press bureau, Kitchener also banned all journalists from visiting the front. Such severe methods failed utterly to recognise the growing power and influence of the press. Even Asquith, the Prime Minister, thought the restrictions went too far and complained to a friend that 'K[itchener] ... has an undisguised contempt for the "public" in all its moods and manifestations'.[4] Kitchener defended himself in the House of Lords in November, saying, 'It is not always easy to decide what information may or may not be dangerous, and whenever there is any doubt, we do not hesitate to prevent publication.'[5] Such an attitude could not be sustained for long in a twentieth-century war.

The legal framework for the press bureau, as for many other restrictions brought in at the beginning of the war, was established in the Defence of the Realm Act (known as DORA), passed in the first weeks of conflict. DORA made it an offence to 'collect, record, publish or communicate' any information that might be 'directly or indirectly useful to the enemy'. Specific clauses prohibited the reporting of information about the movement of troops, ships or aircraft. The Act also made it an offence to publish news 'likely to cause disaffection' among the civil population or the armed services of Britain and her Allies.[6] DORA was not primarily aimed at controlling public opinion, but was intended to enable the government to prevent the flow of sensitive military information and to censor cables and correspondence going overseas. However, it now provided the legal basis on which to control the output of the press, and few newspapers showed any inclination to challenge it. It was clear that with people like Kitchener in charge, an obsession with national security would be paramount and secrecy would be the order of the day.

The result was that for some weeks the British people had little idea of what the war was like. There were no photographs, no reporters and only bland communiqués from the front. For the war pictorials launched at the start of the conflict, artists had to imagine what the war looked like. They looked back to previous conflicts and imagined great cavalry charges and the heroic stance of the noble BEF against the dreadful Germans. As a result, while trying to be serious and accurate, the various illustrated magazines looked more like boys' comics.

When a real story did leak out there was near panic. By the end of August, the British public knew only the barest outlines of the French retreat and their army's action in engaging the Germans at Mons. When *The Times* reporter, Arthur Moore, managed to get hold of the story that the British Army was also in full retreat after the engagement, Lord Northcliffe, the paper's owner, went ahead and published it, reporting the 'terrible defeat of British troops' and of 'broken British regiments'.[7] It was a bombshell to the British public. People read reports gathered by Moore from soldiers who had spent days in a chaotic withdrawal. No one had been remotely aware that the situation was so black. Churchill complained to Northcliffe, 'I think you ought to realise the harm that has been done ... I never saw such panic-stricken stuff written by any war correspondent before.'[8] This was deeply ironic coming from a man who had made his own reputation as a war correspondent on the North West Frontier, in Sudan and during the Boer War, and who was not at all averse to sensationalist writing.

Northcliffe justified the piece by saying it had been approved by the press bureau, and that since it revealed the need for more recruits it was therefore in the public interest. It was true that the report included the words, 'Is an army of exhaustless valour to be borne down by the sheer weight of [German] numbers, while young Englishmen at home play golf and cricket? We want men and we want them now.'[9] And the immediate response was a big rise in the number of volunteers coming forward. The press bureau ought to have learned the lesson of this incident: that they should open up the channels of information so the public were better informed about what was happening with their army in the field. In fact, they drew the opposite conclusion, that what was needed was more control and a greater restriction on the leaking of news.

The irony was that in the first months, the government and the military had little to worry about as public opinion was firmly supportive of the war. There was no need to justify the war other than in the most basic of terms. Two days after war was declared, Prime Minister Asquith addressed a passionate and supportive House of Commons in a rousing speech: 'We are fighting to vindicate the principle that small nationalities are not to be crushed [cheers] in defiance of international good faith, by the arbitrary will of a strong and overmastering power.' Asserting that 'This war has been forced upon us [cheers],' he continued that Britain was 'unsheathing its sword in a just cause [more cheers].'[10] Britain had declared war to protect Belgium, and 'Remember Little Belgium' would become a catchphrase in the early phases of the war.

Britons saw themselves as fighting a moral war, to defend the weak against the strong, to defend what was good against the forces of evil aggression. Political dissension ceased overnight. All the major political parties came together in support of the government. And most of the groups threatening discontent before August, such as the trade unions and the Irish nationalists, declared their support for the war. Even the main body of suffragettes who before the war had threatened the fabric of British society now expressed their support. Christabel Pankhurst wrote in the *Daily Telegraph*, 'The Germans are playing the part of savages, overriding every principle of humanity and morality.'[11] The divisions of the pre-war months were patched up the moment that British soldiers marched off to war.

With the absence of real news, stories of dreadful atrocities and barbarism flourished, fanned by a hysterical popular press. Lurid stories abounded of the murder of civilians, the burning of homes and the mutilation of children. It was a classic case of rumours, unchecked, getting out of hand. When Belgian refugees began to arrive in Britain the stories grew wilder. Everyone seemed to know someone who knew someone who had witnessed an atrocity. Women and children had apparently been gathered and used as human shields by cowardly German soldiers. Young children who had courageously stood in the way of advancing troops had supposedly had their hands amputated. A nurse had her breast cut off and bled to death. Convents had been emptied and the nuns raped. When Rheims Cathedral was shelled on 22 September, it was presented once again as the ultimate infamy of the German war machine.

Such stories could only grow and multiply, of course, if they landed on fertile ground. And the sad fact is that the British people were all too willing in 1914 to believe the worst of their one-time friend and ally. Years of naval, industrial and imperial competition and a fear of all-pervading German *Kultur* was quickly fanned into the belief that the Germans were barbarians committed to an inhuman and overbearing Prussian military culture. Racist cartoons appeared in the press showing obese, swaggering, grotesque German soldiers in pin-helmets banishing everything from their path. The most obscene depicted German soldiers laughing as they paraded babies on the end of their bayonets. Northcliffe's newspapers, which had long been anti-German, led the way. But a belief in German barbarity went right to the top. In October, Asquith spoke of Germans as 'the hordes who leave behind them at every stage of their progress a dismal trail of savagery, of devastation and of destruction worthy of the blackest annals of the history of barbarism'.[12] When even the Prime Minister indulged in talk like this, it was not surprising that very few people were ready to stand up and challenge the prevailing mood of anti-German hysteria.

The fact was that the German army had on occasions behaved harshly in its march across Belgium and northern France. Civilians had frequently been rounded up although only on few occasions had they been maltreated. The Belgian government had, however, encouraged its citizens to conduct a guerrilla campaign against the occupying forces. The Germans found this contrary to the rules of war and as a consequence felt justified in detaining civilians who resisted them. Other attacks on the Germans sometimes had at their root a story that was never reported in Britain. For instance, the press whipped itself into a frenzy over the execution of a British nurse, Edith Cavell, who had stayed behind in occupied Brussels. In the British press she was portrayed as an innocent victim, a martyr to German barbarism. In fact, she had been involved in an underground network to smuggle Allied prisoners out of Belgium, for which she knew the penalty was death. When the French executed two nurses for helping Germans to escape it was not reported.

Other, more generalised stories were also found to have little or no foundation. The story of the nurse having her breast cut off had been entirely made up by the woman's younger sister back home in England. Shelling churches or cathedrals was commonplace on both sides as they could be used as artillery lookouts. A photograph in the *Daily Mirror* of German cavalrymen supposedly holding up stolen booty was actually a pre-war photo of the troops showing off the cups they had won in an army steeplechase tournament.[13] But people almost everywhere believed the stories of atrocities and the accepted image of the German soldier became that of the 'Wicked' or 'Evil' Hun.

The most tragic aspect of this hysteria was that thousands of people with German origins, many of whom had been resident in Britain for decades if not for generations, were suddenly seen as the enemy within. The *Daily Mail* told its readers to refuse service from German waiters and demanded a boycott of German goods. People were told to fear the presence of German spies, particularly near naval bases where it was believed they were cunning enough to be sending signals to ships out at sea. Vigilante committees were formed in towns around the country to protest against what was seen as the lenient attitude of officialdom to enemy aliens. Horatio Bottomley, before the war a convicted swindler, established a newspaper, *John Bull*, that became the ultimate expression of this patriotic fervour, whipping up emotions even further. In its most extreme manifestation, this hysteria led to the banning of German music and even the stoning of dachshunds in the street. Thousands of harmless people with German-sounding names were thrown out of work and made destitute. In October 1914 a form of internment was introduced, although it had to be suspended after a few weeks due to the lack of accommodation for the thousands of internees. The hysteria percolated to the top of society. Prince Louis of Battenberg, the popular First Sea Lord, was accused of being sympathetic to the Kaiser, and was assailed with anonymous letters accusing him of being on the side of Germany. On 24 October, *John Bull* wrote, 'Blood is said to be thicker than water; and we doubt whether all the water in the North Sea would obliterate the ties between the Battenbergs and the Hohenzollerns [the German royal family].'[14] Prince Louis resigned his post four days later.

Lord Haldane, the reforming Secretary of War who by 1914 was Lord Chancellor, was another leading figure suspected of being in league with Germany. It was known that he had gone to a German university, studied German philosophy and admired German values. Now the finger was pointed at him for his army reforms which, it was absurdly claimed had really been intended to wreck the British Army and undermine its ability to fight. Again rumours abounded. It was said that he was in secret correspondence with the German government, even that he was an illegitimate brother of the Kaiser. In 1915, the Conservatives made it a condition that he should leave government (along with Churchill) when they entered the coalition. Haldane, a great believer that logic would always prevail, was deeply scarred by these accusations.[15]

Even the royal family became conscious of its German origins. They sat out this first outburst of anti-German sentiment, but in 1917, on the advice of his advisers, King George V announced that the royals would no longer go by the surname Saxe-Coburg. A committee set up to consider alternatives settled on a suitably solid and appropriate British surname. The royal family would in future be known by the name of Windsor.

In the frenzy of pro-war, anti-German hysteria in these first weeks and months, it was not only journalists but most of the literary establishment who rallied to the cause. Even those who would later come to question the value of continuing the war saw things in simple black and white terms at this stage. Rudyard Kipling wrote in the *Morning Post*, 'However the world pretends to divide itself, there are only two divisions in the world today – human beings and Germans.'[16]

It was not just in Britain that intellectuals and writers came out in support of the war. In France, the leading philosopher Henri Bergson, president of the *Académie des sciences morales et politiques*, was just one of many to enthusiastically take up verbal arms. He could not have put it more simply when as early as 8 August 1914 he declared, 'The fight against Germany is the fight of civilisation against barbarism.' In Germany itself, a group of ninety-three leading intellectuals and professors came together and issued a public declaration entitled an 'Appeal to World Culture' (*Aufruf an die Kulturwelt!*). The signatories were a who's who of German arts and sciences, and included just about every leading academic in the country. They refuted every charge made by the Allies against Germany, from claims that the country was responsible for the war to claims of abuse committed by German soldiers in Belgium. Arguing that because Germany was encircled by its enemies, the war was an act of self-defence and was consequently entirely justified, they denounced the French and the British.[17] Those who took a pacifist position and spoke out against the war – such as George Bernard Shaw and Bertrand Russell in Britain, or Romain Rolland in France – encountered intense personal hostility for their stance. With very few exceptions it was clear that most intellectuals and writers willingly signed up to support their national position.

If the British government had little to worry about in terms of promoting the war at home, it was quite another thing when it came to countries overseas. The Germans bombarded neutral countries from the start with posters, leaflets and pamphlets explaining the reasons for Germany's entry into the war and attempting to discredit the motives and behaviour of its enemies. The British government had no plans to mount a propaganda campaign in neutral countries but grew concerned about the possible growth of anti-British feeling, especially in the United States where German publicity was thought to be particularly virulent. Asquith agreed to ask Charles Masterman, Chancellor of the Duchy of Lancaster, well known as a writer as well as a Liberal politician, to set up a new secret organisation to inform and influence public opinion in the Dominions and in neutral countries. Its official title was the War Propaganda Bureau, but it would be far better known from the name of the building in which it was located, Wellington House.

Masterman was one of a new breed of propagandists (or information controllers) who slowly began to transform the government's attempt to influence public opinion, both abroad and later at home. From a well-established Quaker family, he had grown up in Tunbridge Wells before going to Christ's College, Cambridge. There he became president of the Union, the student debating society, and a leading figure in the University Liberal Club. While still at university he began writing articles for Liberal papers like the *Daily News* and edited the student magazine *Granta*. His two passions were reform, as he came under the influence of the Christian Socialist movement, and literature. After graduating he moved into a tenement block in Camberwell, to live among the poor working class of south London. The experiences he gained there went into various books, including *From the Abyss* (1902) and *The Condition of England* (1909). In 1906 he was elected a Liberal MP and played a central administrative role in the great reforms of 1910–11. He was both Financial Secretary to the Treasury and chair of the National Insurance Commission. Unusually for the time, the latter body employed a series of speakers to travel around the country in order to explain to both workers and employers the benefits of the new National Insurance scheme, a rare example of a government publicity campaign to promote its own policies. At the outbreak of war, then, Masterman was a rare creature, a Cabinet minister with excellent literary and journalistic connections, and with a background in trying to sell the government's message. He was not exactly a man of the people but he had a crusading zeal for causes he took up and could be persuasive and charming when he wished. He was not only an obvious but also a very good choice to take control of propaganda intended for countries overseas.

Masterman had a completely blank canvas. There was no precedent for undertaking the sort of task he now faced. The Foreign Office had a high and mighty view on attempts to influence opinion overseas. According to its traditionalist outlook, diplomacy was for diplomats to conduct between themselves; the less the outside world knew about what they were up to, the better. Any revelations might threaten what was perceived to be the national interest. Often treaties were negotiated in secret, or at least had secret addendums to their public clauses. There was no tradition of transparency at all.

If awkward questions were asked in the House of Commons, the Foreign Secretary simply put a finger to his lips and that was the end of the matter. No further questions would be asked.[18] In this view of international relations there was absolutely no need to consider foreign public opinion, while the gentlemen of the diplomatic corps felt nothing but contempt for what they called the 'gutter press' in Britain. It was a closed, arrogant world. But the war was about to blow it apart.

Masterman started by consulting both the literary and the journalistic worlds. At a conference at the beginning of September 1914 a host of the great literary figures of the day came together, including J.M. Barrie, Arnold Bennett, G.K. Chesterton, Sir Arthur Conan Doyle, John Galsworthy, Thomas Hardy, Rudyard Kipling, John Masefield, Gilbert Murray, G.M. Trevelyan and H.G. Wells. They pledged their full support for the Propaganda Bureau and many would do work for Masterman over the next few years. George Bernard Shaw, the Irish playwright and Fabian, was the only prominent figure who refused to participate. In a pamphlet titled 'Commonsense and the War' he argued that defending Belgium was simply a pretext for going to war to defeat a major rival. The government considered banning the pamphlet, although it never did.

A few days later a group of top newspaper editors and leading journalists met with Masterman. They proposed sending messages outlining the country's case through embassies abroad. Still smarting from the restrictions of the 'suppress bureau', they also demanded that 'unnecessary obstacles to the speedy and unfettered transmission of news should be done away with.'[19] Under pressure, the War Office made a tiny concession to the newspaper editors' demand for access to the front; it agreed to set up a small group of its own writers who would operate as quasi-correspondents at GHQ in France, sending back reports that would be credited as coming from 'Eye-Witness'. The first Eye-Witness was Colonel Ernest Swinton, assistant secretary to the Committee of Imperial Defence, the man whose observations on trench warfare would later prompt him to play a leading role in the development of the tank. Swinton had a more open attitude than his superiors to what could be made available to the press and he soon added a much-needed level of realism to the reporting of military events.

Swinton was the first Eye-Witness but others would soon follow. One of these was Max Aitken, a Canadian millionaire who went to GHQ to write reports about the Canadian troops on the Western Front. Aitken was one of the few relatively classless individuals who had thrived in class-ridden Edwardian Britain. Given his country of origin, he was regarded as something of an outsider. It was also an advantage that he had risen from being a poor country boy to a wealthy Montreal industrialist. The son of a Scottish-born Presbyterian minister, he grew up in Ontario, making his first fortune in securities in Halifax, Nova Scotia before going to Montreal where he acquired a monopoly in the production of cement and made a second fortune. In the wake of a minor scandal about the way the shares had been handled, Aitken left for Britain, taking his money with him.

On arrival he became Unionist MP for Ashton-under-Lyne in 1910. That same year, he bought up Charles Rolls' shares in Rolls-Royce after Rolls' death. But although immensely wealthy, Aitken held radical views on a variety of subjects. For instance, although a committed imperialist he believed in freedom for Ireland as long as Ulster remained in the Union. By 1914 he had acquired a part share in the *Daily Express*, and his attitude to the paper perhaps summed up his attitude to Britain: he wanted it to appeal not only to the lower middle classes, like Northcliffe's *Daily Mail*, but to people of all classes. At the time he was largely successful in expanding its readership. As a well-known financier, he took it upon himself on the declaration of war to try to generate confidence in financial institutions. So on the day war was declared he went to Ashton-under-Lyne and flamboyantly made a deposit of £5000 in the local savings bank. The gesture was much appreciated locally.[20]

In early 1915, the Canadian government appointed Aitken as Eye-Witness at the front. Not only did he send back regular reports on the Canadian troops who fought so bravely in the Second Battle of Ypres, but he wrote the first two volumes of a major history entitled *Canada in Flanders*, following the story of the Canadian troops in battle. He also formed the Canadian War Records Office and assiduously started to collect orders, war diaries and letters, storing them initially in his own office in London. His reporting of events on the battlefield had a great impact in Canada, from where many of his articles were passed on to the United States. When criticism was voiced in London that many Americans seemed to have the impression that Canada was bearing the brunt of the fighting in France, Aitken responded bluntly that the British government should therefore start up its own publicity campaign.[21] That, of course, was precisely what Charles Masterman was trying to do.

Masterman opened shop for the War Propaganda Bureau in the old offices of the National Insurance Commission, which were located in a block of flats in Buckingham Gate. Wellington House would remain headquarters of the bureau for the next three years. Many of the leading figures in the Commission transferred to the new organisation. Masterman took to his task with relish, and the system he created was to set the tone for British official propaganda overseas during the early years of the war. Standing aside from the anti-German hysteria that infected the British press, Masterman wanted to base his propaganda only on reports that could be accurately substantiated. When criticised for not passing on rumours that the Germans had cut the hands off a Belgian baby, he said, 'Find me the name of the hospital where the baby is and get me a signed statement from the doctor and I'll listen.'[22]

Masterman saw his first task as to inform those who helped to create public opinion abroad, including newspaper editors, writers or politicians. He wanted to offer them an accurate, reliable source of information that if challenged could be shown to be based on fact. As all British propaganda was to be carried out in secret, Masterman decided as his second task to use commercial publishers or news agencies to spread the message he wanted to get across.

Consequently, the books and pamphlets produced by Wellington House always appeared to be independent publications, as Masterman thought they would have more impact than propaganda directly issued by the British government. Often, this literature was not distributed freely but was sold for a small sum, as Masterman thought this would give it more value and credibility; he calculated that people were more likely to believe something if they had paid for it. The publications of scholars or experts would also be more likely than blatant British propaganda to impress opinion makers abroad. So Masterman began to set up a network of contacts with publishing houses in Britain and America. By June 1915, Wellington House had distributed two and a half million copies of books, pamphlets and speeches in seventeen different languages, either sending them to influential people, selling them in leading bookshops or placing them in public libraries. This had all been done in total secrecy, with no apparent link to the work of the British government.[23]

Among the academics who worked in secret for Wellington House were two young historians, Arnold Toynbee from Balliol College, Oxford, and Lewis Namier from the London School of Economics. Both men, then in their mid-twenties, would work on propaganda throughout the war, later becoming distinguished figures in the field of British history. The laboratory of war was casting an ever-widening net to draw in expertise and scholarly support.

All British propaganda at this stage of the war was directed at opinion makers, not at the public themselves. Another way of trying to reach newspaper editors, politicians and writers was through the established news agencies, of which Reuters, based in London, was the biggest; its cables went to newspapers throughout the world and its news reports were highly respected. So the government approached Reuters, who agreed to send a certain amount of material on its cables in return for a small subsidy. The Reuters management was concerned at first that this deal would threaten the company's reputation for impartiality, on which much of its credibility as a global news agency relied. But when Roderick Jones, an experienced news reporter and a loyal Reuters man, took over as general manager after the suicide of Baron Herbert de Reuter in April 1915, he justified passing on British official propaganda as straight news by arguing that with British public opinion so firmly behind the war, it was acceptable to present the government line as one that was completely objective. Furthermore, as a patriotic Englishman, he had no qualms about representing the British perspective. Jones also had a further motive. Reuters was in desperate financial straits because of the war and needed the financial help offered by Wellington House to survive.[24]

In addition, Reuters set up a new service to complement its existing news bureau. Known as the Agence Service Reuter, it would supply news in both English and French to Allied and neutral countries in Europe, the Middle East and the Far East. It was entirely funded by the British government, an agreement which served both parties well. Reuters received much-needed funding to ease its financial difficulties and the government could send its propaganda under the cloak of a commercial news agency. As one official noted, the news sent out 'is that of an independent news agency of an objective character with propaganda secretly infused ... it is essential that independence should be preserved.'[25] This was how Reuters reconciled patriotism and impartiality. But the Agence Service Reuter regularly sent out 8000 words a day, and at its peak as much as 60,000 words a day.[26] Many writers had to be found to produce all these words.

One of the writers mobilised by Masterman to fight the war of words was novelist John Buchan, then just forty. Buchan was tall and suave, his good looks marred only by a deep scar on his forehead from a childhood accident. Very sociable, he was much liked for his dry Scottish humour, and was one of the more remarkable characters of the Edwardian era. The son of a Scottish minister who had served a parish in the Gorbals district of Glasgow, Buchan had grown up as a 'son of the Manse', imbibing a strong Calvinist tradition that emphasised the need for hard work. Alongside this he developed a deep love of the Scottish countryside, especially the Borders region around Peebles where relatives lived and where he spent every summer. Buchan attended Glasgow University at the age of just seventeen, but really began to flourish when three years later he went on to Brasenose College, Oxford. While still an undergraduate he published his first novel, became president of the Union and won several literary and history prizes. He was a popular figure, admired for his brilliant, witty conversation. After Oxford, he became a lawyer and was called to the Bar, but he continued to write prolifically, mostly novels but also some popular history. He moved in the top literary and political circles, where he enjoyed the glittering company of many individuals on whom he based characters in his novels. But he still spent as much leisure time as possible rambling in the Scottish highlands. He would think nothing of making a two-day, sixty-mile hike across the mountains of Scotland.

In the aftermath of the Boer War, Buchan had taken a job in South Africa where he served for two years as private secretary to Lord Milner, the High Commissioner, charged with the task of rebuilding the country after the destruction of the war years. This work involved resettling tens of thousands of refugees whom the British had gathered in what were called 'concentration camps'. Milner was particularly impressed by the abilities of his young assistant, who demonstrated not only great administrative skills but also a prodigious ability for hard work. While in South Africa, Buchan was introduced to the world of intelligence and became aware of the work of several remarkable men. On returning to London, he became a director of the publishing house Thomas Nelson & Sons, for whom he edited a set of popular encyclopaedias and started commissioning contemporary writers for a revolutionary new series called Nelson's Sixpenny Classics. Like the popular newspapers of the day, these handy-sized books catered for the tastes of the rapidly growing and increasingly literate lower middle classes, offering good contemporary fiction in an accessible format at a reasonable price.

In the summer of 1914, as the diplomats argued and Europe slid rapidly towards war, Buchan rented a house in Broadstairs, Kent with his wife and young family and spent his spare time writing another novel, what he called a 'shocker'. In this spy thriller he created a new character, Richard Hannay, who would feature in several of his later novels and who was based on the intelligence agents he knew. The storyline played on the idea of a German conspiracy against Britain, a popular theme in fiction since Erskine Childers' bestseller *The Riddle of the Sands* ten years before. But Buchan's story was also very modern in its style and anticipated aspects of the thriller genre of the future. Much of the novel was taken up with a long chase from London to the Highlands of Scotland, in which the innocent Hannay finds himself being pursued by both the adversaries in the plot: suspected by the British police of being a murderer, he is at the same time being hunted by German agents. Published the following year, *The Thirty Nine Steps* was to become probably his most famous novel.[27]

In late 1914, before the book had come out, Buchan suggested to his partners at Nelsons that they should publish a history of the war. He felt that although the full story would not be known for years, it was possible to write a first draft of history, to come out in several volumes roughly three to six months after the events it described. Sir Arthur Conan Doyle was asked to write the *History*. But he was too busy. Nelsons then approached Hilaire Belloc, but at the last minute that deal fell through and Buchan ended up writing the history himself. It would be a massive undertaking. The first of no less than twenty-four volumes appeared in February 1915. Buchan eventually wrote more than a million words.[28] The *History* displayed all Buchan's talents. He summarised large amounts of material and conveyed a real sense of the fighting while at the same time providing a cracking narrative pace.

The venture was independent, with all royalties going to war charities, but it was just the sort of project Masterman approved of. He approached Buchan and started secretly to subsidise some of

Nelsons' output. *Britain's War by Land,* which Buchan wrote in 1915, was translated into several different languages. He wrote several pamphlets for Masterman that were likewise translated and distributed around the world. Masterman's support helped Nelsons survive the difficult times of war and the high costs of paper. And for Wellington House it was the perfect cover. Buchan was a well-respected author writing for an independent publisher. No one would suspect that he was producing British-sponsored propaganda.

In the press, Northcliffe was among those who continued to feel that the war was being run by amateurs, and the criticism began to hurt the government. Then, in the spring of 1915, a major crisis shocked the nation when the press sensationally revealed the shell shortage endured by British forces at the Battle of Neuve Chapelle in March. The *Daily Mail* and *The Times* were accused of being unpatriotic by reporting failings in the army. Copies of the papers were burned in the street, and their circulations plummeted. But Northcliffe regarded his task as one of 'awakening' the British people to the reality of the conflict. He said he intended 'to continue his policy of criticising the Government till such time as we apply ourselves as scientifically as Germany to carrying on the war'. And he dismissed the loss of sales, saying, 'Better to lose circulation than to lose the war.'[29]

However, it was none other than Sir John French, commander-in-chief of the BEF, who had leaked news of the devastating effects of the shortage to Colonel Repington, *The Times* military correspondent in London. When it became clear that the rumours were true, the ensuing political crisis led to the collapse of the Liberal government and the formation of the coalition. The circulation of *The Times* and the *Daily Mail* quickly recovered and Northcliffe's personal influence was restored, alongside that of his papers. He was widely credited with bringing down the old government. It was tremendous power for a press baron to yield.

Moreover, the political upheavals resulted in a change of leadership at the press bureau. Sir Edward Cook and Sir Frank

Swettenham, who now took charge, had a more realistic view of the need to inform the press and, through the newspapers, to influence public opinion at home. By now government ministers were beginning to realise that this was not a limited war of short duration. fought between professional armies. Britain was in for the long haul. And there was a growing awareness of the concept of a 'Home Front', in which public opinion and morale would play an important role. The French had allowed war correspondents to the front for some time, and Northcliffe demanded to know, 'If the French, why not the British?'[30] Eventually the pressure became so great that the War Office decided to allow five correspondents and a small number of photographers to visit the British lines. Later in the year, the Admiralty too relented, permitting selected reporters to visit the Grand Fleet moored at Scapa Flow. At last the War Office and the Admiralty had finally reversed their prohibition on correspondents reporting from the front.

One of the first five journalists sent to the front in France was none other than John Buchan. Having been engaged as a writer by *The Times* and the *Daily News*, he arrived in time to report on the final stages of the Second Battle of Ypres, in which the Germans had first used gas. Later in the year he sent back reports on the Battle of Loos, when Haig first deployed gas. His reports were well regarded. Leo Amery, a junior minister in the coalition government, wrote to his wife saying that the *Times* articles were 'excellent' and that Buchan 'can sense a situation quickly and can with the minimum of effort make a vivid story of it'.[31] Buchan's subsequent career, however, revealed how close these early war correspondents were to the military establishment. When Sir Douglas Haig became commander-in-chief of the BEF in December, GHQ asked Buchan to, as it were, switch sides and write official communiqués for the new chief. Haig, a fellow Scot also from the Borders who had attended the same Oxford college, knew Buchan and liked his highly readable but discreet style. Poacher Buchan had no hesitation in turning gamekeeper; indeed he would not even have seen it in these terms.

Whether he was working for Haig or *The Times*, or writing *Nelson's History of the War*, he saw his role as to cover the war fairly, honestly and tactfully, and to be read by as many people as possible.

Buchan was commissioned as a lieutenant in the Intelligence Corps and joined GHQ just in time for the start of the Battle of the Somme. He prepared communiqués and wrote weekly summaries of the battle that were sent out by Wellington House to British embassies around the world. Later, he wrote up the full story of the battle not only in *Nelson's History of the War* but in two separate volumes also published by Nelsons and simply titled *The Battle of the Somme First Phase* and *Second Phase*. Subsidised by Wellington House, these were then translated into Dutch, Danish, Portuguese, Spanish and Swedish.[32]

The books walk a fine line but are neither simple propaganda nor hard-hitting war reporting. Far from the comic-book approach of the early literature, created when little or no information was forthcoming from the front, they are still a good read today, conveying a first-hand sense of the adventure of war while accurately capturing the landscape of the modern battlefield. They are critical neither of the commander-in-chief nor of the divisional generals who repeatedly sent men in human waves against well-entrenched enemy machine guns. Buchan's reliance on GHQ to supply casualty figures means that British losses are played down and the Germans' built up. Nevertheless, after the ignorance of the early stages of the war, by 1916 Buchan had access to much greater information and detail. His books filled a gap in knowledge about what was happening at the front and tried to explain to readers at home and abroad the scale of the British war effort and the nature of the real fighting. The books did not contain outright criticism. That would have to wait until many years after the guns had fallen silent.

However, by this point of the war, a new, emerging medium would also bring professionals into conflict with the War Office and the authorities. Again, it was the Battle of the Somme that would prove to be the turning point.

13

The War in Pictures

The moving picture or cinematograph had arrived just twenty years before the beginning of the Great War. But a great deal had happened to the medium in the two decades between 1895 and 1914. At first, film had been nothing more than a fairground attraction, a sideshow gimmick where audiences could pay a couple of pennies to see a few minutes' worth of visual jokes or filmic tricks, local scenes of interest that might feature workers leaving a factory, or a sporting event. But the ten years before the war saw the emergence of something resembling a full-scale cinema industry. Special buildings were constructed where the films could be shown, all of course silent but nearly always with some form of live musical accompaniment. By 1914, it was estimated that there were 4500 film theatres across Britain. And the newer ones were large 'electric picture palaces' which the public would enter through marble foyers decked out with palms and floral displays and lined with plush hangings. The new Majestic in Clapham could seat 3000, and by 1917 there were as many as twenty million visits to the cinema every week. Five thousand new films were released every year with many studios producing a film each week, and the industry employed between 80 and 100,000 workers.[1]

A process of consolidation had already begun whereby the

hundreds of early cinema pioneers merged into a few large companies that had their own trade organisations to represent them. Moreover, the early cinema industry was already an international business. There were cinemas around the world, at least 20,000 in the cities of America and more than 1000 even in Russia, despite its relative poverty. French companies such as Pathé and Gaumont were already beginning to dominate the European production market and were well represented in the British industry, which was centred in the West End of London.

Cinema itself had been transformed in the years before the war. Films had grown longer and more sophisticated and a grammar of film narrative had developed. From the beginning, there had been two types of film. Fictional films were emerging by 1914 into a substantial art form in themselves. Dramas were popular all over the world, and comedies too were developing internationally – visual gags and slapstick could be enjoyed just as much in Moscow or Bombay as in Manchester or Baltimore. Already some artists were at work who would take narrative films to great heights. D.W. Griffith was about to start production on the cinema's first great blockbuster, *The Birth of a Nation*, and help establish a new American industry in Hollywood on the west coast. Charlie Chaplin, the British comedian, had already started working there with Mack Sennett and his Keystone Company and had just created the Tramp, the character that would go on to make him world famous.

On the other hand, factual films had also developed from simple scenes that showed workers walking past a camera or taking a ride on the top deck of a tram, to impressive feature documentaries that recorded major events such as polar explorations or imperial pageants. In addition, weekly newsreels containing five or six separate stories, known as 'Topicals', had been produced and shown for some years. By 1914 there were three principal newsreel companies in Britain: Pathé Gazette, Gaumont Graphic and the Topical Budget. 'Topicals' contained a set of stories that might include anything from a royal visit to a few seconds of a football match, from an aviation fair to the launch of a ship. Already the Topical companies had inter-national links, so that film shot in Britain could be seen anywhere around the world, while footage shot abroad of, say, the Russian imperial family or an American cowboy pageant, could also be shown in Britain.[2]

Most educated people looked down on film as a vulgar, cheap and mindless form of entertainment for the working classes. But the industry made high claims for itself, calling the cinema 'the greatest social force in existence' and claiming that film was capable of 'sat-isfying an indispensable human requirement'.[3] So when war was declared the young film industry too was keen to do its bit. The cinema trade argued that films could be used to 'arouse patriotism', to help recruitment and to keep the public informed with a film record that would show 'the actual likeness of events' at the front.[4] The trade further asserted that it had the basic production and dis-tribution infrastructure in place to offer access to a worldwide cinema audience. The War Office and the Admiralty responded exactly as they had with the press: they banned all photographers and cinematographers (film cameramen) from going anywhere near the front. The same knee-jerk reaction that barred newspapers from reporting events also prevented the cinema industry from showing any kind of military activity. Any cameraman who ventured within about thirty miles of the front was arrested and faced being shot as a spy.

The cinema-going public was denied any view of what was hap-pening at the front and instead had to put up with simple recruiting pictures in which portraits of Wellington, Nelson and Gordon of Khartoum would be screened alongside girls in khaki singing patri-otic songs and pointing out the location of the nearest recruiting office. By early 1915, however, both the Germans and the French were beginning to allow film cameramen access to the front, and their films were being shown around the world. Film of advancing columns of German infantry, of German generals benevolently pat-ting the heads of little children, or of the Kaiser at the front began to 'conquer the screen' in picture palaces worldwide.[5] Charles Masterman, busy setting up his literary propaganda operation at Wellington House, argued that film should also be exploited as an international propaganda medium. But he came up against a barrier. Thanks to their sense of contempt for the cinema, his peers were unable to comprehend that this trivial form of amusement might have anything to do with the serious business of winning the war.

However, when it wanted to, the cinema trade could act as an effective lobby. Joseph Brooke Wilkinson, secretary both of the Kinematograph Manufacturers' Association and the British Board of Film Censors, was deputed by the Topical companies in the spring of 1915 to negotiate with the War Office on their behalf. Brooke Wilkinson was from Manchester, where after leaving technical college he had worked in a firm of photographic chemists. He came to London just before the turn of the century and was spotted by Cecil Hepworth, one of the pioneers of the British film industry. Wilkinson was a dapper little man with endless patience and energy, and both of these qualities were called upon in his negotiations with the War Office. Fortunately, it seemed that the initial hostility of the military authorities was beginning to change. After arguing the case for several months, Brooke Wilkinson persuaded the War Office to allow film cameramen to visit the Western Front as long as the army had control of the material shot, in case the images gave away vital information to the enemy. The authorities also wanted the cinema industry to donate to military charities a share of the profits made by screening war films. The industry was reluctant to agree at first, but eventually gave in and conceded the concept of profit participation. It was agreed that only a small number of film cameramen would be allowed to visit the front and that the footage they took would be shared or 'pooled' by all the Topical news companies. A War Office Topical Film Committee was formed to supervise the whole operation. After months of negotiations, the first two cameramen finally set off for the front on 2 November 1915. They were Edward 'Teddy' Tong of Jury's Imperial Pictures and Geoffrey Malins of Gaumont.

In the next year they would bring about a revolution in British film propaganda.

While these negotiations took place, Masterman was also at work, pushing as it were from the inside. In the summer of 1915 he formed within Wellington House a Cinema Committee consisting of T. Welsh of Gaumont, Charles Urban, a producer of factual films, and William Jury, a prominent cinema distributor, along with the ubiquitous Brooke Wilkinson. Masterman managed to convince the First Lord of the Admiralty, Sir Arthur Balfour, that German propaganda films were having it all their own way and that the Royal Navy ought also to be seen on cinema screens in Britain and around the world. Balfour presented the case to Admiral Sir John Jellicoe, commander-in-chief of the Grand Fleet, stating rather contemptuously that because film 'required no reading' and 'threw no strain upon the spectators' powers of realisation' it had the ability to reach 'the intelligence of the least intelligent'. Although he said he personally hated the cinema, Balfour argued that 'the reality and magnitude of the Fleet' should be shown to audiences in Britain, France and Russia, as well as in America.[6] Even though he probably shared the same disdain for moving pictures, Jellicoe agreed that a group of cameraman could visit the Fleet in Scotland. The result was *Britain Prepared*, the first ever official British propaganda documentary.

The producer of the film was to be Charles Urban. He was a pushy young American from Ohio who had originally come to Britain in 1897 as a salesman to sort out an American-owned company that dabbled in cinema distribution. He became fascinated with film and settled in Britain and so enjoyed mixing with British high society that he bought a large country estate where he raised orchids and kept birds. In 1903 he formed his own company, the Charles Urban Trading Company. Although, because of their entertainment value, early films had prospered in the music hall and the fairground, Urban's passionate mission was to use film to educate, inform and show audiences a new perspective on the world. He produced some of the first scientific and medical films, his cameramen pioneering the use of micro-photography and stop-motion effects to record sequences such as a bud opening into a blossom. Audiences were stunned. He sold short films of the Russo-Japanese War and the Balkan wars, while his operators travelled to Africa, Asia and the Far East, bringing back hours of material which was edited into short travel and adventure films. The motto of his company was 'We Put the World Before You'.[7]

Urban's assistant, George Albert Smith, developed for him the first ever natural film colour system, called Kinemacolor. It became a sensation before the war. In addition to shooting bird films in colour at the aviary on his own estate, he made a spectacular colour film of the visit of the King and Queen to India in 1911 and of the dazzling Delhi Durbar, at which thousands of Indian dignitaries gathered to pay homage to the British sovereign. Screenings of the Durbar film featured a lecturer reading a narration and an orchestra playing a specially written accompaniment, and soon became the talk of the town. By the First World War Urban was a leading figure in the British film business, running what today would be called an independent factual production company with a big, plush office in Wardour Street, Soho.

With his go-getting, no-nonsense attitude, Urban was appalled when the War Office initially prohibited filming at the front. He worked closely with Masterman to persuade the Admiralty to lift their ban on filming with the navy, and as soon as permission finally came through he went with three cameramen to Scapa Flow in the Orkneys to film the fleet on exercises.

The small film crew were put up on HMS *Queen Elizabeth*, one of the newest battleships in the Royal Navy. Right from the start, Urban was horrified by the behaviour of his navy 'minders'. He was allowed to film nothing without their approval and they proved to be supremely cautious. Whenever a glimpse of shoreline in the background of a shot might allow a viewer to identify where the fleet had assembled, a naval officer put his hand over the lens or simply stopped the cameraman from operating the camera. Urban was furious but somehow held his temper. Finally he secured a meeting on his flagship with Admiral Jellicoe, a man of whom he was greatly in awe; 'When you caught his eye you realised that here was a *man*,' Urban wrote later.[8] The meeting went well, and Jellicoe agreed that Urban and his team must carry on and were to be given full support.

Still there were difficulties. Urban was keen to film the great fifteen-inch heavy guns of the battleship firing broadside, but when they opened fire the blast was such that the camera tripod lifted several inches into the air and the shot was ruined. Urban's cameramen were eventually allowed to board a sister ship, from which he was able to film the heavy guns firing from a safe distance. They also managed to film destroyers on patrol and the firing of a torpedo, much of the footage being taken in Kinemacolor. Even when the material was sent back to London for developing, however, the navy were still nervous and insisted on having a representative present throughout to ensure that extra copies were not made. They were also obsessed that 'foreign' technicians should not be able to see any of the footage before it had been viewed and approved by the naval censor.

Eventually, the naval footage was edited together with film taken at Aldershot that showed a group of recruits being trained for the army, a sequence showing Royal Flying Corps aircraft practice-bombing (and missing) a farmhouse on Salisbury Plain, and the return of a photo reconnaissance aircraft. To these were added at the last minute shots of women war workers in a Vickers factory, as it was thought this would impress audiences abroad as to the extent to which Britain was already mobilised for war. The film, which originally ran for more than three hours with parts of the newly shot material in colour, was premiered on 29 December 1915 at the Empire, Leicester Square. Foreign ambassadors, members of the diplomatic corps, MPs and members of the House of Lords, senior representatives of the armed services and many other dignitaries were present. Balfour himself came on to the stage before the screening to introduce and commend the film as an 'entertainment [that] is something more than an entertainment'. The film was shown with a specially prepared musical accompaniment of marches, airs and songs selected to match each scene, and it ended with the national anthem. The audience was apparently thrilled by the climax of the film, when the battleship fired its broadside of salvoes and flashes of orange flame and black smoke lipped out of the huge gun barrels.[9]

Britain Prepared is rather quaint by later standards.[10] It primarily presents scenes of the navy, with shots of the army and air force added as an afterthought. But the film makes no claims as to the superiority of British weapons or the invincibility of the armed services. Nor does it pretend that any of the shots are of actual fighting; it simply shows the army and the navy preparing for war. In this sense the film uses no fakes and is totally honest, in line with Masterman's view of what propaganda should be.

Indeed, Masterman was delighted with the film and wrote, just after Bulgaria had joined the war on the German side, 'If we had only had permission to arrange this six or seven months ago I believe we could have shown places like Bulgaria pictures of the fleet and they would never have gone over to the enemy.'[11] The film was shown all over Britain and throughout the world. Audiences in Britain were especially enthralled with the scenes of the Royal Navy's massive warships, which were often cheered and applauded as they appeared on the screen. In Europe too the film went down very well. The French were impressed by the scenes of the vast munitions in production, and even a German newspaper reporter who saw the film in neutral Switzerland wrote, 'We must admit, a more clever advertisement could hardly be made by the English Ministry of War for its Army and Fleet and its war services in general.'[12]

In the United States, however, the film failed to excite audiences. In part this was down to German sympathies among cinema owners and film distributors who did not want to show it. Urban, who had taken the film to America himself, struggled hard to find distributors. This he justified to Wellington House by saying that for many

Americans the film was 'too classy', although some merely thought it was too bland and uninteresting, and lacked punch. 'If it showed troopers being blown to pieces, it would go all right' was one typical comment.[13] At its New York opening at the Lyceum Theatre on 29 May 1916 there was nearly an incident. The film included shots of George V reviewing soldiers, captioned as Irish, on their way to relieve others at the front. It was only a few weeks since the Easter Rising in Ireland and, with an audience that had great sympathy for the Irish cause, Urban feared the worst. However, to his delight, there was spontaneous cheering as the scenes were shown, and when the orchestra struck up 'It's a Long Way to Tipperary' a famous opera singer led the whole audience in singing along. Eventually, the film was shown to audiences in all the major cities of the east coast. One of those who saw it in Washington was the Assistant Secretary of the Navy, young Franklin D. Roosevelt. He was impressed, and believed it demonstrated the need for the American government to begin preparations in case it found itself in the war before long.[14]

Meanwhile, footage sent back by the two cameramen on the Western Front was edited into occasional short films seven or eight minutes in length. But none had the impact of a major compilation like *Britain Prepared*. Furthermore, the endless sparring between the War Office and the cinema trade continued. The public officials from the War Office found the cinema men grasping and focused on profit. The trade representatives still thought the army was being unhelpful and restrictive, and had failed to realise the full potential that film had to offer.

The cameramen were officially a part of GHQ under Brigadier Charteris, whose intelligence responsibilities also covered the press. Their day-to-day care was in the hands of Captain Faunthorpe, who would give them instructions each morning as to where to film and what of interest was happening. With the film cameramen there sometimes went a stills photographer whose photos would be released to the press.[15] They were given an honorary rank and dressed as officers, wearing the green armbands of the Press Corps rather than unit badges. Filming was limited by the parameters of how the cameramen worked and what their cameras were capable of. Film cameras of the day were heavy and cumbersome, and were hand-cranked rather than motorised. An experienced cameraman could crank at exactly two complete turns per second, which would pull sixteen frames through the camera gate for exposure, and as film was usually projected at exactly this speed it made for a smooth and precise moving sequence. However, as most projectors were still hand cranked as well, the process was far from an exact science.[16]

The heavy hand-cranked cameras had to be mounted on large wooden tripods to provide stability. It was possible to move the camera only by turning or panning it sideways, or tilting it slightly up and down. This involved cranking the pan or tilt handle in the opposite direction with the left hand at the same time as cranking the camera with the right. Although this sounds complex, an experienced cameraman could do it quite smoothly even under shellfire. The final problem was the result of the standard 50mm lens used on the Western Front. This is the lens most commonly carried by photographers and it made for a good wide shot and a fine panorama, but it was not much good at detail or on close-ups in the battle zone. The army did not allow the use of longer, telephoto lenses in case sensitive detail was picked up, and the zoom lens had not then been invented. So most shots were either static wide shots or slow panoramas. The film itself was slow and the lenses usually operated at an aperture of about f 3.5, so filming was possible only in good light and never at dusk or in the dark. A cameraman could usually carry with him only 1200 feet of film stock, enough for about fifteen minutes of filming, although this was usually quite sufficient for most days' work.[17]

The weight of all this equipment was more than 100 lb, and so the cameramen were reliant upon GHQ to provide transport to the point at which they could set up the camera and start filming. It was easy to set up and shoot film of troops moving up to the front or of artillery pieces firing, and such sequences contained their own movement; easy too to film groups of soldiers resting behind the lines, or groups of prisoners, or the wounded being transported. However, walking up through the reserve trenches to the front line carrying a camera and tripod would be slow and arduous. Moreover, a cameramen filming near the front trenches would not only be in immense danger, risking almost instant death if he showed himself above the trench parapet, but would also be unpopular with the troops for blocking the trench. So, in practice, there were severe restrictions on what it was possible to film. In today's era of lightweight, hand-carried, highly portable cameras, it's difficult to imagine the challenges facing First World War cameramen. But it is essential to do so in order to understand their output.

Of the two cameramen who had gone to the Western Front in November 1915, Teddy Tong was soon invalided out with influenza, leaving only Geoffrey Malins to film in the first months of 1916. Malins, then aged twenty-nine, was forceful and ambitious. Born in Hastings, he had worked as an assistant at a photographer's portrait studio in Weymouth. The demand for formal portraits had declined with the spread of portable box brownie cameras, so Malins moved into the growing cinema industry instead. In 1910 he got a job as principal cameraman for a south London production company, Clarendon Films of Croydon, for whom he shot a run of highly successful melodramas. But, itching for action, in August 1914 he transferred to the Gaumont film company and became a cameraman for their twice-weekly Graphic newsreel, even though to transfer to news filming was almost certainly a step down for the equivalent of a feature film cameraman. Malins filmed for a few months with the Belgian army, the only military force that allowed cameramen near the front, and then went back to London to work for Gaumont Graphic. It was thanks to his experience with the Belgians that he was selected to go to the front as one of the first two official British cameramen. In the deal negotiated between Brooke Wilkinson and the War Office, the cinema trade paid Malins £1 per day, but he had to live in army accommodation and on military rations in France.

The seven-to-eight-minute films into which for several months his footage was edited by what was now called the Topical Film Committee were made up of little more than a few scenes shot near the front, and had titles like *Ypres – The Shell Shattered City in Flanders*, *The Destruction of a German Blockhouse by 9.2 Howitzer*, *HRH The Prince of Wales with the Guards* and *In Action with our Canadian Troops*. Although the close supervision of Captain Faunthorpe, who acted as his 'minder', could be frustrating, a rapport built up between the two men, and Faunthorpe paid tribute to Malins in a report to the War Office, saying that he was often 'constantly under fire' when filming and that shells sometimes landed within twenty yards, 'ploughing up the ground almost at their feet and showering missiles round the camera'.[18] Malins showed great ingenuity on one occasion. rigging a camera in a BE2c aeroplane and filming the trenches from the air. And he invented a variety of imaginative ways of filming near the front under canvas sheets and through mock sandbags. Back in Britain the reviews of the films edited from his material were very favourable.

In June 1916, with the approach of General Haig's Big Push along the Somme, GHQ decided to ask for a second film cameraman to come out to France in order to record what they hoped would be an historic victory. John Benjamin McDowell, thirty-eight-year-old head of the British and Colonial Film Company, volunteered for the role. In the last week of June, Malins started filming troops moving up towards the front and shot scenes of the heavy artillery barrage that preceded the launch of Battle of the Somme. McDowell arrived in France a couple of days later to join him. On 1 July, the bloody and catastrophic first day of the battle, Malins was allocated to film with the 29th Division in the vicinity of Beaumont Hamel, where the attack proved to be a disaster; McDowell was sent to film alongside the 7th Division further south near Mametz where it was relatively successful. Malins filmed a memorable sequence of Lancashire

Fusiliers in a sunken road in no man's land waiting for the order to advance. The men's faces are lined with anxiety and tension, capturing a rare glimpse of what it must have felt like just before going over the top.

A few minutes later, at 7.20, Malins prepared to film a huge mine set to go off at the Hawthorn Redoubt. Malins later gave an account of the incident in his book *How I Filmed the War*. Sadly, the book is largely unrealistic; Malins presents himself in it like the hero of a trashy novel, rushing hither and thither as explosions rage around him, but always miraculously pulling through. By contrast, his description of filming the mine explosion is totally believable. He set up his camera and tripod in the midst of a group of engineers who would rush into the crater once the mine had gone off, describing the next few moments as follows:

Time: 7.19. My hand grasped the handle of the camera ... Another thirty seconds passed. I started turning the handle of the camera, two revolutions per second, no more, no less ... I fixed my eyes on the Redoubt. Any second now. Surely it was time. It seemed to me as if I had been turning for hours. Surely it had not misfired ... I looked at my exposure dial. I had used over a thousand feet. The horrible thought flashed through my mind that my film might run out before the mine blew. Would it go up before I had time to reload? ... I had to keep on. Then it happened. The ground where I stood gave a mighty convulsion. It rocked and swayed. I gripped hold of my tripod to steady myself. Then, for all the world like a gigantic sponge, the earth rose in the air to the height of hundreds of feet ... From the moment the mine went up my feelings changed. The crisis was over and from that second I was cold, cool and calculating. I looked upon all that followed from the purely pictorial point of view.[19]

Many combat cameramen in the next hundred years would experience similar tensions: waiting, anxiety about running out of stock, and then, blocking out everything else, total concentration on the job to be done.

Malins and McDowell spent a week filming events the length of the battlefield. According to Malins, a shell landed so close to him at one point that it ripped his tripod in half and threw him backwards. Both cameramen filmed the arrival and treatment of the wounded at Casualty Clearing Stations. They filmed prisoners being brought in and reinforcements marching up to the front. Whenever men saw Malins with his camera and tripod, they asked to be in the picture and waved at the camera. He and McDowell also filmed in German trenches captured on the southern flank of the offensive.

There was one problem, however. The critical moment of a First World War infantry assault came at zero hour, when the officers blew their whistles and led the men 'over the top', climbing out of their trenches to advance across no man's land. John Buchan wrote at the time of the Somme, 'The crossing of the parapet is the supreme moment in modern war. The troops are outside defences, moving across the open to investigate the unknown. It is the culmination of months of training for officers and men, and the least sensitive feels the drama.'[20] It had proved impossible to film such a scene in the confined space of the front trench with the cumbersome film cameras available, and it would have been suicidal for a cameraman to get up before or behind the troops as they clambered forward. However, Malins later said that he was very aware of the need to provide 'thrills' in the footage he shot, material that people 'had never seen before, and had never dreamed possible', and at the War Office there was no doubt the expectation that this central moment should be recorded.[21] So Malins went to a trench mortar battery training school in the rear at Rollencourt, where a set of trenches had been laid out quite authentically. There he set up his camera behind a practice trench and filmed a squad of men going over the top.[22] Malins' previous experience in shooting dramas at Clarendon Films clearly came into play here. He laid down a smoke screen in the barbed wire in front of the trench into which the men were to advance. Not satisfied with this, he asked two men to fall as though shot as they climbed out of the trench, and directed two more to fall as they advanced into the smoke. It looked realistic. But he had faked the whole scene.[23]

On 10 July, Malins and McDowell returned to London with several hours of footage. When it was screened for members of the Topical Committee for War Films (as it was now called) there was general agreement that it contained some outstanding material. William Jury, the chair of the committee, argued that because of the strength of the images and the importance of the continuing battle in France, they should not edit the footage into a series of short films but instead turn it into one long, feature-length documentary. One of the first exhibitors of film in Britain, Jury ran one of the country's major distribution companies, Jury's Imperial Pictures. He had a good sense of what the public would want and persuaded the rest of the committee and the War Office to agree to produce and distribute a single feature-length film on the Battle of the Somme. After the army censor in London had viewed the footage, Jury supervised the editing of the film, which took place quickly over two weeks.

The resultant film is structured chronologically and very simply, showing three phases of the battle. First there are the preparations, the build-up of supplies, the artillery bombardment, and endless columns of men marching to the front. Second comes a short sequence depicting the launch of the attack. Finally, the third phase depicts the aftermath, in which long lines of the wounded and prisoners are brought back and bodies are seen scattered across the battlefield. The film ends by showing preparations for the next attack. Once inter-titles had been written and shot to provide a basic guide to the action, the edited film was taken back to GHQ in France and shown again to the high command, who approved it for public screening.

By early August the film was complete. It ran for one hour and seventeen minutes, and the main trade magazine, *The Bioscope*, published a list of music recommended for playing as an accompaniment to each scene.[24] David Lloyd George, who had recently become Secretary of State for War, wrote a short introduction, recommending that every cinema manager should read it before the film was screened. He called the film 'an epic of self-sacrifice and gallantry' and 'the most important and imposing picture of the war that our staff has yet procured'.[25] While the battle continued to rage in France, and barely six weeks after the worst day in the history of Britain's armed forces, the documentary feature had its trade preview on 10 August at the Scala cinema in London's West End. Simply titled *The Battle of the Somme*, not only was it one of the first, but as it proved, the most successful official propaganda film ever produced in Britain.[26]

Crowds flocked to see the film. Queues formed outside cinemas, and at first more people were turned away than managed to buy tickets. By the end of the first week more than one million people had seen it in the thirty-four cinemas screening it in London alone. Across the country it was equally popular. In the industrial cities of the north, cinemas were packed for continuous screenings from morning to late evening. In Glasgow, one cinema manager arranged an extra screening, free for wounded servicemen at home on leave. Other films that had been booked were cancelled to make space for it. Everywhere it broke box office records. And the press response was equally positive. The *Evening News* wrote that *The Battle of the Somme* was 'the greatest moving picture in the world' and compared it favourably to D.W. Griffith's *Birth of a Nation*.[27] *The Star* wrote that 'Somebody in the War Office or at GHQ has at last ... grasped the power of the moving picture to carry the war on to British soil.'[28] The *Manchester Guardian* wrote that it was 'the real thing at last'. The trade press was even more ecstatic. *Kine Weekly* claimed it was 'the most wonderful battle picture that has ever been taken', and *The Cinema* affirmed that 'There is no make believe. This is the real thing.'[29]

Audiences were stunned at some of the sequences. For the shot of the huge mine exploding at the Hawthorn Redoubt, the musical instructions were that a drummer should play a quiet roll, building up to a crash and roll at the explosion itself. In cinemas up and down the country people would cry out with a spontaneous 'Ooh' in awe at this moment. But it was the staged scene of the men going over the top that attracted most attention. Almost everyone who saw the film was affected by the sequence. In one London cinema, the orchestra went silent during the scene, and as the soldiers fell a young woman screamed out in the audience, 'Oh god, they're dead.' The sixty-year-old novelist Henry Rider Haggard wrote in his diary after seeing the film, 'There is something appalling about the instantaneous change from fierce activity to supine death ... War has always been dreadful, but never, I suppose, more dreadful than today.'[30] Frances Stevenson, Lloyd George's secretary and mistress, was in her twenties and had just lost her brother, Paul. After seeing the film she wrote, 'I am glad to have seen the sort of thing our men have to go through, even to the sortie from the trench, and the falling in the barbed wire ... It reminded me of what Paul's last hours were. I have often tried to imagine to myself what he went through, but now I know: and I shall never forget.'[31] A man wrote to *The Times*, 'I have lost one son, and have a second home wounded ... I have been twice to see these films and was profoundly struck by the emotion and almost reverence, with which they were followed.'[32]

Later in the film there is a shot held for some time in which men can actually be seen moving forward across no man's land in a genuine attack. It was filmed from a trench to one side of the advance, probably through a periscope raised above the parapet. Limited to the 50mm lens, however, Malins could reveal no detail, and although it is fascinating to see an authentic 1 July assault today, it had at the time no impact on audiences who probably could not discern clearly what was happening on the screen. In addition to the poignancy of the comments about the 'over the top' scene, therefore, there is great irony as well. Although most of the footage in the film is totally authentic, the scene that caused the greatest impact was one that had been faked.

The Battle of the Somme raised ethical questions for viewers in 1916 that are as relevant to combat filming today as they were one hundred years ago. These include questions not only as to what is real and how legitimate it is to stage sequences, but also about how much death it is fair to show. It was quite likely that certain members of audiences would have recognised in the film their own husband, son, brother or boyfriend marching up to the front, but by the time they saw the film, they would already have received a telegram informing them of his death. Another issue the film raises is the extent to which it is reasonable to show images of the dead and the dying, even if, as in this film, those images are of the enemy.

These issues were keenly debated. 'Is it right to let us see brave men dying?' asked the reviewer in *The Star*, though he answered his own question with a 'Yes' and went on, 'Is it a sacrilege? No. These pictures are good for us.' Even the pages of *The Times* were filled with debate about the morality of showing the film, until a leader concluded, 'The War Office has set the public an example. It had given the cinematograph a golden opportunity and the cinematograph has used it well ... In the ages before the invention of this machine it was impossible to convey so clear and full an idea of warfare as anyone may now see for a few pence.'[33] However, the manager of the Broadway in Hammersmith took a very different view and, unusually, decided not to show the film at all. He put up a notice, 'This is a Place of Amusement, not a Chamber of Horrors.'[34]

The Battle of the Somme is an extraordinary film document. Its success at the time was unprecedented. No one knows how many millions saw it during its run, but it was still being screened in November 1917, so it is possible that in the region of 20 million Britons (out of a population of 43 million) saw the film.[35] Even a majority of children polled in two schools in Holborn and St Pancras the year after the film's release named it as their favourite.[36] But the film does not have the makings of an obvious hit. It has a structure but no narrative. There is no sense of a plan for the battle and no footage of Haig or Rawlinson or GHQ. The only general who appears is an anonymous figure on a horse riding past his men. Only occasionally can officers be picked out. There are scenes of cheerful men waving at the camera marching to the front and of wounded men hobbling back. The climax of the film, the advance over the top, comes just after its halfway point. How the film should end must have exercised those who put it together, but after many scenes of the wounded, of prisoners and of German dead, it closes with a group of British soldiers cheerfully going up the line for the next phase of the battle.

One or two scenes have obvious propaganda intent. Shots showing the supplies of shells going forward to keep the guns blazing, for instance, would have helped to motivate the munitions workers at home who saw the film. And yet the inter-titles create no drama or uplift, occasionally naming the men who are about to be seen simply as 'Lancashire Fusiliers', 'The Manchesters' or 'The Royal Field Artillery', but creating no closer affinity with any particular group. The film lacked the sense of jolly patriotic optimism so characteristic of later film propaganda. Nor is any part of it particularly anti-German. The extraordinary success of *The Battle of the Somme* was no doubt at the time very much due to the fact that although the war had been going on for two years, people at home had never seen such images before. Audiences and critics believed it was so real because they had nothing else to judge it by. And, without doubt, its presentation from the point of view of the ordinary soldier, going about his everyday duties and taking part in a great battle, strengthened its impact. There was no obvious message or explanation of the strategy of the battle, and that very much contributed to its appeal.

Today, we can be far more critical of the film's production and clear about what is real and what is faked than could audiences at the time. But it still has a mesmeric, haunting effect. Somehow its images have come to epitomise our understanding of what the Western Front looked and felt like. Its footage, one hundred years later, has attained almost iconic status. In 1916 it was recommended that a jaunty cavalry march should be played during the 'over the top' sequence. After all, the battle was still being fought and, as far as most people were concerned, being won. Today, there appears nothing heroic about these scenes and the images of the front line are melancholic in the extreme. They have become enduring symbols of the sacrifice and futility of the Great War.

The Battle of the Somme had been edited with contemporary overseas audiences as well the home audience in mind. And it was shown in eighteen countries – in western Europe, in all the Dominions and as far away as China, Iceland and Peru. In the United States there were problems, as there had been with *Britain Prepared*. The International Film Corporation offered to buy the rights. However, as the company was owned by William Randolph Hearst, who was known to be pro-German, it was feared that he intended to purchase the film simply in order to suppress it and that he would never show it. Instead another pro-British group distributed the film, while Charles Urban re-edited it for America and it was shown among a variety of other attractions within cinema programmes. Urban later claimed that by the end of 1917, 65 million Americans had seen the film, although this figure seems highly exaggerated, probably to build up Urban's success back in London.[37]

The film was also taken to Russia. Captain Bromhead, working for Wellington House, had successfully shown *Britain Prepared* to the imperial royal family and to an estimated 100,000 Russian soldiers, and he now took *The Battle of the Somme* on tour, showing it to huge crowds of soldiers in mass open-air screenings behind the lines. Sometimes more than 5000 men saw it at a single showing. At first Bromhead wrote back saying how successful the screenings had been in showing the Russians how much pressure Britain was putting on Germany on the Western Front. However, as Russia went through its first revolution in February 1917, the soldiers' view seemed to change, with unexpected consequences. As the poorly equipped and provisioned Russian troops watched images of well-armed and well-fed British soldiers, the film began to encourage disaffection within the ranks. Robert Bruce Lockhart, British vice consul in St Petersburg, noted that screenings were being followed by mass desertions, which had clearly not been the film-makers' intention. Events were running out of the propagandists' control and after the second revolution in November, led by the Bolsheviks, Bromhead and his cinema team were called home.[38]

The German cinema industry jealously followed the huge success of *The Battle of the Somme* and soon attempted to emulate it. The trouble was that no cameraman had been filming on the German side of the lines until much later in 1916. So a half-hour documentary was put together using authentic material of British prisoners, coupled with a lot of faked material of German soldiers under bombardment from the British, who were clearly painted as the aggressors. Released in January 1917, *Bei unseren Helden an der Somme* (*With Our Heroes on the Somme*) was billed as 'the greatest cinematic document of this terrible war'. However, the public were not fooled and the film had no significant impact in Germany, nor was it ever shown abroad.[39]

Production of *The Battle of the Somme* had marked a new phase in British film propaganda. In November 1916, partly as result of the film's success and partly as a consequence of further pressure from the War Office, a new organisation called the War Office Cinematographic Committee took over from the Topical committee. Sir Max Aitken (created Lord Beaverbrook in January 1917), who had been running the Canadian military propaganda operation since early in the war, became chairman and there were only two other members: Sir Reginald Brade represented the War Office and William Jury the cinema industry. The new committee continued to produce feature-length films. *The Battle of the Ancre and the Advance of the Tanks*, which followed the second part of the Battle of the Somme in September–October, was released at the beginning of 1917. Both the *Daily Mail* and the *Evening News* preferred it even to the Somme film and called it 'all real, all unrehearsed' and 'the greatest war picture yet produced'.[40] *The German Retreat and the Battle of*

Arras, the third and last of the 'Battle' films, was released in June 1917.

Beaverbrook was keen to move into new areas of film propaganda and launched a twice-weekly official newsreel, produced by the Topical Company. From May 1917 it was called the *War Office Topical Budget*, changing its name later to the *Pictorial News (Official)*. There followed a rapid increase in the amount of film-making. Hundreds of short films were made for the Home Front in 1918, many featuring not just scenes from the front line but about work and attitudes at home, illustrating how important the concept of the Home Front had become. *A Day in the Life of a Munitions Worker* provided a glamorised view of working in a munitions factory, intended to encourage more women to sign up to the war factories. *Mrs John Bull Prepared* was aimed at changing the attitude of men who were reluctant to allow their wives or daughters to do war work, and showed the multitude of roles taken by women, from driving trams to running post offices, from working as welders to toiling on the land.[41] There were short films intended to encourage healthier eating, to avoid waste and to show how the new state bureaucracies were bringing a fairer distribution of resources to everyone. Much of this set a precedent for what would be achieved with tremendous effect in the official films of the Second World War.

Film was also shot and edited into documentaries of the campaigns in Italy, Egypt, Palestine and Mesopotamia (Iraq). Frank Hurley, for instance, who had filmed Sir Ernest Shackleton's epic expedition to Antarctica, worked as an official Australian film cameraman, taking some powerful images of the Battle of Passchendaele and then recording General Allenby's entry into Jerusalem in December 1917. But there was never more than a handful of film cameramen on the Western Front. Geoffrey Malins went on filming in 1917 until he had to withdraw, suffering from battle fatigue and a form of shell shock. He later returned to the front to work for Beaverbrook and the Canadians, but finally opted out in early 1918. He was awarded the OBE for his services. J.B. McDowell stayed on as an official film cameraman for the rest of the war and was awarded an MC for his outstanding work. William Jury, who had given up commercial work in order to concentrate on the war effort, eventually became head of the Cinema Division in 1918 and was knighted. Charles Urban, the American who had helped to start it all, received nothing and resented Jury's elevation. The film propaganda campaign had ultimately shown how outside professionals could successfully be brought into the war effort. But this had only been possible after the opposition by the armed services to any sort of joint venture with the commercial 'picture men' had been overcome.

When the United States entered the war, American film-makers brought their own unique approach and skills to the conflict. Unlike the simple recruiting films shown in Britain, often little more than screen versions of the successful posters, stars turned out en masse in the American films to exhort audiences to join up, to buy war bonds or to give to the Red Cross. Douglas Fairbanks, Mary Pickford, Charlie Chaplin, William S. Hart and producers like Cecil B. DeMille and D.W. Griffith all contributed. The war offered an opportunity for the development of new ideas in film advertising. With the soldiers arriving in Europe came a special film unit of the US Army Signal Corps, and for some months they concentrated on producing short films about training and collaboration with the armies of France and Britain. But by 1918, the American cameramen were trying to create heroes by following small groups of soldiers in actions supposedly recorded at the front – although most were of necessity shot well behind the lines. And in their captions, the American films were far more overtly political and anti-German. After the end of the war, with much of the European film industry in ruins, the American industry based in Hollywood was to emerge as the leading cinematic and cultural force, selling its output throughout the world.

Visual coverage of the war was not limited to film and photographs. In Britain, Charles Masterman at Wellington House had also come up with the idea of sending war artists to the front. The first to go, in May 1916, was sketcher Muirhead Bone, who was given permission to travel up and down the front making sketches. The results were published in December 1916 in an anthology called *The Western Front* and formed part of a very successful exhibition the following year. Portraitist Francis Dodd was then asked to paint a series of portraits of 'Generals of the British Army' and 'Admirals of the British Navy'. Although this early work had some propagandist intent, it seems that the briefs given to the artists allowed them considerable freedom. One artist, Christopher Nevinson, asked before leaving if there was any subject he should avoid. 'No, no,' said Masterman with a wave of his hand. 'Paint anything you please.'[42]

Max Aitken was also keen to encourage war artists to visit the front and make their own personal interpretations of what they saw. And established and 'safe' artists were not the only ones to paint for the propaganda bureau; several radical young artists of the day were sent to France as well. Paul Nash had been invalided home from the infantry in early 1917, but when Masterman asked him to return to the Western Front later that year he produced many memorable paintings, including *The Menin Road* and *The Ypres Salient at Night*. Stanley Spencer, then part of the avant-garde Vorticist movement, had served in an ambulance unit in Macedonia when he was asked to work as an official artist. Commissioned to produce work that would in some way symbolise Anglo-American sacrifice, American painter John Singer Sargent painted one of the most famous artworks of the war, *Gassed*, in which a long and tragic line of blindfolded soldiers with their hands on the shoulder of the man in front are depicted being led along a series of duckboards by a medical orderly while other wounded soldiers lie in agony around the line of wounded men. Several of these paintings were exhibited in the final year of the war and caused much controversy. However, the work of these artists has helped shape our vision of mud, barbed wire, trenches, and life and death on the Western Front.[43] And, the precedent having been set, official war artists have accompanied British units on most subsequent military campaigns.

Many of these images have since contributed to our perception of the Great War. However, nothing compares to the success of *The Battle of the Somme*. Truly pioneering and completely exceptional at the time, it lives on today in that almost every television documentary made about the First World War uses clips from it. The footage of Malins and McDowell, the editing of William Jury, is to be seen somewhere on television most nights of the week. And the men who marched, cheering, up to the front, who went over the top and were carried back on stretchers, are ghosts on the screen, reminders of the horrors of battle and the power of film propaganda.

The summer of 1916 marked an even deeper transformation of Britain. Day after day, appallingly long lists of the dead and wounded appeared in the newspapers. In some cities whole districts went into mourning if a local regiment or Pals Battalion had gone over the top. Accrington in Lancashire, Belfast in Northern Ireland, Sheffield, Barnsley, Bradford and many other towns and cities were hit hard by the losses. But, even when city dwellers across the country witnessed the packed trains bringing the horribly maimed and wounded back from France to hospitals at home, the support of the British people for the war did not collapse. The Battle of the Somme had proved a turning point in many ways. It had led to the slaughter of the men who eagerly flocked to join up in 1914–15, 'Kitchener's volunteers', and would leave a scar across an entire generation. But it had marked an acceptance that strict censorship was no longer a possibility. It had created the realisation that in order to run a modern, industrial war, the government and the military would need the support of workers and civilians at home. It helped to define what was needed to sustain morale on the Home Front. And it turned the conflict into the first ever mass media war.[44]

14

Masters of Information

The long suffering and sacrifice of the Battle of the Somme left a general sense in Britain that the war was not being run efficiently, and that something needed to be done. Most politicians could not decide what was required. David Lloyd George, on the other hand, felt that he alone possessed the ideas and leadership to win the war. He had made the Ministry of Munitions work. He had run the War Office since Kitchener's death in 1916, when the ship in which he was sailing to Russia was sunk off the Orkneys. Unlike many of the old-style grandees who were running the country, Lloyd George was a man of the people. In December 1916, barely a few weeks after the guns ceased firing on the Somme, he outmanoeuvred Herbert Asquith in Parliament and became Prime Minister of a new coalition. He had the support of the Conservatives and Unionists under Andrew Bonar Law, while the Labour Party agreed to back him and entered the coalition. The Liberal Party, split between support for Asquith and Lloyd George, would remain divided for the next twenty years.

Lloyd George's entry to Number 10 marked a victory for those who wanted a more ruthless prosecution of the war. He created a new War Cabinet to run the war, a sort of political-strategic politbureau consisting at first of only five members. For the first time, Cabinet meetings were minuted, agendas were prepared and action points circulated, while Maurice Hankey moved from the Committee of Imperial Defence to become Cabinet Secretary. In a move that gave him almost presidential status, Lloyd George created a new private staff, a sort of think tank, to report directly to him; housed in temporary huts erected in St James's Park behind Downing Street, they became known as the 'Garden Suburb'. Four new departments of state – the ministries of Shipping, Labour, Food and National Service – were established almost overnight. There was now a sense that the whole nation must be mobilised to win the war and that the Home Front would play an essential role in final victory.

In the very first meeting of the new War Cabinet on 9 December, it was decided that the whole question of propaganda 'would require consideration at an early date'.[1] By this time three separate government departments had a hand in propaganda; each guarded its own ground jealously and was suspicious of the others. The War Office had accepted the need to send propaganda to foreign nations. Its press bureau had opened up its operations and John Buchan was one of several authors writing material to be sent abroad weekly. The Admiralty too had come around and was issuing material through naval attachés abroad. Even the Foreign Office had overcome its original scruples and accepted that propaganda was a necessary evil in wartime; it was using the Agence Service Reuter as well as distributing material supplied from Wellington House to opinion makers abroad. The situation had become a complete muddle. Now all these bodies submitted their arguments to the War Cabinet, claiming that they should take control of any new organisation. However, Lloyd George decided not to listen to any of them. He turned instead to an old friend and supporter, Robert Donald, to investigate and make recommendations. The editor of the Liberal-leaning *Daily Chronicle*, Donald had been involved with all the different groups who now set out their case. He was an independent arbiter who, at least in Lloyd George's eyes, was well suited to his task. And having the Prime Minister's backing was all that mattered.

Donald spent a week exploring the current arrangements for the management of propaganda. It was not a lengthy investigation, but Donald concluded the situation was a shambles. He thought British propaganda too defensive, in contrast to the more offensive German output; too reactive to foreign criticisms rather than proactive in promoting Britain's interests. And Britain was far too slow to get its messages out. Alarmed to find that news of a recent important speech by Lloyd George had not been sent abroad, he came down hard and said, 'It is through news that public opinion in neutral countries is most easily influenced. At present our news propaganda department seems to be asleep for more than ten hours out of twenty-four (it does not work at night) and it is not alert during the rest of the time.'[2] This was just what Lloyd George wanted to hear. He saw his role as to wake up a sleepy establishment and make it fit for total war.

With the aim of establishing a single straightforward message that would be sent out clearly, both at home and abroad, Donald suggested the creation of a centralised organisation that would direct the output of all the existing groups with an interest in propaganda. He probably hoped to run it himself. However, when the War Cabinet discussed the proposal, Lord Milner suggested John Buchan. Not only was Buchan already doing good work, but Milner still admired the administrative abilities he had displayed when he was on his staff in South Africa a decade before. So, on 9 February 1917, Buchan became director of a new Department of Information. He would report directly to the Prime Minister.

Almost immediately, however, the new department lost some of its teeth. Arthur Balfour, Lloyd George's new Foreign Secretary, insisted that information should still be distributed abroad through conventional Foreign Office channels. Many of the other officials who had been organising propaganda for the last two years also argued their corner forcefully. Buchan, skilled though he was as a writer, had no experience in settling bureaucratic rivalries and rather than abolish any of the existing organisations he kept most of them, albeit with slightly different names. His changes were more of a rationalisation than a root and branch reform: for instance, Charles Masterman at Wellington House now reported to Buchan, rather than the other way around. And Buchan's decision to base himself in the Foreign Office led many outsiders to think he was under the heel of this ancient department of state.

Buchan did bring in as advisers some of his own new men, including explorer Sir Ernest Shackleton, cricketer Pelham Warner, and Roderick Jones, the head of Reuters, who was put in charge of sending news cables. The latter worked part time, in a voluntary capacity, while still managing Reuters as a neutral and objective news agency.[3] And a distinction was drawn between news propaganda and literary publications. News was to take priority, with a renewed emphasis on the cinema and on sending war artists to the front. An office was opened in still-neutral America, and American journalists were invited over for tours of the front and of war factories. In addition, propaganda was to be directed at enemy countries as well as to neutrals. These were all important steps forward. But Buchan found the going difficult, especially when he came up against the domineering position of the Foreign Office. The core of the dispute was that the news-oriented propagandists wanted to gather information from all the relevant departments and then feed appropriate versions to the public in each country by getting sympathetic or pro-British stories out into the press. The Foreign Office view was still that the diplomatic corps abroad, with their links to the great and the good in each country, were better placed to influence the key decision-makers.

The clash between these two points of view would rumble on. Meanwhile, Military Intelligence was drawn into propaganda: the War Office set up its own unit, MI7, which jealously retained control of the release of military information from the front to the extent of censoring every photograph in case it revealed details that might be helpful to the enemy. Buchan was losing control over many aspects of his operation. And to add to his frustration, he found it impossible to access Lloyd George, who was always too busy to see him.

When the Cabinet also criticised Buchan for not doing enough to direct and organise propaganda for the Home Front, he complained that he did not have the finances or resources to handle this huge responsibility as well. In the absence of any coherent plan for domestic propaganda, therefore, Parliament set up the National War Aims Committee (NWAC), an all-party organisation run by the three party whips which opened offices in constituencies up and down Britain. The NWAC presented its work as purely voluntary, thereby maintaining the pretence that government was not directly involved with propaganda at home. By the end of the year it was represented in one-half of the constituencies in the country. These local branches encouraged and briefed speakers to address meetings in order to remind audiences about the causes for which Britain had gone to war, and of the objectives in winning. This presented speakers with a problem: the government had not yet defined what its war aims were. It was not until the following January that Lloyd George, largely under pressure from President Woodrow Wilson of America, by this time of course an ally, formulated Britain's official war aims.

The speakers of the NWAC were instructed to talk on a specific range of subjects. Mostly they were to combat the population's increasing war-weariness. This often came down to attacking the occasional voices of pacifists, whom many people were beginning to take more seriously as the war entered its fourth year. Pacifism was equated by many with socialism, which in the light of the Russian Revolution was seen as a growing threat. So speakers were encouraged to denounce pacifism and call upon British workers to make further sacrifices for the war effort. The committees specifically targeted areas that were thought to be experiencing particular industrial troubles. So, for instance, speakers were sent to Wigan and Hull, where it was said workers were willing to 'down tools' at the slightest provocation. This offensive was by no means always successful; in Brighton the local trade unions voted to reject any speakers and not only to burn the literature that had been sent to them but also to send back the ashes to the NWAC head office.[4] In the light of these concerns, it is perhaps not surprising that when Lloyd George finally made a speech outlining the nation's war aims, it was addressed to an audience of trade unionists at Caxton Hall in London.

By the summer of 1917, the conflicting interests within Buchan's department were once again pulling propaganda in many different directions. Desperately needing someone at Cabinet level to argue his case, Buchan appealed to Milner, and the War Cabinet appointed Sir Edward Carson to supervise the work of the Department of Information, NWAC and MI7. Carson was a strange choice. A fiery Ulster orator, he was almost a one-man propaganda machine in his own right. In the two years before the Great War he had mobilised the people of Ulster into resisting all calls for Home Rule and had formed the Ulster Volunteer Force, overseeing its growth into a powerful separatist militia with its simple slogan of 'No Surrender'. But in 1917 he did almost nothing to support Buchan or to focus the various arms of propaganda. He was in any case preoccupied with Irish matters for much of the time, obsessed with guarding the interests of Ulster that were always his principal passion. Only a month after Carson's appointment, in October 1917, Lloyd George turned back to Robert Donald and asked him to review the whole system of propaganda once more.

For the second time, Donald came up with damning criticism, this time of the very department that he had helped to create. He felt that the work of the news propaganda division was still amateurish, despite the presence of Roderick Jones at its helm. He thought there was still far too much literary work coming out of Wellington House and that it was not having enough impact. He argued that experienced journalists like himself were not being consulted enough. And with a staff of 300 people, he believed the department had grown too big and too wasteful.

Both Masterman and Buchan felt they had to defend their record. Masterman claimed that his pamphlets and books had indeed had a major impact in neutral countries. He pointed out that '19 countries have declared war against Germany and 10 have broken off relationships with her. At the same time she is continually complaining of the poor results of her own propaganda contrasted with that of the British and announcing that "malignant British lies" have turned all the world against her.' How could this be a sign of failure? Masterman asked.[5] The problem that he could not escape, however, was that his work was done in total secrecy; not only was it impossible for the public to know the details, it was also extremely difficult to accurately assess the effect.

Buchan, who was now increasingly under attack in the press for failing to make the propaganda system work, made his own defence and argued that 'We welcome such criticism when it is not merely ignorant gossip, for propaganda is not an occult science, but a matter on which every citizen has a right to judge ... Moreover, there is no finality to it; it may be improved but it can never be perfect.'[6] He claimed that it was unreasonable to say not enough was being done as most of the department's work was surrounded by strict secrecy and so was invisible to most observers. He put the case that what was needed was leadership at Cabinet level, not by the constantly distracted Carson but by a more prominent figure. Buchan was not resisting change, but he was in effect asking for a new supremo to seize control.

The end of 1917 was a particularly black period for the British people and their Allies. The war had been going on for well over three years. The grinding, gruesome struggle at Passchendaele that ended only in November had achieved little, but had filled the newspapers once again with long lists of the dead and missing. The triumph at Cambrai had been only brief and short-lived. The French army had mutinied after the Nivelle Offensive, threatening to undermine France's position on the Western Front. The defeat of the Italian allies at the Battle of Caporetto in October had nearly knocked Italy out of the war. Russia meanwhile was negotiating a treaty with Germany after the Bolshevik Revolution and so had completely removed itself from the conflict, freeing dozens of German divisions to leave the Eastern Front and head west. The U-boat campaign had caused massive destruction to shipping and the lack of supplies was becoming evident in the shops; almost everyone felt they did not have enough to eat. Meanwhile workers were being called upon to work ever harder. The one bright spot was that America had joined the war – though it had so far contributed nothing to the titanic struggle on the Western Front. Across Britain there was a profound sense of war-weariness. Some people began to wonder if the war would ever end. It was clear that Britain needed a powerful tonic to lift morale.

On 23 January 1918, Sir Edward Carson resigned from the Cabinet to concentrate on Irish affairs. This provided the opportunity for Lloyd George to act. His chief whip suggested that Lord Beaverbrook, the former Max Aitken, should be appointed to the propaganda job. Beaverbrook had by now taken control of the *Daily Express* and the *Sunday Express* and had also acquired the *Evening Standard*. Lloyd George enjoyed a close relationship with many press men and editors. He did not fear their hostility, rather, he embraced their support. And both Beaverbrook's and Northcliffe's papers had supported him when he challenged Asquith for the leadership of the country a year before. Sensing the nation's black mood and the need for dramatic action, Lloyd George decided to go one step further than his chief whip had suggested and created what would be the last new ministry of the war, the Ministry of Information, appointing Beaverbrook as minister with a seat in Cabinet as Chancellor of the Duchy of Lancaster.[7] He also offered a role to Northcliffe, who had just returned from running the British Mission in the United States.

Northcliffe was at first reluctant to accept office, as he prized his journalistic independence and did not want to be put in a position in which his papers could not attack the government. And he refused to play second fiddle to Beaverbrook, whom he regarded as his junior. It was Beaverbrook who came up with a clever compromise. Northcliffe would take charge of propaganda against enemy countries and report directly to Lloyd George. Beaverbrook would himself take over the rest and the Cabinet position. After a few days of bargaining a deal was done.[8] On 4 March 1918 the new Ministry of Information opened for business at offices in Norfolk Street.

The story of propaganda in Britain had turned full circle. At first it had been a case of the utmost secrecy and of withholding information from the press, who were seen as little more than a nuisance when it came to fighting a war. Now the two top press barons in the country had entered government, bringing their professionalism and expertise to help buck the nation up, lift morale and sustain it for the long fight that was to come, and to lead the efforts to undermine the enemy's fighting spirit. It was a sign of how far the political leadership had come in accepting the concept of total war. Propaganda, whether in words or pictures, would now be as important as the fleet at sea or the armies on land in bringing victory.

With his experience as Canadian Eye-Witness on the Western Front, his role in setting up the Canadian War Records Service, and his responsibility for commissioning photographers, war artists and film-makers, Beaverbrook was well suited to take charge of propaganda. But his arrival was not universally welcomed. Many of the staff of the political intelligence section, including Arnold Toynbee, Lewis Namier and other academics, refused to serve under a newspaper man, and the section was quietly transferred to the Foreign Office to avoid a confrontation. Others in Whitehall were equally horrified at the thought of press men being appointed to senior positions; it was almost an affront to their dignity as civil servants and gentlemen. Even today, the idea of bringing media moguls into the heart of government would be difficult for the establishment to accept. But Lloyd George was not bothered by this reaction, and neither were Beaverbrook and Northcliffe. Desperate times required novel solutions. As if to rub the new situation in, Beaverbrook was made answerable not to Parliament, the Treasury or the War Cabinet, but directly to the Prime Minister himself.

Many feared, possibly correctly, that by bringing the press barons inside government Lloyd George intended to create his own fiefdom in order to muzzle press criticism of his premiership. Rumours abounded that Northcliffe and Beaverbrook were moving into offices in Downing Street. Some believed that in the post-war world Lloyd George would use them to build up his own reputation and bolster his popularity. He was acting like a dictator, they felt, riding roughshod over political conventions. Maurice Hankey, Secretary to the War Cabinet, wrote that he thought the reason for Northcliffe's appointment was because Lloyd George wanted to raise funds for a new party organisation, describing it as a 'shady business'.[9] Many senior officials, especially in the Foreign Office, felt it was inappropriate to allow the 'press gang' access to confidential information. A confrontation was inevitable.

Beaverbrook brought in a new group of advisers, as his predecessors had done, forever widening the group who were to influence the development of propaganda. Many of the new intake were from the City – men like Sir Harold Snagge, a director of Barclays Bank; Sir Hugo Cunliffe-Owen, a tobacco magnate; and Sir Eric Hambro, another banker (whose wife was already working in Naval Intelligence at the Admiralty). Writers such as Arnold Bennett and Hugh Walpole were also given senior positions. Walpole was another author who was able to litter his novels with characters based on individuals he met while carrying out his secret war duties. He had worked for the Red Cross in Russia and was charged with co-ordinating the flow of propaganda to the newly born Soviet state. In the new ministry, John Buchan was made Director of Intelligence, charged with processing information for the propaganda departments.

Buchan was also to be responsible for inviting foreign journalists to tour British war factories, and it was in this capacity that he met Lowell Thomas, an American journalist searching for an individual whose dramatic wartime story could be promoted to help lift morale. Buchan put Thomas in touch with Allenby's headquarters in Palestine, where stories were beginning to spread about a young intelligence officer who was leading an Arab revolt in the desert. Thomas went with a cameraman to the Middle East, where he met

Colonel T.E. Lawrence, writing up the story as a series of illustrated lectures and films that had the titles *With Allenby in Palestine* and *With Lawrence in Arabia*. They proved hugely popular and made both Thomas's reputation and that of 'Lawrence of Arabia'. This phase of the propagandists' work gave birth to at least one enduring legend of the war.

Beaverbrook continued to do battle with the Foreign Office, just as Buchan had before him. Balfour, the Foreign Secretary, and Lord Hardinge, his Permanent Secretary, were formidable opponents, continuing to insist on Foreign Office control over the distribution of propaganda overseas. Used to getting things his own way, Beaverbrook was not good at the patient game of point scoring and jockeying for position in the hierarchy of government. While Balfour pretended to respect the position of the young new ministry, he then carried on exactly as before. Feeling that there was no point in having a ministry to control information if he was unable to manage its various elements, on 13 June he wrote to Lloyd George, 'I am nearly worn out with my effort to put this Ministry on its legs.'[10] But the situation on the Western Front was critical and the Prime Minister did not reply. The dispute with Balfour escalated further. On 1 July, Beaverbrook offered his resignation, but Lloyd George was again too busy to see him and did not respond. Later in the month, Lloyd George tried to mediate and came up with a compromise. Balfour did not accept it. Beaverbrook had lost his battle.

Beaverbrook did not make a good minister.[11] Unsuited to a situation in which he had to play a subordinate role, he was too direct, too pushy and too much of an outsider, both as a Canadian and as a press baron, to progress within the corridors of established power in Britain. He had great skills and a brilliant ability to find the popular way of approaching a subject, but he was unable to use these in government. He could rage and shout, and offer his resignation. But it made no impression on the mandarins of Whitehall. As Stanley Baldwin said of him at the time, 'The Minister of Information is a man of very strong personality. Men with strong personalities have

this in common, that the magnetism that comes with that personality either attracts or repels.'[12] In October Beaverbrook successfully
resigned, giving ill health as the reason. His health was indeed
poor – he had an operation for a glandular swelling in his neck – but
more than anything he had been worn down by the process of government. He would return to writing, journalism and running his
newspapers.

By contrast, Northcliffe, appointed by Lloyd George to run the
Enemy Propaganda Department at Crewe House in Mayfair, did a
masterly job. The operation he led has been described as bringing
'the greatest victory ever achieved by war propaganda'.[13] Lloyd
George wanted to launch an offensive directly against enemy morale
and Northcliffe did this on a vast scale. Once again he brought in
outsiders, including writers, journalists and historians, who managed the policy for him.

His department's first target was the Austro-Hungarian Empire,
which Northcliffe identified as the weak link among the countries at
war with the Allies. Henry Wickham-Steed, the foreign editor of *The
Times*, ran the campaign with the assistance of young historian
Robert Seton-Watson, an expert on the various peoples ruled by the
Hapsburg monarchy. Northcliffe quickly realised that if the Allies
could make it clear that they favoured independence for the many
nationalities within Austria-Hungary, such as the southern Slavs (or
Jugo-Slavs), the Croats, Czechs, Poles and Slovaks, then they could
weaken the military effort of the Austro-Hungarian armies by entic-
ing soldiers to desert. If successful, this might hasten the
disintegration of the Austro-Hungarian Empire and fundamentally
weaken Germany's war effort.

The problem once again lay with the Foreign Office, which had
not yet decided its post-war policy towards the ethnic groups of central Europe. Balfour refused to be steamrollered into making policy
in order to meet the demands of propaganda, which he saw as putting the cart before the horse. Northcliffe, however, wanted the
go-ahead as quickly as possible. In April, a congress in Rome, held

to clarify Allied policy, declared support for the many different nationalist groups governed by the Hapsburgs. The War Cabinet told Northcliffe to go ahead with his offensive as long as 'no promise ... [was] made to the subject peoples of Austria that we could not redeem'.[14] Northcliffe's department employed a variety of techniques to achieve its objective. Leaflets were dropped over the lines where Austro-Hungarian troops were stationed. A weekly newspaper was produced and smuggled into central Europe. On the Italian Front, gramophone discs of patriotic, nationalist songs were played in no man's land, directed towards the Austrian lines. Colour pictures of religious and nationalist symbols were distributed. The result was that thousands, possibly tens of thousands of Slav soldiers deserted. Many of them crossed to the Italian lines carrying copies of the leaflets and said to their captors, 'I have come because you invited me.'[15] Many front-line units had to be replaced with more trustworthy troops.

The Austro-Hungarian offensive against the Italian lines in June was a failure, and as the military situation in the west deteriorated for the enemy powers, the Austro-Hungarian Empire finally began to crumble. In October, new independent nations were proclaimed in Prague, Warsaw and Zagreb. Even the Hungarians declared they were now a separate state. The Austrian armies fell apart. As the Italian army cautiously moved forward, more than three hundred thousand men deserted. The Austro-Hungarian navy surrendered to the Yugoslav National Council at Pula. In early November, what was left of Austria-Hungary signed an armistice. The collapse of the Austro-Hungarian empire had not all been down to the propaganda campaign, but it certainly helped.

When it came to directing propaganda at Germany, Northcliffe called in novelist H.G. Wells to help define policy, appointing alongside him John Headlam-Morley, a specialist on German history. A supporter of socialism in his younger days – although he had always approved of Britain going to war in 1914 – Wells had also been a great believer in world government and advocated the creation of a global state in which the emphasis would be on the improvement of science and not the development of nationalism. Wells realised that the defeat of Germany was tied to the establishment of a 'world peace that shall preclude the resumption of war' in some sort of international organisation like a 'League of Free Nations'.[16] Wells said that propaganda against Germany must make it clear first, that the Allies would continue with the war until they had secured complete victory; second, that the Allies had nothing against the German people, but were determined to destroy the militaristic leadership that had led Germany to war and sustained it ever since. In other words, to avoid financial ruin, utter destruction, starvation and military humiliation, the German people must throw off their current imperial government and negotiate with the Allies. It was a clever line to pursue, as it encouraged moderate Germans who would want their country to participate fully in the systems, such as the League of Free Nations, created after the war. Again, Balfour felt he was being pressured into a formulating a foreign policy simply in order to suit the propagandists, another case of being asked to proclaim his war aims before ready to do so. However, on this occasion he declared himself to be 'in general agreement with the line of thought'.[17]

From July 1918, literature was smuggled into Germany from neutral Holland and Switzerland calling on the German people, exhausted by four years of war, to demand peace. In addition, vast numbers of leaflets were printed and dropped from aircraft over German lines. Some were carried in high-flying balloons, sent to float east towards Germany in the hope of reaching the civilian population as well. The total numbers of leaflets dropped amounted to 1.7 million in June, 2.2 million in July and 4 million in August.[18] Some leaflets claimed the Allies were making huge advances along the Western Front, others, listing the numbers of submarines sunk or captured, made it clear that the U-boat campaign had failed. Most reminded the Germans that tens of thousands of American troops were arriving daily in France. Another regular feature was to point out the gains Allied armies were making in Greece, in Italy, against Bulgaria and in Palestine. Every leaflet strongly emphasised the heavy German losses and the futility of making further sacrifices for a lost cause.

As the German army went into retreat in the west, even more leaflets were dropped: 3.7 million in September, 5.4 million in October. The Germans punished those who read the publications and offered three marks to every soldier who handed in a leaflet to an officer, eight marks for a book. But German officers collected only about one-tenth of the leaflets dropped, so it can be deduced that most of them were read and kept. One German writer described the leaflets as 'English poison raining down from God's clear sky'. The German military chief, Marshal von Hindenburg, described how the propaganda leaflets intensified the demoralisation of German soldiers, writing, 'The enemy seeks to poison our spirit ... He bombards our Front, not only with a drumfire of artillery, but also with a drumfire of printed paper. Besides bombs which kill the body, his airmen throw down leaflets which are intended to kill the soul.' Hindenburg recorded that soldiers sometimes took the leaflets home with them on leave, where they were passed 'from hand to hand' and widely discussed 'at the beer table, in families, in the sewing room, in factories, in the street. Unsuspectingly many thousands consume the poison.'[19] Once again, as with Austria-Hungary, it was military defeat in the field that caused the Germans to sign an armistice. But it is clear that the Allied propaganda campaign in the last months of the war made a major contribution to the collapse of morale that led to this defeat.

Northcliffe had proposed an active propaganda campaign of dropping leaflets as long ago as September 1914. His proposal had been turned down at the time, the Chief of the Imperial General Staff saying that propaganda was 'a minor matter – the thing was to kill Germans'.[20] It was yet another indication of how far propaganda had come during the war years. In November, on the eve of the armistice, Northcliffe resigned as Director of Enemy

Propaganda, his job done. Now he could return to his favourite pastime of leading his newspapers as they attacked the government and its policies.

Despite their obvious skills in managing public opinion, Northcliffe and Beaverbrook were, strangely, denied control over one vital area, propaganda at home. That was still controlled by the NWAC. During 1918, the NWAC organised large patriotic rallies across the country, often with music hall stars – the celebrities of the day – turning up to provide entertainment. Soldiers gave positive talks on their experiences at the front in an effort to bolster support for the war factories. Five mobile film projection vans were sent to areas where there were no cinemas. In Birmingham, on what was called 'Win the War Day', a big procession passed through the city led by a tank. Much of this was fairly crude, and some of it harked back to the old claims of German atrocities: a 'Crimes Calendar' was produced in which each month highlighted a different enemy atrocity.[21] More than anything, however, an exhausted and war-weary nation was kept going by the growing certainty that victory was not far off.

Two days after the Armistice, the War Cabinet decided to close down the Ministry of Information. Maybe they had sensed the general feeling that Lloyd George should not be allowed to use the experts gathered in Norfolk Street to boost his own personal standing. The notion that propaganda, if a necessary evil in war, was not suited to peacetime government also prevailed. Whatever the reasoning, Information was the first of the wartime ministries to be shut down. Masterman said, 'They're getting rid of me. I dare say that's right. But if I were them I'd keep the department going, to work in favour of a reasonable peace.'[22] Buchan was asked to liquidate the ministry and had the task of dismissing staff and reallocating its assets. There was a general sense that with the war won it was time to go, and he told Masterman, 'I am not sorry to get quit of the business.'[23]

Somehow, despite the pressures of war, of writing his mammoth *History* and countless weekly news digests, as well as running the

Department of Information and then playing a senior role in the ministry, Buchan had still found time to write two novels. The more successful of these, *Greenmantle*, published in 1916, again centred on the exploits of Richard Hannay.[24] Convalescing after the Battle of Loos, Hannay is asked to take on a mission to track down Muslim extremists organising a jihad against British rule in the Middle East and India. The book was informed by much that Buchan had learned in office and was so successful that it outsold *The Thirty Nine Steps*. After the war, however, Buchan collapsed in exhaustion for some weeks. He eventually rejected a life in public affairs and bought a house in Oxfordshire, moving there to concentrate on his writing. He had lost several of his best friends during the war, along with his youngest brother. For years he lived in the shadow of the war and wrote many books of remembrance and tribute. Like many of his generation in the 1920s, he felt a sense of guilt that he had survived when others had not.

Of the assets that Buchan had to dispose of there were all the films, photographs and paintings made and recorded under the auspices of the Ministry of Information. In March 1917, the War Cabinet had agreed to establish as a memorial to the wartime sacrifice a museum to record all aspects of the war relating to Britain and her empire. Sir Alfred Mond was appointed its director and Charles Ffoulkes, Master of the Armouries at the Tower of London, was made curator. Mond visited Haig at GHQ, and in the last year of the war, with the help of senior army figures, Ffoulkes began to collect artefacts such as trench signboards, disused field artillery pieces, war relics and military machines of every sort. Lorries visited waste dumps, gathering up huge collections of rifles, machine guns, flags, posters, maps and endless spoils of war. No doubt suitably puzzled, soldiers still fighting the war described these early museum curators as 'souvenir collectors', but as Ffoulkes later wrote, 'one had to consider that while such things might be of transitory value for the moment our business was to make history, or rather to record history.'[25] After some debate, as the memorial was intended to reflect the work of all the Commonwealth armies, it was decided to call the new enterprise the Imperial War Museum.

On 1 January 1919, about 42,000 photographs, along with the paintings of more than thirty official war artists, were transferred from the defunct Ministry of Information to the new museum. The ministry had opened a shop in Coventry Street in London at which members of the public could buy copies of photographs, and this had proved immensely popular. The shop too now became a part of the museum.[26] Ffoulkes wrote, 'What a collection it was ... War conditions on all fronts and at sea, an almost complete photographic survey of the whole of the British front ... in fact all aspects of what was then the greatest war in history, and miles and miles of cinematograph films.'[27]

King George V opened the Imperial War Museum at its first home at the Crystal Palace in south London in June 1920. A crowd of 40,000 people gathered for the occasion, at which the director, Sir Alfred Mond, made it clear that the museum was 'not a monument of military glory but a record of toil and sacrifice' in which 'everyone who took part in the war' would find 'an example or illustration of the sacrifice that he or she made'.[28] From the beginning, the museum included objects relating to women's war work, war factories and industrial production as well as to the activities of the army, navy and air force. It interpreted its brief widely and even collected letters and songs containing examples of war slang, some of which were thought to be so profane that they had to be kept under lock and key and could only be read by 'very discreet members of the staff'.[29] In November of that year the museum began to acquire the official films made for the War Office, the Admiralty and the Air Force, in order that they 'might be classified and preserved as records'. Six hundred and thirty separate films were collected.

The museum proved a huge success from the day of its opening, with more than 1.4 million visitors in its first nine months.[30] In 1936 it moved to its permanent home in Southwark at the Bethlem Royal Hospital, a lunatic asylum once known by a shortened version of its name as 'Bedlam'. The outstanding work of the photographers, film-makers and war artists, along with the records of the men and women who had fought or otherwise taken part in the war, or who had toiled on the Home Front, found a permanent home. One hundred years later these unique records can still be studied, assessed, reviewed and copied.

In terms of propaganda, however, there was to be a sea-change in attitude within ten years of the Armistice. Many of the claims of German atrocities and barbarism were rapidly discredited. Arthur Ponsonby argued in *Falsehood in Wartime*, a highly influential book published in 1924, that British propaganda had repeatedly lied to the people, especially about the atrocities. Ex-soldiers wrote about the contrast they felt between life at the front, where many of them grew to respect their enemy, and the hysterical anti-German mood they found when they came home on leave.[31] But, elsewhere, very different conclusions were drawn from the use of propaganda in the First World War.

The Germans continued to argue that British propaganda had been used very successfully against them. The German War Minister, General von Stein, said that 'In propaganda the enemy is undoubtedly our superior.' General Ludendorff, commander-in-chief on the Western Front in 1918, wrote in his memoirs:

We were hypnotised by Allied propaganda as a rabbit is by a snake. It was exceptionally clever and conceived on a grand scale ... many people were no longer able to distinguish their own impressions from what enemy propaganda had told them ... with the disappearance of our moral readiness to fight everything changed completely. We no longer battled to the last drop of our blood. Many Germans were no longer willing to die for their country ... The attack on our home front and on the spirit of the Army was the chief weapon with which the Entente intended to conquer us, after it had lost all hope of a military victory.[32]

What this demonstrates is the beginning of the argument that the German army was stabbed in the back in November 1918 by the collapse of morale at home. This was a theme later picked up and exploited powerfully by Adolf Hitler, who wrote in *Mein Kampf* of 'the very real genius of British propaganda'. Part of Hitler's appeal derived from his claim that Germany had not lost the war on the battlefield but the German people had been betrayed by the demoralised peacemakers at home. In Nazi Germany these lessons would not be forgotten. Joseph Goebbels at the Propaganda Ministry in the late 1930s would take the science of propaganda to new heights, introducing radio and sound films to create one of the most powerful propaganda campaigns of the century.

On the other hand, in Britain, propaganda became a dirty word. The falsehoods that had been put out, and the lies that had been believed, greatly discredited its use. There was a continuing belief in some of the methods used successfully in 1918, so that for several months at the start of the Second World War the RAF dropped on Germany nothing but millions of leaflets. But there was a big difference between the mood of imperial Germany at the end of a long, exhausting war, and that of Nazi Germany where people felt buoyant and supreme at the start of another. This time the leaflets were completely ignored. Equally, the Second World War Ministry of Information put a lot of emphasis on film, as it had been so effective in the First. As a consequence, film-makers like Humphrey Jennings and Harry Watt produced some great film classics. But more profoundly, the pendulum had swung. If they thought it was propaganda, British people no longer believed what they were being told. When genuine atrocities, such as the Nazis' gassing and burning of millions of innocent civilians in the extermination camps, were revealed, many therefore dismissed them as alarmist propaganda. One of the cruel ironies of propaganda in the twentieth century was that people readily believed what was false and sometimes disregarded what was dreadful but true.

Epilogue:

The First Boffins

The work done by the scientists and mavericks called on to contribute to the progress of the war between 1914 and 1918 would help lay the foundations for much scientific and technological progress in the following decades. Some of these changes would no doubt have happened anyway. Others were very much a consequence of a war which, as so often in history, brought about immense technological developments.

For instance, the world of aviation had been transformed. In 1914, the armed services possessed 272 aircraft. By 1918, the RAF had over 22,000 machines. Aircraft powered by engines of only a few horsepower that were blown backwards in the face of a heavy wind in 1914, had developed by 1918 into formidable flying machines with engines of 250–350 hp. The four-engined Handley Page bombers designed to bomb Berlin drew on a total of 1500 hp, had a wingspan of 126ft, could carry 7500 lb of bombs and remain airborne for seventeen hours.

After the war, the heavy bombers were adapted for civilian use and helped to create a new industry. In June 1919, Captain John Alcock and navigator Lieutenant Arthur Whitten Brown made the first pioneering flight across the Atlantic in a converted Vickers bomber. After sixteen hours struggling with bad weather and the difficulty of navigating across a featureless sea, they came down in an Irish bog near Clifden on the coast of County Galway. In that same year the first regular aviation service in Europe started between London and Paris using a converted Handley Page bomber carrying ten passengers. Later in the year direct flights to Brussels began. Soon a host of small airlines were flying converted bombers around Britain and Europe.

In 1924 the government brought several of these airlines together in Imperial Airways, which was tasked with developing air links between Britain and the empire. Five years later the first passenger service from London to Karachi was inaugurated in a flying boat built by Short Brothers, a company that had been in at the beginning of flying in Britain, at their base in Rochester, Kent. Before long, small groups of passengers were flying on routes across the Mediterranean, over the Pyramids, then either south across the savannahs of Kenya to South Africa, or east across the desert of Iraq to India and all the way to Australia. By the 1930s, Imperial Airways had become a by-word for luxury. The cabins of their Short flying boats were beautifully furnished, while barmen from top London clubs would mix cocktails for passengers about to be served meals freshly cooked in a galley kitchen in true silver service on laundered linen tablecloths. It was a long way from John Moore-Brabazon bouncing across a field in Kent at fifty feet, followed by his dog, only thirty years before.

Radio technology had improved consistently through the war. Although there had been no revolutionary advances, there were considerable developments in the use of thermionic valve transmitters; more importantly, a huge number of servicemen in the army, navy and air force had become proficient in the use of wireless technology. When peace came, a group of enthusiasts argued that this technology could be put to new uses to communicate music and information to audiences across large areas. As a result, in October 1922, an alliance of manufacturing companies including Marconi, Metropolitan Vickers and the General Electric Company came together with the General Post Office to form a new organisation with the purpose of broadcasting an experimental service. Transmitting from a station known as 2LO, based in Marconi House on the Strand in London, it was called the British Broadcasting Company. As transmissions improved, so listeners around the country began to tune in, many of them ex-servicemen who built their own receivers from crystal kits. Meanwhile, the first large 'wireless' sets were mass produced for the domestic market. It was soon clear that the experiment had been a success and that there was a national market for a broadcast service. A series of government committees explored how to bring this about and under the guiding hand of the first powerful and dominating general manager, John Reith, the government decided to re-form the organisation without the commercial interests. Renamed the British Broadcasting Corporation, it went on air on 1 January 1927 as a public broadcaster established under a Royal Charter and funded by a licence fee to be paid by everyone who had a wireless set.

A completely new medium of mass communication thus came into being. Within three years the BBC was broadcasting a national programme from London and a regional service from other cities around the country, and by the 1930s the BBC had become a much-loved public institution. Whole families would gather around the wireless to listen to popular broadcasts of music programmes, talks and entertainment, or, at times of crisis, to the nine o'clock evening news. By the Second World War, 23 million Britons listened regularly to the BBC radio news. A new cultural force had been created out of the remnants of the electronics technology of the Great War.

In Britain, the foundations of a new chemical industry had been established during the war from the need to develop antiseptics, home-grown explosives, poison gases and compounds such as acetone. In 1926 four large companies including the descendant of British Dyes, created by the Ministry of Munitions during the war, came together in a merger to form Imperial Chemical Industries (ICI). For many decades this was to be the largest manufacturing company in Britain, and in addition to its continuing association with explosives, it produced a broad range of products including fertilisers, insecticides, dyes, artificial fibres and paints. The company continued to come up with new products into the 1930s; among these innovations were perspex, polyethylene and, at the end of the decade, nylon. From 1931, in association with DuPont, ICI produced the alkyd and polyester based paint known as Dulux and created a new advertising jingle, 'Say Dulux to your decorator'.

In the United States, developments in these parallel fields were determined entirely by entrepreneurs in the marketplace. Juan Trippe founded Pan American Airways as a private company and built up a network first linking Latin and South America and then extending around the globe. In broadcasting, advertising revenues funded vast new communication empires like the National Broadcasting Company (NBC) and the Columbia Broadcasting System (CBS), which were soon making huge profits and were quite different organisations to the BBC. In the chemical industry, American companies eclipsed the German pre-war giants and emerged as the biggest global players. But in post-war Britain the state played a role in encouraging developments in the country's newest industries: civil aviation, broadcasting and electronic goods, and chemicals and pharmaceuticals. Wartime controls and attitudes were becoming a permanent feature of government. William Beveridge, then a newly appointed official at the Board of Trade, wrote, 'We have ... under the stress of war made practical discoveries in the art of government almost comparable to the immense discoveries made at the same time in the art of flying.'[1]

Meanwhile, a generation of doctors and nurses had seen more injuries and treated more wounds and trauma in four years than had their predecessors in a lifetime. The process of blood transfusion developed on the battlefield by Canadian and Australian doctors came into regular use, while the chemical and pharmaceutical industries started marketing new drugs. Alexander Fleming's experience with antiseptics at the Base Hospital in Boulogne helped him with the discovery of penicillin. Psychology became a mainstream subject of study, and work was done on exploring the subconscious and the libido. Having advanced the new form of treatment known as plastic surgery, Harold Gillies in the inter-war years developed this further to make available for wealthy clients cosmetic surgery of various types; he even carried out the first gender reassignment surgery.[2] A report of the Medical Research Committee concluded that the war had been 'a great stimulus' providing 'unequalled opportunities for study and research, of which the outcome may bring lasting benefits to the whole future population'.[3]

Behind the closed doors of Whitehall, the code breaking performed under the auspices of the Admiralty in Room 40 was transferred in 1922 to the Foreign Office, which somewhat reluctantly picked up the mantle of reading other nations' cables and signals. The Government Code and Cypher School created by the Admiralty at the end of the war continued to monitor the activities of the nation's enemies, and also those of many of its friends. With war clouds gathering in the late 1930s it was decided to relocate the school from central London to a larger and more suitable base outside the capital, and officials found a country house near a railway line with good links into London. In 1939, the school moved to Bletchley Park and thereby into a new and even more distinguished phase of intelligence gathering, code breaking and technological development that ultimately ushered in the computer age.

After the Second World War, the Code and Cypher School continued its top secret work of monitoring communications but was renamed the Government Communications Headquarters or GCHQ. During the Cold War a close link was formally established between GCHQ and the National Security Agency in the USA, a link that was only publicly disclosed in 2011. The incredible revelations in the internet age of the scale of government surveillance on both sides of the Atlantic thus have a direct lineage from the pioneering code breaking work in Room 40 of the Admiralty, through the Government Code and Cypher School to Bletchley Park and then to

GCHQ. The legacy of the signals intelligence work that produced such extraordinary breakthroughs in the First World War is the technology behind today's surveillance state.

In all of these fields, the role of science would gain a higher profile in the Second World War, when the public became more aware of the work done by the 'boffins', a word unknown in the Great War but one that became popular in the Second. The term 'boffin' initially appeared in RAF slang as an affectionate word for the scientists who worked quietly in the background to develop new military technologies. They were also known as the 'backroom boys'. The word boffin was first used to refer to the civilian scientists who came up with the concept of radar, but came to have a much broader meaning and to include servicemen as well as civilians. The word also carried with it associations of eccentricity and unusual behaviour, of someone labouring long and hard in secret to invent devices incomprehensible to most people but which had revolutionary military applications.

The origins of the word 'boffin' are obscure. There was a strange character in the Dickens novel *Our Mutual Friend* called Nicodemus Boffin, while jokes abounded that the word represented the offspring of two unusual creatures, a puffin and a baffin. However, it was more likely to have begun as an acronym for the words 'back office intern'. During the Second World War most of the boffins were unknown, working quietly behind the scenes and recognised only by their peers. But three boffins did achieve fame during that war, becoming household names and popularising the role of the wartime scientist to a wider public. And all three men had done important pioneering work during the First World War, early in their careers. Their stories provide an illustration of the link between the little-known work of the Great War scientists and the better-known achievements of the Second World War boffins.

Frederick Lindemann was probably the most famous scientist of the Second World War, and also the most controversial. In true eccentric form, he always appeared wearing a bowler hat and a long black Melton coat, carrying an umbrella whenever he went outside, in winter or summer, rain or shine. Universally known simply as 'the Prof', he was a close friend and confidant of Winston Churchill and in May 1940, when Churchill entered Downing Street, he appointed Lindemann as head of his Statistical Office. Lindemann became what today would be called a 'special adviser', although the term was not then in use. Acting for Churchill like the head of an independent think tank, Lindemann had a roving commission to dig into any aspect of the wartime government and administration. He met with Churchill almost daily, advising him on scientific matters, military issues of all sorts, logistical problems, and the economy. Most weekends, Lindemann joined Churchill and his entourage at Chequers. And he sometimes accompanied Churchill when he travelled abroad.

Lindemann had been born in Baden-Baden to a German father from the Alsace border region and a half-American mother. But his parents lived in England, where his father, a successful businessman and part-time scientist of distinction, had settled and was later naturalised. Lindemann grew up in Sidmouth, Devon where he learned to love the English countryside. From his early years he developed a formidable memory for numbers and was supposedly able to recite the number pi (π) to 300 decimal points. In 1900, aged thirteen, he and his brother were sent to Darmstadt in Germany to finish their education and he went on to enter one of Germany's superb technical universities, a *Technische Hochschule*, where he began to excel at science. While there he also became a tennis champion; mixing in high social circles, he played tennis with both the Kaiser and, on one of his visits to the town, with the Tsar. In 1908, the young Lindemann went to Berlin, the scientific capital of the world at the time, as research student to the great German scientist Walther Nernst. In Berlin, Nernst introduced Lindemann to many of the scientific luminaries of his day, including Albert Einstein, Max Planck and Marie Curie. He performed important research for Nernst on specific heat, calculating the amount of energy needed to increase one gram of several different substances by one degree centigrade. From the results it was possible to gain an important understanding of the substances' atomic structures. His work for Nernst brought him international acclaim and invitations to lecture in Britain, France and America.

Lindemann had spent fourteen years in Germany when the prospect of war suddenly interrupted his life in the summer of 1914. He was playing in an international tennis championships at Zoppot on the Baltic and had just qualified for the finals when, to avoid being interned, he returned to England immediately in order to reclaim the British side of his character. As Europe divided he seems to have had no hesitation about asserting his British identity. Within weeks he had offered his services to the War Office and the Admiralty but received no reply from either. In March 1915, Mervyn O'Gorman at the Royal Aircraft Factory in Farnborough asked him to join a group of young scientists to continue the scientific work on aeronautics started by the Advisory Committee before the war. But since he had spent so many years in Germany, questions were raised about Lindemann's nationality and his loyalty to Britain. Amid the prevailing hysteria, some of his colleagues even privately feared at first that Lindemann was a German spy in their midst. Shaken by this, he was thereafter rather cagey about his background and was always keen to prove himself as more British than the next man.

However, Lindemann soon demonstrated his loyalty and became a central member of the extraordinary group of young civilian scientists O'Gorman had recruited to Farnborough, called the 'H Department' or 'Physics Department'. The team lived together in a house outside the town in an atmosphere like that of an Oxbridge college, arguing late into the night about key questions of science. Soon nicknamed the 'Chudleigh Mess', the group included three future Nobel Prize winners, three future peers, five knights and several professors and Fellows of the Royal Society.[4] The purpose of their work was to further improve the scientific understanding of aeronautics, and they provided a model for many such groups that would exist in the next war. Lindemann himself worked on a variety of tasks that included detecting aircraft by sound, designing and making rate-of-climb meters, turn indicators and range finders, and the use of infra-red rays to detect the approach of enemy aircraft at a distance. Elements of this research would be picked up and continued in the next war.[5]

By 1916, it was clear to many of the young scientists that their work would be improved if they understood the practicalities and not just the theories of aviation, and so they put in a request to learn how to fly. The War Office approved and Lindemann, along with a few others, went through a flight training programme, eventually qualifying to fly for the RFC in October 1916. Maintaining the eccentricity that was his trademark, Lindemann would turn up at an RFC airfield wearing his bowler hat and long black overcoat, a stiff wing collar, and carrying his umbrella. Before boarding his plane he changed into a flying suit, rolled up his civilian clothes and packed them under his seat. Having taken off, piloted his aircraft and carried out his mission, he would change again on landing and walk away from the aircraft wearing his overcoat, bowler hat and wing collar and carrying his umbrella, while the RFC ground crews looked on in amazement.

It was as a scientific test pilot that Lindemann carried out, in the summer of 1917, the research that would make him famous. Many pilots in this period of the war were being lost when their engine stalled and the aircraft went into a spin. Since no one understood how to pull out of a spin, such incidents almost invariably proved fatal. If a pilot miraculously survived he would usually say he had no idea what he had done to save himself. Having observed a few fatal spins, Lindemann noticed that the plane's rate of turn did not increase as it spun to the ground. He concluded that the lift on both wings must therefore be equal, as the outer wing which was beating against the air turned to become the inner wing which was not. He realised that in order to escape from a spin the pilot should avoid doing the obvious thing, to pull the joystick towards himself to try to make the aircraft climb, and instead counter-intuitively push the joystick forwards to try to make the aircraft dive more steeply towards the earth, picking up speed so both wings would generate lift. His theory would however remain no more than a mathematical formula if somebody were not willing to try it out. Lindemann decided he must do so himself.

With remarkable courage he flew to 14,000 feet in a BE2 biplane, stalled the engine and put it into a spin. His colleagues watched in horror as the aircraft went spinning earthwards at about forty feet per second. In the cockpit Lindemann remained calm and used his remarkable memory for numbers to store in his mind the readings of the air speed indicator, the angle of incidence of the two wings (measured by tapes attached to the struts), the height at the beginning and end of the spin, the time taken and the number of turns. Furthermore, his method of getting out of the spin worked. In his experiments he gathered enough data to prove the mathematics of his theory. Once new instructions based on his findings had been issued to all pilots, the number of fatalities from spins began rapidly to diminish. Lindemann's bravery, his good memory and his mathematical theorising were to save hundreds and possibly thousands of lives.[6]

Lindemann went on to do other important experimental work, including research on stabilising bomb-sights for high altitude bombing in tests at Orford Ness on the Suffolk coast, in which he again acted as pilot and bomb aimer. He had overall what could be described as a 'good war', helping to demonstrate that the application of science could have immensely valuable practical results and displaying great bravery in his flight testing, even though he did not have to endure the horrors of the trenches.

After the war, Oxford University appointed Lindemann Professor of Experimental Philosophy, their term for physics. Science at Oxford was very much the poor relation to the humanities, a state of affairs Lindemann was determined to change. In order to fulfil his vision of a new start to Oxford science, he slowly turned the

Clarendon Laboratory into a powerhouse of scientific research to rival the Cavendish Laboratory at Cambridge and the laboratories at Imperial College, London. He devoted himself to administration, recruited a new generation of young scientists and raised funds from government and industry. But despite his outward success he was a prickly personality to deal with. He remained a very private person who took offence easily and, although quick to bear a grudge himself, seemed to have no hesitation about offending others. Though supremely confident, he was abrasive, short tempered and distant to many of those around him. He was keen to become a public figure and stood for Parliament as Member for Oxford University, but failed to win the seat. He did better by cultivating his relationship with Winston Churchill; the two men became close, Lindemann providing the politician with an insight into many of the scientific developments taking place at the time.

It was in the Second World War that Lindemann achieved the peak of his public success. From May 1940, when Churchill asked him to become his chief scientific adviser, he sent Churchill about 2000 memos (or minutes), often two or three per day but averaging about one per day for six years. They covered every conceivable scientific subject – from an explanation of what it meant to split the atom, to the workings of the Mills grenade; from how the Germans used beams to help their bombers navigate at night, to the principles of a gas turbine jet engine. There were also memos relating to economic matters – anything where a quantitative or analytical approach could make a fresh contribution. Lindemann had a knack of expressing complex scientific ideas simply and clearly. Churchill demanded brevity and he would often forward to Lindemann a long, complex civil service paper with a note, 'Prof – summarise in 10 lines pls', So the papers Lindemann prepared were rarely more than two pages in length, in large type and double spaced, bite-sized digests of science for a man who was too busy to absorb long and detailed papers and probably would not have understood them anyway.[7] Churchill created Lindemann Viscount Cherwell in 1942, and his official biographer described him as having 'power greater than that exercised by any scientist in history'.[8]

Lindemann was to gain notoriety in the Second World War from his dubious use of research into the effects of bombing to 'prove' that a full-scale bombing campaign against German towns and cities would eventually undermine the German war machine. Although the bombing offensive would entail the ruthless destruction of civilian centres in the country where he had spent his youth, Lindemann became one of the greatest exponents of area bombing within the senior echelons of government. And he was among the first senior government scientists to acknowledge the possibility of an atom bomb.

One of those with whom Lindemann fell out badly was another of the three most famous scientists of the Second World War, Sir Henry Tizard, regarded as one of the greatest defence scientists in Britain. Born in Gillingham, Kent, the son of a Royal Navy captain and a mother from an engineering family, Tizard went in 1899 to Westminster School, where he excelled as a Scholar. In 1905 he went up to Magdalen College, Oxford, graduating three years later with a First in mathematics and chemistry. In 1908 he moved to Berlin to study with Walther Nernst in the year in which Lindemann went to study with the same scientist.. Lindemann and Tizard got to know each other well, although Tizard carried out no research of any distinction and left Berlin after just a year. He then spent a year working in the Faraday Laboratory of the Royal Institution, where he carried out original research into the colour changes of indicators such as litmus and began to acquire a reputation as a research scientist. In 1911, he was offered a fellowship at Oriel College, Oxford along with a lectureship in Natural Sciences. Tizard settled into a new, comfortable life as an Oxford don and might have continued happily there for many years had not his career been transformed by the coming of war.

In August 1914, Tizard was attending the British Association's annual meeting in Australia, but on his return he immediately joined up and was commissioned into the Royal Garrison Artillery in Portsmouth. The following year he transferred to the RFC as a scientific experimental officer based at the Central Flying School at Upavon on the edge of Salisbury Plain. This was the era in which many pilots still regarded flying as an empirical activity reliant upon trial and error, and were suspicious of those who wanted to investigate, measure and analyse the actions they took by sheer instinct. During the early months of 1916, Tizard learned to fly. He found the experience exhilarating but was told by the flight commander, in a perfect demonstration of the prevailing prejudice against science, that he 'would never make a pilot' if he had to rely on instruments when flying.[9]

Tizard carried out important work on the development of bombsights. Until now a bomb had simply been thrown over the side of the fuselage or released from under the wing at the moment the pilot or bombardier thought was about right. Tizard set up a system for photographing a test aircraft on a camera obscura, monitoring the angle at which a bomb dropped from different heights and speeds and gradually calculating the mathematics on which an effective aiming system could be based. He also experimented with new radio equipment, new weapons and new cameras, taking one of the first aerial pictures of Stonehenge – which clearly showed the existence of the outer ring of the ancient stone circle. In autumn 1916 the Aircraft Testing Flight needed to find a new site. Tizard and Bertram Hopkinson, the Professor of Applied Mechanics at Cambridge, who was carrying out war work with the Royal Engineers, prospected a location for an airfield on heathland at Martlesham, near Woodbridge in Suffolk, convenient for flying over the test area at Orford Ness on the coast nearby. A new landing ground was constructed here, and for forty years Martlesham Heath was to be at the centre of experimentation first for the RFC and then for the RAF.

Here, Tizard set up procedures for checking and double checking the specifications of every new mark of aircraft that came off the factory lines, including the confirmation of maximum speeds and rates of climb. He brought a level of academic accuracy to the process which was essential to build a complete understanding of the science of aircraft performance. He checked the rate of climb in different atmospheric conditions; he recorded speeds not just once but at least five times over several days in different winds in order to calculate an average maximum over time. He experimented with flying in and out of clouds and thunderstorms and carried out tests designed to improve fuel efficiency collaborating on some of this research with his old friend Lindemann. Tizard's career closely paralleled Lindemann's, with the difference that Tizard was a commissioned officer operating within the RFC and Lindemann a civilian scientist working out of Farnborough. Tizard also had his moment of heroic daring. He was trying out a new Sopwith Camel in July 1917 when a squadron of twenty giant Gotha bombers passed by heading home after a bombing raid on London. Heavily outnumbered, Tizard climbed to 17,000 feet and attacked the rearmost aircraft, but his brand new and untested guns soon jammed. Instead of giving up and returning to base, Tizard continued to fly alongside the German bombers and made detailed observations of their speed and performance, information that was unknown at the time. He later wrote, 'The German crews were all looking up at me, wondering I suppose what I was doing. I then waved goodbye to them, they waved back, and I went home to lunch.'[10]

In 1918, Tizard joined Hopkinson in the new Air Ministry as administrator in the department of Research and Experiments with the rank of major. Looking back later on his Great War experience, he wrote, 'The war did me a great deal of good. It pulled me out of the ruck at Oxford ... It brought me into close and friendly contact with all sorts and conditions of men, and it made me realise that a purely scientific education *was* of value for men who had to deal with the practical affairs of life.'[11] After the war, Tizard briefly returned to Oxford, but it was clear to him that his future lay not in academic research but in the practical application of science, which despite the war he still felt to be 'a neglected field in England'.[12] He worked part time as consultant to an oil company and helped devise the system for allocating octane numbers to different types of fuel. Then in 1920 he decided to leave Oxford to join the Department of Scientific and Industrial Research, the body set up by Sir Frank Heath at the Board of Education in 1916 to mobilise scientific support for the war effort, alongside the committees of the Royal Society. Dedicated to advising the armed forces on different aspects of scientific research, the department was in effect a sort of meeting point for defence officials and academic scientists. With his service and academic backgrounds, Tizard was well suited for the job, and in 1927 he became secretary of the department. Two years later he was appointed rector of Imperial College, London, while in 1933 he added to this the role of chairman of the Aeronautical Research Committee (the latest name for Haldane's Advisory Committee, originally formed in 1909). Throughout the post-war years he continued to cultivate a series of relationships with scientifically minded civil servants and leading academic scientists.

During 1918, by one of those serendipitous twists of fate, both Lindemann and Tizard were to meet at Farnborough the third boffin who would achieve fame and even some fortune in the Second World War. Robert Watson-Watt had been born in Brechin in northeast Scotland. Both the Watsons and the Watts were from Aberdeenshire, and his most famous if distant ancestor was James Watt, the inventor of the condensing steam engine. Watson-Watt's father was a carpenter and Robert grew up as part of a strict Scottish Presbyterian family. He went to the local grammar school and won a bursary to University College, Dundee, then part of the University of St Andrews, where he was a star student and graduated in engineering in 1912. His professor, William Peddie, not only offered him an assistant lectureship on graduation but also personally tutored him in a course on radio waves. He was the only postgraduate on the course and the one-to-one tuition changed the direction of his career. In the summer of 1914 Watson-Watt offered his services to the War Office but heard nothing. Then, in the autumn of 1915, he was offered a job as a meteorologist at the Royal Aircraft Factory at Farnborough. His daily task was to take records of weather conditions by measuring in great detail the progress of a balloon as it ascended skywards, and from this to produce a daily forecast. In addition, in his spare time Watson-Watt applied his own knowledge of radio waves to try to predict the approach of thunderstorms from the 'atmospherics' they generated. Thunderstorms were then a real menace for aviators, and any ability to forecast their approach and their strength would be a great aid to flying. Watson-Watt identified at this early stage that the cathode ray oscilloscope could be used to display the information reflected back by radio signals, but such a device did not become widely available for another decade and much of his research remained of theoretical rather than practical use during the war. Watson-Watt's meteorological station was located on the roof of the Royal Aircraft Factory at Farnborough where Lindemann and the galaxy of talents from the Chudleigh Mess were based. This enabled Watson-Watt to write later, 'I was fortunate to sit on the roof which covered such distinguished heads.'[13]

During the 1920s, Watson-Watt continued his pioneering work on the use of radio waves to predict the weather, and began using radio to explore the ionosphere, the ionised upper layer of the earth's atmosphere. He became Superintendent of a Radio Research Station at Slough, initially reporting to the Department of Scientific and Industrial Research, where he got to know Tizard well. In a restructuring in the 1930s, the station came under the auspices of the National Physical Laboratory.

Watson-Watt was short, chubby, hugely enthusiastic and endlessly verbose. It was written of him that 'He never said in one word what could be said in a thousand.'[14] Partly because of this and partly thanks to his humble Scottish background, Watson-Watt remained something of an outsider in the world of English science in the post-war years. But all that would change in 1935.

The careers of these three Second World War boffins who had started their pioneering scientific research during the previous war became intertwined in the second half of the 1930s. They were all involved in one of the most famous twentieth-century examples of science coming to the aid of the military. Inside the Scientific Section of the Air Ministry a committee of prominent scientists was formed to explore how to provide effective early warning of the approach of enemy bombers towards the British coast. Tizard was asked to chair the committee, which began investigating possibilities that ranged from the use of huge audio detectors to the employment of infra-red rays. Quite separately, Lindemann denounced in a letter to *The Times* the fatalism of the prevailing view that 'the bomber will always get through' and suggested that scientists should be called upon to find a way of detecting the approach of bombers at long distance. Lindemann took great personal offence when he discovered that Tizard rather than he had already been asked to chair a secret committee to pursue the matter. This led to a dramatic falling out with his old friend that was to last for the next decade.

The committee consulted Watson-Watt, who recalled stories of GPO engineers recording that their high-frequency radio signals were disrupted when an aircraft flew overhead. Using the cathode ray oscilloscope as a visual display, Watson-Watt laid on a demonstration to show how radio signals bounced back after they had hit a metal object like an aeroplane and how this could be measured to identify the position of the distant aircraft. In a single four-page memo to the committee in February 1935, he outlined the basis of radio direction finding, later to be known by the American term radar.

A group of talented young scientists working for him, first at Orford Ness and then at Bawdsey Manor on the Suffolk coast, began to develop the increasingly efficient use of radar, which was later called 'the invention that changed the world'.[15] By 1940 the radar shield known as the Chain Home system had been constructed along the east coast of Britain from the Orkneys to the Isle of Wight. Put in place in the nick of time, the system enabled the RAF to defend the country in what became a turning point of the Second World War, the Battle of Britain. Watson-Watt has been described as the 'father of radar' and after the war he was given a substantial financial award for his invention. To twist Churchill's well-known phrase, never in the field of human history has so much been owed by so many to such a tiny number of boffins.

There were other fields in which research carried out in the First World War laid the foundations for the innovations of the Second. One of the most interesting is what later became known as operational research (known in America as operations research). Archibald Vivian Hill was a Fellow of Trinity College, Cambridge in 1914, a young scientist who in the four years before the war had done important work in the university's Physiological Laboratory on heat production in nerves and muscles during contractions. Using mathematical formulae for analysing his research, he helped to promote the new science called biophysics. He became world famous for this and after the war, in 1922, was awarded a Nobel Prize for his work. In 1914 he joined the army and became a captain in the Cambridgeshire Regiment, but the following year he was put in charge of an Anti-Aircraft Experimental Station under the aegis of the Ministry of Munitions. He recruited several other young scientists to work with him, among them R.H. Fowler and E.A. Milne, both of whom went on to become distinguished Fellows of the Royal Society. Analysing a variety of different techniques for targeting anti-aircraft guns in order to assess which had greater accuracy, and looking at the optimum layout for the guns to maximise the concentration of fire, they used mathematics to observe, assess, review and ultimately to improve military operations.

In the next war a group of boffins would build on this use of mathematics in the field of military problem solving. Among them was Professor Patrick Blackett, a seventeen-year-old naval cadet in 1914 who went on to serve in the navy during the Great War, taking part in the Battle of Jutland and later inventing a new gunnery device for the navy. After the war he went up to Cambridge where he worked with Rutherford at the Cavendish Laboratory and, specialising in quantum mechanics, helped to build the new science of nuclear physics. During the Second World War he was to establish the discipline of operational research as a tool for improving military techniques. Initially called in to help improve the accuracy of anti-aircraft guns in the Battle of Britain, he worked afterwards for both the RAF and the Admiralty. Tall, striking and with a natural flair for inspiring those with whom he worked, he surrounded himself with a team of young physicists and mathematicians who became known as 'Blackett's Circus'. He defined his role as one of bringing scientific techniques and numerical assessments to the improvement of traditional military operations in order 'to avoid running the war on gusts of emotion'.[16] In the small world of interwar science, Blackett had like Hill been a member of Tizard's committee in 1935, and had worked with Watson-Watt on the introduction of radar and the building of the Chain Home network. Together these men formed the nucleus of boffins during the later war.

The conventional view for many years has been that the armed services went into a spectacular decline after the First World War and that, in an era of massive cuts in defence budgets followed by the policy of appeasement in the 1930s, much of the progress made during the war went into reverse. It is argued that this entailed a substantial decline not only in numbers of personnel in the three services but also in the scientific research that brought innovation and new technologies to the military. The notion is usually illustrated by reference to the fact that the RAF still had fighters in the mid-1930s that looked like the biplane fighters of the Great War; that the army ignored the development of tank warfare even though it had played a major role in the victory of autumn 1918; that the Royal Navy failed to develop anti-submarine weaponry and seek improvements in radio and navigation; and that the armaments industry as a whole and the aircraft industry in particular turned its back on research and development and found itself 'in a position of chronic penury and sometimes on the very verge of bankruptcy'.[17]

David Edgerton has argued in recent years that this 'declinist' view of inter-war Britain is entirely wrong and that the country still had a powerful armaments industry, a strong navy, a small but efficient army and one of the best air forces in the world.[18] He also shows how the influence of the Great War scientists lived on in the public sector during the 1920s and 1930s. Whereas in 1914 military men had frequently questioned the value of consulting outside experts – engineers, chemists, physicians, doctors, psychiatrists and propagandists – the Great War banished this attitude for ever. Science had arrived in the military and in state planning, and it had come to stay.

We have seen how after the war the Department of Scientific and Industrial Research, under the leadership of men like Sir Frank Heath and Sir Henry Tizard, brought together defence scientists and industrial chiefs. In addition, each of the armed services in the post-war era created its own Director of Scientific Research (DSR). In the Admiralty, the DSR was supported by a team of naval advisers who reviewed new techniques for ship construction. The gunnery school at HMS *Excellent* and the torpedo school at HMS *Vernon* in Portsmouth became centres for naval research and development. A new Admiralty Research Laboratory was established alongside the National Physical Laboratory in Teddington where work was done on improving underwater sonar, Asdic. The Army DSR oversaw developments in explosives, gunnery, ballistics and signalling at Woolwich. A new chemical warfare research station was established at Porton Down in Wiltshire, in 1920, on the site where experiments had taken place with chlorine, phosgene and mustard gas during the war. The government was determined not to be caught off guard again and by the mid-1920s at least sixty scientists were permanently based there.[19]

In the Air Ministry, the DSR from 1925 to 1937 was Henry Wimperis, a Cambridge engineer who had spent the war in the Royal Naval Air Service. In 1935 he brought together the committee that oversaw the development of radar and its introduction as a practical tool to guide RAF fighters to approaching fleets of enemy bombers. During the 1920s and 1930s there was much movement between scientists in uniform who would take time out to work in the universities, and civilian scientists who would consult with the armed services. Moreover, the public sector employed some 6500 technicians and scientists on a regular basis during the inter-war years.[20]

With hindsight, knowing the strength of the enemy Britain would come up against in Europe from 1939 and in Asia from 1941, the research divisions in the pre-war army and navy might be judged to have been inadequate for the purpose. However, that cannot be said of the vast pool of scientific research that continued to contribute to the world of aviation. The Air Ministry was the largest publicly funded organisation for research and development in Britain. In 1932 the budget for the Royal Aircraft Establishment at Farnborough (the post-war name of the Royal Aircraft Factory) amounted to £430,000 (roughly equivalent to £30 million in 2014). This compares to a total spend on research and development across the board in ICI of only a little more, at £500,000.[21] Under its chair, the ubiquitous Sir Henry Tizard, the Aeronautical Research Committee brought some of the best scientific brains in the country to bear on the challenges of aviation. Not only did such scientists develop the revolutionary concept of radar, but Frank Whittle, a young RAF apprentice who was sent to Cambridge on a RAF studentship, patented the concept of the gas turbine jet engine in 1930. Although it took many years to produce a functioning model of such an engine, Whittle and the RAF had laid down a marker for the future of aviation technology into the second half of the twentieth century and beyond.[22]

Of the three services, the RAF was most open to new scientific ideas. Obviously, an aircraft flies through the sky only by the continuous application of science to power the engines, provide the lift and control the aeronautics, and to allow navigation in three-dimensional space. And, being the youngest of the three services, the RAF was still creating new traditions and not living off old ones. So it was the Air Ministry and the RAF that did most to bring to the fore the boffins of the Second World War.[23] Speaking of the war in general, but of aviation in particular, Tizard reported to a parliamentary committee in 1942 that 'you could hardly walk in any direction in this war without tumbling over a scientist.'[24]

The links between the scientists of the First World War and the boffins of the Second were considerable. War was becoming increasingly technical through the twentieth century, and the experience of fighting the Great War had, despite initial opposition, put scientists at the centre of planning for future technologies. There they would remain for the rest of the century and beyond.

The technological changes to which the scientists of the Great War made such a contribution, helped to change the world in which most people lived. Consider the life of a boy born towards the end of the nineteenth century, say in 1890. He would have been born into an era dominated by the horse and carriage, in which city streets often swirling with sulphurous fog were lit by gaslight, and in which only a tiny minority of homes had been touched by electricity. He might have had a bicycle, but unless he was from a wealthy family he would be unlikely to travel far from the parish or town of his birth. While still a boy he might have seen the first motor car drive through his town; and, in great excitement, as he watched it pass he would probably have been splashed by the muck and mud thrown up from its wheels. Still a boy, he might have seen the first flickering moving picture shows projected in a penny arcade. His parents probably bought cheap newspapers for the first time and he would have seen the advertisements for new brands of food and drinks.

As a teenager he might have become more aware of some of the technical changes offered by the growing use of electricity. Streets and even a few houses were now lit by electric light bulbs and if he visited London he would have seen the first electric trains crossing the city. The internal combustion engine was everywhere by now, powering vehicles on the road and buses in the cities. He might also have stood in awe, looking up to see and hear a pioneer aviator splutter across the sky in a canvas and wire box kite.

With the surprise coming of war in his twenties he might have rushed to volunteer, or later have been conscripted into the armed forces. Perhaps he was taught how to use wireless communications while his wife or girlfriend filled shells with liquid chemicals. If he had been wounded in the trenches he might have seen a modern operating theatre for the first time in a Casualty Clearing Station. He could have been treated with drugs that would have been rare before the war. And assuming he survived to return home, he would have experienced a series of tumultuous changes. If moderately wealthy, his home might have a vacuum cleaner, a telephone, a gramophone; it would almost certainly possess one of the wireless sets that would soon take pride of place in most living rooms. He probably went to the cinema at least once a week to watch films featuring glamorous stars from Hollywood.

During yet another war he would have been too old to fight, but he would have heard about radio beams that patrolled the skies, and he might have experienced bombs and then missiles falling from above. In his mid-fifties he would have seen in the newsreels pictures of the first atom bombs being dropped, and he would have heard about the harnessing of an entirely new source of energy. As he approached retirement he might have bought a television set to watch moving images of the coronation of the new Queen, live from Westminster Abbey. He would have heard about jet planes crossing the Atlantic in a few hours and satellites circulating the globe in space.

That boy's life would have spanned the era from the horse and carriage to a world living in the shadow of nuclear weapons. It was a period in which science and war proved to be fine bedfellows and, for good or ill, had transformed the landscape of the twentieth-century world.

Who's Who of the Secret Warriors: key scientists and scholars who contributed to the war effort, 1914–1918

Details of professional careers pre- and post-Great War that are relevant to work carried out during the war

Abbreviations:

BMA	British Medical Association
CH	Companion of Honour
Dir	Director
Exec	Executive
FBA	Fellow of British Academy (date created a Fellow)
FRAS	Fellow of Royal Aeronautical Society (date created a Fellow)
FRS	Fellow of Royal Society (date created a Fellow)
GHQ	General Headquarters of the British Army in France, 1914–18
Inst	Institute
Lieut	Lieutenant
Min	Ministry
Pres	President
RAMC	Royal Army Medical Corps
Regt	Regiment
RCS	Royal Chemical Society
RS	Royal Society

Max Aitken, Lord Beaverbrook (1879–1964)
Businessman, newspaper baron and propagandist

Born: Ontario, Canada, son of Scottish Presbyterian minister who had migrated to Canada in 1864.

Educ: local schools, Newcastle, New Brunswick; failed entrance to Dalhousie University, Halifax, Nova Scotia.

1900–10: business interests in Canada, including buying and selling bonds and assets; dabbled in newspapers; became very wealthy.

1910: emigrated to England following rumours about dodgy dealing in creation of the Canada Cement Company; became Liberal MP for Ashton-under-Lyne. 1911: bought country estate near Leatherhead, Surrey.

1915: appointed Canadian 'Eye-Witness' reporter on Western Front; set up Canadian Records Office. 1916: acquired the *Daily Express*. 1918: Feb–Oct Minister of Information.

Post-war: leading press baron; circulation of the *Daily Express* increased from 0.4 million (1919) to 2.3 million (1937). 1923: bought *Evening Standard*. 1928: started *Scottish Daily Express*; led several political crusades including those for Empire Free Trade and appeasement.

1940: appointed Minister of Aircraft Production and member of War Cabinet by Churchill; galvanised aircraft industry. 1941: Minister of Supply; resigned (1942).

Post-WW2: took up several controversial causes, suggesting more distant relationship with America and opposing European union.

Knighted 1911. Created Baron Beaverbrook, 1916.

Patrick Blackett (1897–1974)
Physicist and founder of operational research

Born: Kensington, London, son of a stockbroker.

Educ: Royal Naval Colleges at Osborne and Dartmouth (1907–14); Magdalene College, Cambridge (1919–21).

1914: naval cadet, took part in battles of Falkland Islands (1914) and Jutland (1916); invented new form of gunnery device.

Post-war: 1919: finished his education at Cambridge. 1921: researcher for Rutherford at Cavendish Laboratory. 1923: Fellow, Kings College, Cambridge. 1924–5: Gottingen University; studied quantum mechanics and helped Rutherford to establish nuclear physics. 1933: Birkbeck College, London. 1937: succeeded W.L. Bragg [q.v.] as Langworthy Prof of Physics, Manchester University.

1935: member of Tizard [q.v.] Committee that encouraged the development of radar.

1940: adviser to head of Anti-Aircraft Command; devised more efficient and accurate use of anti-aircraft weapons. 1940–1: member of the Maud Committee investigating the possibility of developing an atomic bomb.

1941: adviser to RAF Coastal Command; applied mathematical techniques to improving anti-submarine tactics. 1942–5: adviser at Admiralty; devised more efficient forms of hunting U-boats; with group of mathematicians known as 'Blackett's Circus' established techniques of operational research to observe, assess and improve military tactics.

Post-WW2: awarded Nobel Prize for pre-war work (1948); hostile to nuclear weapons; believer in expansion of universities and application of science and new technology; encouraged development of computer industry; scientific adviser to the Labour Party; Pres of RS 1965–70.

FRS 1933. Created life peer, Baron Blackett, 1969.

Sir Anthony Bowlby (1855–1929) Surgeon

Born: Namur, Belgium, where his father was correspondent for *The Times*.

Educ: Durham School; St Bartholomew's Hospital, London, qualified 1879. 1880: house surgeon; for twenty years successful London surgeon.

1899–1901: volunteer army surgeon in South Africa during Boer War.

1908: commissioned in the new Territorial Medical Service.

1914–16: Consulting Surgeon to army. 1916–18: Consulting Surgeon to the British Army in France and Belgium with rank of Maj-General; made changes to practice of military surgery and major improvements in organisation of Casualty Clearing Stations. Known as a fine administrator.

Post-war: did not return to medical practice.

Member of Exec Committee of British Red Cross.

Co-wrote *Official History – Medical Services: Surgery of the War* (1922).

Knighted 1911. Created Baronet 1923.

Sir William Lawrence Bragg (1890–1971) Physicist

Born: Adelaide. Family moved back to England 1909.

Educ: St Peter's College, Adelaide; Adelaide University at age fifteen, took a First in Mathematics (1905–8); Trinity College, Cambridge (1909–12).

1914: Fellow of Trinity College. 1915: awarded Nobel Prize with his father for work on using X-rays to identify the atomic structure of crystals. 1915: Lieut in Royal Horse Artillery. 1915–18: working on sound ranging for GHQ.

Post-war: 1919–37: Langworthy Prof of Physics at Manchester University, pioneering work on crystallography. 1937–8: Dir National Physical Laboratory. 1938–54: Prof of Physics and Dir Cavendish Laboratory, University of Cambridge; oversaw a flowering of British science; under his leadership, James Watson and Francis Crick discovered the structure of DNA (1953). 1954–66: Dir of Davy-Faraday Laboratory, Royal Institution, London.
OBE 1918. FRS 1921. Knighted 1941.

John Buchan (1875–1940) Author and propagandist

Born: Perth, eldest son of Free Church of Scotland minister.
Educ: Hutcheson's Grammar School, Glasgow (1888–92); Glasgow University (1892–5); Brasenose College, Oxford (1895–9).
Pre-war: wrote several novels, many set in Scotland, and much journalism; studied for the Bar (called 1901); assisted Alfred Milner in the reconstruction of South Africa after the Boer War (1901–3); worked for and became partner in Nelsons publishers.
1915–16: reporter at the Western Front. 1916–17: official press representative for GHQ and Lieut in Intelligence. Feb 1917: Dir of Dept of Information. March 1918: Dir of Intelligence at Min of Information under Beaverbrook [q.v.]. Dec 1918: dismantled the Min of Information.
Post-war: settled into rural life at Elsfield Manor near Oxford. 1927–35: MP for Scottish universities. 1935–40: Governor-General of Canada; visited Washington and was the first Briton to address Congress (1937).
Most famous of his books: *Prester John* (1910), *The Thirty Nine Steps* (1915), *Greenmantle* (1916), biographies of *Sir Walter Raleigh* (1911) and *The Marquis of Montrose* (1913, revised 1928), *Nelson's History of the War* (24 vols, 1915–19), *These for Remembrance* (1919), *Sir Walter Scott* (1932).
Created Baron Tweedsmuir 1935.

Viscount Cherwell – see Frederick Lindemann

Alexander Denniston (1881–1961) Code breaker

Born: Greenock.
Educ: Bowdon College, Cheshire; universities of Bonn and Paris.
1906–14: teacher of foreign languages at Marchiston Castle School and then Royal Naval College, Osborne, Isle of Wight. 1908: played hockey for Scotland in London Olympic Games.
1914: one of first to join Room 40 in Admiralty because of linguistic skills; worked in Room 40 throughout war.

1919–42: remained in code breaking as head of Government Code and Cypher School, which in 1939 moved to Bletchley Park.

1939–42: achieved technical success in deciphering the German Enigma machine and recruited a new generation of mathematician code breakers to Bletchley Park, inc. Alan Turing and Gordon Welchman.

1942: retired from Bletchley Park but remained in intelligence to 1945.

Sir James Alfred Ewing (1855–1935)
Engineer and cryptographer

Born: Dundee.

Educ: Dundee High School; University of Edinburgh (1871–8).

Pre-war: worked with Great Western Telegraph Company laying undersea cables in the south Atlantic. 1878: Prof Mechanical Engineering at Imperial University of Tokyo and worked on earthquakes. 1883: Prof Engineering at newly endowed University College of Dundee and worked on the effects of stress on metals. 1890: Prof Mechanism and Applied Mechanics at Cambridge, where he set up new laboratory and oversaw huge growth in the engineering school.

1903: Dir Naval Education in Admiralty, introduced programme of scientific and engineering training for naval officers.

1914: set up Room 40 to decipher German wireless intercepts.

1916: left to become Principal and Vice-Chancellor of Edinburgh University, supervised a period of immense growth with creation of 13 new chairs and several new depts.

1929: retired to Cambridge.

FRS 1887. Knighted 1911.

Sir John Ambrose Fleming (1849–1945)
Electrical engineer and inventor

Born: Lancaster, son of a Congregational minister.

Educ: University College School, London; University College, London (1867–70); St John's College, Cambridge (1877–80).

1870s: science teacher at various schools. 1882: consultant to the Edison Lighting Company, London, and involved in production of first electric light bulbs; helped to develop alternating current for power transmission. 1885: Prof of Electrical Technology at University College, London. 1899: scientific adviser to Marconi Company.

1904: invented the thermionic valve. 1904 onwards: lectured extensively on wireless telegraphy, telephony, mathematics and mechanics, many lectures for popular, working and not academic audiences.

1915: lectured on the importance of science in the war.

1926: retired to Sidmouth, Devon where his house had a basement laboratory. FRS 1892. Knighted 1929.

Sir Harold Gillies (1882–1960) Plastic surgeon

Born: Dunedin, New Zealand, son of contractor.

Educ: Wanganui College, NZ; Gonville and Caius College, Cambridge; St Bartholomew's Hospital, London, qualified 1906. Champion golfer.

1915: joined RAMC; set up specialist dept in facial surgery at Aldershot. 1917: unit moved to Queen's Hospital, Sidcup. Pioneered development of plastic surgery and facial reconstruction.

Post-war: set up private practice in plastic surgery; pioneered cosmetic surgery for the wealthy; became plastic surgeon at St Bartholomew's Hospital, London; continued with reconstruction work on deformities cause by terrible accidents.

WW2: with his cousin, Archibald McIndoe, trained more plastic surgeons and treated wounded at Rooksdown House, Basingstoke; McIndoe worked at Queen Victoria Hospital, East Grinstead, and his patients formed the 'Guinea Pig' club.

Post-WW2: pioneered sexual reassignment surgery.

CBE 1920. Knighted 1930. First Pres of British Association of Plastic Surgeons 1946.

Nigel de Grey (1886–1951) Publisher and code breaker

Born: Copdock, Suffolk, son of the rector. Aristocratic parents.

Educ: Eton.

1907: failed language exam to join Foreign Office and joined publishers William Heinemann; joined Royal Naval Volunteer Reserve.

1914: posted to Belgium. 1915: joined the code breaking team in Room 40.

1917: deciphered the Zimmerman Telegram with Rev. William Montgomery.

Post-war: ran the Medici Society specialising in publishing prints of Old Masters.

1939: started work at Bletchley Park code breaking centre. 1941: picked up first evidence of genocide against the Jews in the occupied Soviet Union, passing reports on to Churchill.

1945: deputy director of GCHQ, intercepting Soviet cables.

John Scott Haldane (1860–1936) Physiologist

Born: Edinburgh into a Scottish aristocratic family; younger brother of Richard
    Burdon Haldane, reforming Secretary of State for War (1906–10).
Educ: Edinburgh Academy and Edinburgh University (1876–9).
Pre-war: career in medical research specialising in analysis of foul air in schools
    and slums, and of poisonous gases in coal mines; devoted much time to
    improving industrial health and the ventilation of factories, workshops
    and mines. 1887–1913: member of Physiology Dept Oxford University
    and Fellow of New College.
1914–18: member of RS War Committee. 1915: consulted by Kitchener on the
    best form of respiratory masks for protection from poison gases;
    advised on medical aspects of gas poisoning.
Post-war: continued with work in industrial welfare. 1921: member of Safety
    in Mines Research Board; lectured widely on science of industrial dis-
    eases.
FRS 1897.

Hugh Hamshaw-Thomas (1885–1962)
Palaeo-botanist and photo interpreter

Born: Wrexham, Denbighshire, son of a gentleman's outfitter.
Educ: Grove Park, Wrexham; Downing College, Cambridge.
Pre-war: researched fossil plants, inc. shale plants from the north of England.
1914: volunteered for the army and joined the artillery. 1916: transferred to the
    RFC. 1917–18: worked on mapping Egypt, Palestine and Syria from
    aerial photos.
Post-war: sent to India to consider an aerial survey of the country; returned to
    Downing College and continued to research and publish in palaeobotany.
WW2: went back to aerial photography at RAF Medmenham and became a
    specialist in interpreting photos of German industry.
Post-WW2: returned to Cambridge and the broader study of botany.
MBE 1918. FRS 1934.

Sir Frederick Handley Page (1885–1962)
Aircraft designer and manufacturer

Born: Cheltenham, son of a master upholsterer.
Educ: Cheltenham Grammar School; Finsbury Technical College.
1906: chief electrical designer at Johnson & Phillips engineers, Charlton. 1908:

sacked for spending too much time on aviation experiments. 1909: founded Handley Page Ltd to build aeroplanes but only established when contracted by the War Office to produce versions of official designs.

1914: built first twin-engined bomber for Royal Naval Air Service. 1918: built a four-engined bomber to bomb Berlin.

Post-war: 1919: turned to passenger aviation. 1924: part of consortium that founded Imperial Airways. Produced several long-range passenger aircraft.

1940: Handley Page Co produced the four-engined Halifax bomber.

Post-WW2: 1946: helped to establish College of Aeronautics at Cranfield. Handley Page designers produced the Hermes airliner and the Victor bomber. Refused to accept government pressure to merge with other aviation companies to create a stronger, consolidated aviation industry.

CBE 1918. Knighted 1942. Pres RAS 1945–7.

Edward Harrison (1869–1918) Chemist

Born: Camberwell, London.

Educ: United Westminster Schools; apprenticed to pharmaceutical chemist.

Pre-war: pharmacist in Croydon; industrial chemist and researcher; carried out analysis for BMA of medicines to prevent deception of public by quacks and published *Secret Remedies* (1909).

1915: joined Royal Fusiliers as a private; transferred to chemists' corps of Royal Engineers to research anti-gas measures; designed new form of respirator to give protection from range of different gases with lime permanganate capsules developed with Bertram Lambert at Oxford. 1917–18: assistant then controller of Chemical Warfare Dept of Min Munitions responsible for anti-gas section.

Nov 1918: succumbed to Spanish influenza epidemic and died one week before Armistice.

FRCS 1894.

Sir James Headlam-Morley (1863–1929)
Historian of Germany

Born: Barnard Castle, Co Durham, vicar's son.

Educ: Eton; Kings College, Cambridge (1883–7); Gottingen and Berlin, Germany.

Pre-war: historian of Germany and author of *Bismarck and the Foundation of the German Empire* (1899). 1902: joined Board of Education as school inspector.

1914–17: joined Charles Masterman [q.v.] at Wellington House; wrote many pamphlets on different aspects of the war and Germany's guilt.

1918: Min of Information worked with H.G. Wells on propaganda directed against Germany and German troops.

1919: acted as liaison between PM Lloyd George and the Foreign Office at Versailles Peace Conference.

1920: historical adviser to Foreign Office, collator of documents relating to the war; founder of Royal Inst of International Affairs (Chatham House).

Knighted 1929.

Sir Frank Heath (1863–1946)
Educationalist and scientific administrator

Born: London, eldest son of portrait painter.

Educ: Westminster School; University College, London; University of Strasbourg.

1890: Prof of English at Bedford College and King's College, London; 1895: Librarian, London University; 1901: Registrar, London University.

1902: Board of Education, director of special inquiries. 1910: joined new branch of Education Board to allocate funding across the growing university sector, out of his reforms came the University Grants Committee in 1919.

1915: helped design new tin helmet for soldiers and suggested creation of a new group to co-ordinate government funding for scientific research needed for war. 1916: first permanent secretary of new Dept of Scientific and Industrial Research (DSIR).

1925–7: helped set up similar DSIRs in Australia and New Zealand.

1927: retired and Henry Tizard (q.v.) took over DSIR.

1929: secretary Universities Bureau of the British Empire; 1931: governor of Imperial College, London.

Knighted 1917.

Archibald Vivian Hill (1886–1977)
Physiologist and founder of Operational Research

Born: Bristol, only son of a timber merchant.

Educ: Blundell's School, Tiverton; Trinity College, Cambridge, reading maths and natural sciences (1905–9).

1910: Fellow of Trinity College, Cambridge, and researcher in the Cambridge Physiological Laboratory; investigated energy used by nerves and muscles; emphasised the mathematical and quantitative assessment of his research; helped create the science of biophysics.

1914: joined up and became capt in the Cambridgeshire Regt; directed an anti-aircraft experimental station as part of the Min of Munitions to provide mathematical and quantitative assessment of anti-aircraft fire; helped to establish what was later called operational research.

1920: Prof of Physiology, Manchester University. 1922: Nobel Prize for his research on muscles. 1923: Prof at University College, London.

1935: member of the Tizard [q.v.] Committee that led to the introduction of radar.

1940–5: independent MP for Cambridge University; member of 1940 Tizard Mission to USA; 1943–4 mission to India to advise on reconstruction.

1951: retired but kept up voluntary work on many scientific societies.

FRS 1918.

Sir Gordon Holmes (1876–1965)
Surgeon and brain specialist

Educ: Dundalk Educational Institute; Trinity College, Dublin.

1901: National Hospital for Nervous Diseases, Queen Square, London; carried out pioneering work on the functioning of the brain.

1917: took over from Charles Myers as leading psychologist to the British Army in France.

Post-war: wrote extensively; editor of the influential journal *Brain* (1922–37); taught widely. Was bombed out from his home in Wimpole Street in 1940 and retired the following year.

FRS 1933. Knighted 1951.

Bertram Hopkinson (1874–1918)
Mechanical and aeronautical engineer

Born: Woodlea, Birmingham, the son of engineer John Hopkinson.

Educ: St Paul's School, London; Trinity College, Cambridge (1893–6).

Pre-war: successful research engineer; elected to chair in Mechanism and Applied Mechanics at Cambridge in 1903 aged only twenty-nine, taking over from J.A. Ewing (q.v.); did pioneering work on metal fatigue, the magnetic properties of iron, the effects of explosions and the development of internal combustion engines.

1914: commissioned in the Royal Engineers, helped to develop bombs to be dropped from aircraft; 1915: head of Dept of Military Aeronautics supplying new aircraft with armaments; Secretary of RS committee advising government on scientific problems of the war. 1917: Established an experimental flying station at Orford Ness and Martlesham Heath with Henry Tizard (q.v.). August 1918: killed in an air crash while planning to set up a school of aeronautical engineering post-war.

FRS 1910.

Sir Arthur Hurst (1879–1944)
Psychologist and physician

Born: Bradford. Original surname Hertz; his father's cousin was Heinrich Hertz, the discoverer of radio waves. Name changed by deed poll in 1916.

Educ: Bradford and Manchester grammar schools; Magdalen College, Oxford; Guy's Hospital, London; studied at Munich, Paris and USA; qualified 1907.

1909: assistant physician, Guy's Hospital with responsibility for neurological dept.

1916: in charge of neurological section at Netley Military Hospital. 1918–19 worked at Seale Hayne Hospital, Devon, as specialist in removing the symptoms of hysteria. Some of his cases were filmed before and after treatment.

Post-war: returned to Guy's and worked on the alimentary tract and on gastroenterology; founded New Lodge Clinic, Windsor.

WW2: taught at Oxford.

Pres Royal Society of Medicine 1927–9. Knighted 1937.

Sir Roderick Jones (1877–1962)
Director of Reuters News Agency

Born: Dunkinfield, Cheshire, of poor family.

Educ: Parents could not afford a top school or university so taught by grandfather at home.

1894: reporter in Pretoria, South Africa. 1902: joined Reuters news agency, London. 1905: Reuters regional manager, South Africa.

1915: recalled to London to take charge on suicide of Baron de Reuter. 1916: restructured Reuters with covert government backing; gave support to British propaganda output, including Agence Service Reuter in English and French to neutral countries. 1917–18: worked at Dept and Min of Information as head of news division.

Post-war: increasingly autocratic style at Reuters alienated many. 1930: acquired the Press Association, making a personal fortune out of the deal; began to lose confidence of the board and of the British government. 1941: forced to resign.

Knighted 1918.

Sir William Jury (1870–1944) Cinema distributor

Pre-war: one of the first cinema distributors in Britain; by 1914 his company, Jury's Imperial Pictures, was a leading film exhibition company.

1915: chair of the Topical Committee for War Films. 1916: supervised the editing of documentary *The Battle of the Somme*. 1918: chair of the Film Committee of the Ministry of Information.

Post-war: President of the Cinematograph Trade Benevolent Fund.

Knighted 1918.

Dilwyn 'Dilly' Knox (1884–1943)
Classicist and code breaker

Born: Oxford, son of tutor of Merton College, later Bishop of Manchester; one of four famous brothers.

Educ: Eton; King's College, Cambridge (1903–7).

1909: Fellow of King's College and classicist deciphering fragmentary texts of Herodas; lover of puzzles.

1915: code breaker at Room 40; broke the German admirals' flag code.

Post-war: continued at Government Code and Cypher School which moved to Bletchley Park 1939.

WW2: worked on understanding functioning of German Enigma machine; broke the Italian naval code. Died of cancer while still at Bletchley Park.

Frederick William Lanchester (1868–1946)
Engineer, inventor, car and aircraft designer

Born: Lewisham, south-east London, son of an architect.

Educ: Brighton primary and secondary schools; Hartley College (later part of the University of Southampton); Royal College of Science.

1888: before finishing his education offered job of assistant works manager of Forward Gas Engine Company of Birmingham; constantly inventing new mechanical devices and techniques. 1894: built first British motor boat. 1895: built first British four-wheeled petrol driven motor car. 1899: established Lanchester Engine Company, Birmingham, reconstructed as Lanchester Motor Company 1904; worked on technology that led to the first gearbox. 1909: worked for Daimler Motor Company.

1907–8: fascinated by aeronautics, published two books on aerial flight and aircraft design. 1909: joined government advisory committee on aeronautics and predicted major role aircraft would play in war;

encouraged new designs in aircraft production throughout the war; later frustrated by lack of official recognition for his war work.

Post-war: moved back to Birmingham, worked on several pioneering engine designs including those for Malcolm Campbell's record-breaking Bluebird; lost money in several business ventures and ended his life in poverty, with a charity taking over payment of his mortgage.

Registered 426 patents.

Pres Inst of Automobile Engineers 1910. FRAS 1917. FRS 1922.

Alfred Leete (1882–1933) Graphic artist

Born: Northamptonshire, son of a farmer.

Educ: Kingsholme School, Weston-super-Mare, and School of Science and Art.

Pre-war: cartoonist and commercial graphic artist working for several daily papers, *Pick Me Up* magazine, *Punch, Strand Magazine, Tatler* and the *Pall Mall Gazette*; also designed advertising posters.

Sept 1914: designed the Kitchener 'Your Country Needs You' poster later described as 'one of the most famous posters of all time', copied in America ('Uncle Sam wants you') and in many other countries.

1915: joined Artists' Rifles and fought on Western Front.

Post-war: produced advertising designs for London Underground, Guinness, Youngers' Ales, Bovril and Rowntrees, for whom he created the character 'Mr York of York'.

Often returned to Weston-super-Mare and designed adverts for local businesses.

Frederick Lindemann, Viscount Cherwell (1886–1957) Physicist and scientific adviser to Winston Churchill

Born: Baden-Baden, Germany, where his mother was 'taking the cure', of German Alsatian father and American mother.

Educ: Blair Lodge Scotland; from 1900 at Darmstadt, Germany, in Gymnasium and Hochschule. 1908–14: studied in Berlin under Prof Nernst, met some of the leading scientists of the day inc. Albert Einstein and Max Planck.

August 1914: returned to England. 1915: joined Physics Dept of Royal Aircraft Factory, Farnborough where he formed part of the Chudleigh Mess with other young scientists. 1917: purposefully flew aircraft into spins to calculate the mathematics and prove the theory of getting out of a spin.

Post-war: 1919: Prof of Experimental Philosophy (Physics) at Oxford; lived in rooms at Christ Church College; made the Clarendon Laboratory into highly respected institution; recruited new generation of young scientists; became good friends with and informal adviser to Winston Churchill. 1934: argued there should be a scientific committee to explore possibilities of future air defence; took great offence when he discovered Henry Tizard [q.v.] was already chairing such a committee. 1937: defeated in by-election as MP for Oxford Univ.

Sept 1939: to Admiralty with Churchill. May 1940: to Downing Street with Churchill as scientific adviser; at centre of decision making for rest of war. 1942: argued for increase in bombing offensive once navigational aids had been improved; supported development of atomic weapons. 1944: sceptical about intelligence claims the Germans had developed missile technology. 1942–5: Paymaster General.

Post-WW2: 1945–51 returned to Oxford; member of Conservative Shadow Cabinet in House of Lords. 1951–3 Paymaster General. 1956 retired from Oxford professorship.

1920 FRS. Created Baron Cherwell 1941. CH 1953. Created Viscount Cherwell 1956.

Geoffrey Malins (1886–1940) Film cameraman

Born: Hastings, son of a hairdresser.

Educ: unknown but assumed locally.

1906: photographer's assistant in portrait studio, Weymouth. 1910: cameraman with Clarendon Film Company, Croydon; shot several features.

1914–15: news cameraman with Gaumont Graphic; filmed with Belgian army.

Nov 1915–June 1918: Official Kinematograph Operator at the front; filmed many of the scenes for *The Battle of the Somme* (1916). 1918: left due to ill health.

Post-war: wrote *How I Filmed the War* (1919); continued writing and producing fiction films and exploration documentaries. 1932: moved to South Africa.

1918 OBE.

Charles Masterman (1874–1927)
Liberal politician and propagandist

Born: Wimbledon, fourth son of evangelical Quaker parents.

Educ: St Aubyn House, Brighton; Weymouth College and Christ's College, Cambridge.

1900: Fellow of Christ's College, Cambridge; associated mostly with liberals and Christian socialists; lived in a tenement in Camberwell, south

London, and wrote about the contrast between the squalor of the slums and the luxury of the West End. 1906: elected as Liberal MP and became junior minister in the reforming Liberal government. 1911: as leading reformer was appointed chair of the National Insurance Commission with responsibility to introduce NI to employers and workers.

1914: Asquith placed him in charge of official propaganda at Wellington House; brought together a strong team of writers, pamphleteers, film-makers and artists to promote the British line in the war.

1917: demoted and put under John Buchan in Dept and later Min of Information; services dispensed with at the end of the war.

Post-war: continued to write prolifically in the *Guardian* and the *Sunday Express* on liberal subjects but never reconciled with the demise of the Liberal Party, nor did he become a member of the Labour Party.

Sir William Mills (1856–1932)
Engineer and inventor of the Mills grenade

Born: Sunderland, son of a joiner in shipbuilding yard.

Educ: unknown but assumed locally.

1870s and 1880s: marine engineer, repaired underwater telegraph cables, designed lifeboat disengaging gear.

1885: established an aluminium foundry, Sunderland; later established a second foundry in Birmingham; produced castings for motor cars and later aircraft.

1915: took up Belgian design for a hand grenade and adapted it to what became the Mills grenade or bomb; 75 million produced by 1918.

Knighted 1922.

John Moore-Brabazon (1884–1964) Aviator

Born: London, to parents from the Anglo-Irish aristocracy.

Educ: Harrow School; Trinity College, Cambridge (1901–3); left without taking a degree.

Pre-war: racing driver; first Briton to make a powered flight in Britain (Apr 1909); won aviation prizes from the *Daily Mail*. July 1910: gave up flying temporarily on the death of his friend Charles Rolls in an air accident.

1914–18: served in the RFC, mostly in the development of aerial photography and reconnaissance; worked closely with Hugh Trenchard; awarded MC.

Post-war: MP (1918–42). 1919–22: Parliamentary Private Sec to Winston Churchill, Secretary of State for War and Air. Parliamentary Sec to Min of Transport (1923–4 and 1924–7); supporter of Churchill's policy of re-armament in late 1930s.

Oct 1940: Minister of Transport in Churchill's wartime coalition. 1941: Minister of Aircraft Production, replacing Lord Beaverbrook [q.v.]. Feb 1942: resigned after making a speech calling for Russian and German armies to annihilate each other.

Post-WW2: became father figure to British aviation, encouraging new ideas like the jet engine; chaired the committee that planned construction of civil aircraft, inc. the 'Brabazon', a huge airliner that never flew and was cancelled in 1953; Pres Royal Aero Club; Pres Royal Institution.

Created Baron Brabazon of Tara, 1942.

Charles Samuel Myers (1873–1946) Psychologist

Born: London, of wealthy merchant parents.

Educ: City of London School; Gonville and Caius College, Cambridge; St Bartholomew's Hospital, London.

1898: joined a pioneering anthropological expedition to study the natives of New Guinea but returned to Psychology Dept at Cambridge to study the minds of Britons.

1912: established the first Experimental Laboratory for Psychology in Cambridge.

Early 1915: coined the phrase 'shell shock' and believed the condition was treatable; commissioned into the RAMC. 1916: appointed consulting psychologist on Western Front, but eventually disillusioned with the army's harsh response to victims of nervous breakdowns and returned to Cambridge.

Post-war: left Cambridge and set up National Institute of Industrial Psychology (1921), became the first president of British Psychological Society and editor of *British Journal of Psychology*. 1920s and 1930s: wrote and lectured extensively; carried out a great deal of laboratory experimental work contributing to the development of the science of psychology.

FRS 1915

Sir Lewis Namier (1888–1960) Historian

Born: Wola Okrzejska in Russian Poland of Polish-Jewish gentry parents, surname Niemirowski.

Educ: Lvov University (1906); LSE, London (1907); Balliol College, Oxford (1908–11).

1910: changed name by deed poll to Namier.

1914: worked for Charles Masterman [q.v.] at Wellington House as expert on Germany, Austro-Hungary and eastern Europe. 1918: Min of Information, worked on propaganda to assist break up of Austro-Hungarian Empire.

Post-war: worked at Oxford, Vienna and Prague; became committed Zionist after meeting Chaim Weizmann [q.v.]; wrote two major tomes on eighteenth-century British politics (1929–30). 1931–53: Prof of History at Manchester University; during late 1940s and 1950s respected as leading historian and 'Namierism' became shorthand for detailed structural analysis of people and power systems, the opposite of historical narrative.

FBA 1944. Knighted 1952.

Viscount Northcliffe (Alfred Harmsworth) (1865–1922)
Newspaper baron and propagandist

Born: Dublin, grew up in north London, eldest son of a barrister who took to drink and an Anglo-Irish mother.

Educ: Stamford Grammar School; Henley House, Hampstead.

1880s: freelance journalist with gift for popularising. 1886: editor *Bicycling News*. 1888: established his own Amalgamated Press (later Associated Newspapers) with brother Harold, to publish a series of popular papers and journals using new printing technology and new layouts. 1894: bought *Evening News*. 1896: launched *Daily Mail* bringing all his innovations into one new daily paper. 1900: circulation peaked at over 1 million copies. 1903: launched *Daily Mirror*. 1905: bought the *Observer*. 1908: bought *The Times* in a deal that initially kept his ownership secret; his papers campaigned on many issues inc. warning about the threat from Germany and in favour of new technologies, especially aviation; mixed in high political circles.

1914: his papers helped whip up anti-German fever. 1915: exposed the shells scandal. Dec 1916: replacement of Asquith by Lloyd George enhanced his reputation as a king-maker.

1917: head of British War Mission in USA.

March 1918: head of enemy propaganda in Min of Information; led effective campaign against Austria-Hungary and Germany. Nov 1918: resigned.

Post-war: campaigned for a punitive settlement on Germany at Versailles;

Lloyd George denounced his 'diseased vanity'. 1920–1: his behaviour became erratic and rude. Aug 1922: died of a blood infection that was wrongly thought to be syphilis.

Created Baron Northcliffe 1905; Viscount Northcliffe 1917.

Baron Rayleigh (John William Strutt) (1842–1919) Physicist

Born: Langford Grove, Maldon, Essex, son of 2nd Baron Rayleigh.

Educ: Eton, Wimbledon, Harrow and Torquay (moved due to ill health); Trinity College, Cambridge (1861–5).

1866: Fellow of Trinity College, Cambridge; set up a laboratory at his family home, Terling Place, Witham, Essex, where he worked throughout his life.

1870s: studied psychic phenomena, acoustics, optics, physics. 1879: Cavendish Prof of Physics at Cambridge. 1880s: studied electromagnetism, thermodynamics and acoustics for the telephone. 1890s: helped calculate exact values for the ohm, ampere and volt. 1895: announced the discovery of argon, an inert gas (with William Ramsay); won Nobel Prize for this discovery (1904).

1896–1911: chief adviser to Trinity House and helped design foghorns. 1905–9: chair of War Office Explosives Committee. 1909–14: Pres of War Office Advisory Committee on Aeronautics. 1914–18: member War Committee of RS.

Inherited baronetcy 1873. FRS 1873. Pres RS 1905–8. Chancellor Cambridge University 1908–19.

Dr William Halse Rivers (1864–1922) Psychologist and anthropologist

Born: Chatham, Kent, eldest son of a curate.

Educ: Tonbridge School; London University; St Bartholomew's Hospital, London, qualified 1888.

1891: house physician at National Hospital for Paralysed and Epilectic. 1892: assistant at Bethlem Royal Hospital. 1897: lecturer in Psychology at Cambridge University. 1898: part of the Cambridge anthropological expedition to Torres Strait with among others Charles Samuel Myers [q.v.]. 1900–15: took part in anthropological expeditions to South India, Melanesia and New Hebrides in between work at Cambridge.

1915: joined Maghull Military Hospital to work on shell shock with among others Grafton Elliot Smith [q.v.]. 1916: commissioned as capt in RAMC and sent to Craiglockhart Hospital, Edinburgh to help officers recover from shell shock where he developed a close relationship with Siegfried Sassoon. 1918: RFC Central Hospital, Hampstead.

Post-war: wrote on the interpretation of dreams and on social anthropology. FRS 1908. Pres Royal Anthropological Inst 1921–2.

Sir Ernest Rutherford, Baron Rutherford (1871-1937) Nuclear physicist

Born: Brightwater, Nelson, New Zealand, the son of a farmer who had emigrated from Scotland.

Educ: Havelock School; Nelson College; Canterbury College, University of New Zealand, 1890-94.

Pre-war: 1895 to Cavendish Laboratory, Cambridge; worked under Prof J.J. Thomson, carried out early work on sound waves but was overshadowed by Guglielmo Marconi. 1898 to McGill University, Montreal, Canada, carried out pioneering work on radioactivity, understanding how much energy could be released by a single atom. 1907 to Manchester University as Prof of Physics, discovered atomic nuclei.

1915: appointed to the Admiralty Board of Invention, worked on underwater sound waves and the detection of submarines with William Bragg (senior);

1916: joined the Dept of Scientific and Industrial Research under Sir Frank Heath [q.v.]

1917: led Anglo-French delegation to USA to share with Americans Allied scientific advances in war.

Post-war: 1919 to Cambridge as Dir Cavendish Laboratory; split the atom, 1919; discovery of the neutron by James Chadwick, 1931; Rutherford was known as the 'father of nuclear physics' but continued to insist that it was impossible to harness the power released in an atomic bomb.

FRS 1903. Nobel Prize 1908. Knighted 1914. Pres RS 1925-30. Created Baron Rutherford 1931.

Robert William Seton-Watson (1879–1951) Historian and specialist on central Europe

Born: London, to Scottish parents with wealth from India.

Educ: Winchester; New College, Oxford.

Pre-war: travelled to Berlin, Vienna and Budapest; learned Hungarian, Croat and Czech: published *Racial Problems in Hungary* (1908).

1918: called in to become co-director of Austro-Hungarian section of Crewe House in Min of Information, helped organise the highly successful propaganda campaign against Austria-Hungary.

Post-war: helped to establish what is now the School of Slavonic and East European Studies, a faculty of University College, London.

WW2: served briefly in Political Intelligence Bureau of the Foreign Office.

Post-WW2: as a great friend of Czechoslovakia he was upset by that country's absorption into the Communist Eastern bloc in the Cold War.

Sir Grafton Elliot Smith (1871–1937)
Anatomist and doctor

Born: Grafton, New South Wales, Australia.

Educ: Sydney Boys High School; University of Sydney; to Cambridge 1896.

1909: appointed Prof of Anatomy at Manchester University; extensively studied the anatomy of the brain believing that humankind originated in Europe and not Africa or Asia; did much work on early magic and religion. 1912: involved in the 'discovery' of the Piltdown skull (later established to be a forgery).

1916: worked on shell shock at Maghull Hospital; became friend of Dr William Halse Rivers [q.v.].

Post-war: Prof of Anatomy at University College, London. 1924–7: Pres of the Anatomical Society; carried out much work in Egypt; first to X-ray a mummy; reported the discovery of Tutankhamun's tomb (1922). Wrote extensively about human evolution.

FRS 1907. Knighted 1934.

Sir Wilfred Stokes (1860–1927)
Civil engineer and inventor of the Stokes mortar

Born: Liverpool, son of school inspector.

Educ: St Francis Xavier's College, Liverpool; Catholic University College, London.

Pre-war: apprentice with Great Western Railway; bridge designer for Hull and Barnsley Railway; Chairman and Managing Dir of Ransomes & Rapier, engineering company and producer of agricultural machinery, Ipswich.

1915–18: Inventions Branch, Min of Munitions; designed what became known as the Stokes mortar, after modifications a very popular trench weapon.

Knighted 1917.

Sir Ernest Swinton (1868–1951)
Engineer, army officer, writer and inventor

Born: Bangalore, India, son of a judge in Indian civil service.

Educ: various schools in England (moved for reasons of economy); Royal Military Academy, Woolwich, commissioned in Royal Engineers 1888.

1890s: military engineer in India and South Africa during Boer War (1899–1902).

1906: instructor at Royal Military Academy, Woolwich. 1910: wrote official history of Russo-Japanese war and became aware of power of modern weaponry. 1913: assistant secretary to the Committee of Imperial Defence.

1914: appointed official war correspondent by Kitchener, known as Eye Witness, and observed stalemate of Western Front. 1915: involved with the development of the tank. 1916: helped train volunteers for the new Tank Corps.

1917: assistant secretary to War Cabinet.

1919: left army and worked in Air Ministry.

1925: elected Prof of Military History, Oxford. 1934–8: Col commandant of Royal Tank Corps. 1940: broadcast on BBC a series of War Commentaries.

Knighted 1923.

Sir Henry Tizard (1885–1959) Scientific administrator

Born: Gillingham, Kent, son of a naval officer.

Educ: Westminster School; Magdalen College, Oxford; Berlin (1908–9).

1909: researcher at Faraday Laboratory of Royal Institution. 1911: tutor in chemistry at Oriel College, Oxford.

1914: joined Royal Garrison Artillery, Portsmouth. 1915–17: experimental officer for RFC at Upavon and then Martlesham Heath, developing bomb-sights, accurately measuring aircraft performance. 1918: scientist at Air Ministry.

Post-war: 1919: briefly returned to Oxford; then worked on petrol content at Shoreham Laboratory. 1920: assistant secretary to Dept of Scientific and Industrial Research. 1927: head of DSIR. 1929: Rector of Imperial College, London. 1933: chair of Aeronautical Research Committee. 1934: Tizard Committee on aerial defence and was closely involved with the development of radar by Robert Watson-Watt [q.v.] and his team; Frederick Lindemann [q.v.] fell out with him badly over this. 1937–8: led

Biggin Hill trials to create system for the integration of radar into the operation of RAF Fighter Command.

WW2: 1940: dispute with Lindemann reached a peak and Churchill sided with Lindemann; appointed to lead a delegation to share British military secrets with Americans, including the revolutionary cavity magnetron. 1943–7: Pres of Magdalen College, Oxford.

Post-WW2: 1947: returned to Whitehall as chair of defence research policy committee; helped appoint a chief government scientist and argued for recruitment of scientists into government. 1952: retired.

FRS 1926. Knighted 1937. Pres British Association 1948.

Henry Tonks (1862–1937) Anatomist, artist and teacher

Born: Solihull.

Educ: Clifton College, Bristol; Royal Sussex County Hospital, Brighton.

1886: house surgeon at the London Hospital. 1888: Fellow of the Royal College of Surgeons; taught anatomy at the London Hospital.

1892: tutor at Slade School of Fine Art, becoming one of the most renowned art teachers of his generation; greatly influenced by the French impressionists.

1915: served as a medical orderly in France. 1916: Lieutenant in the RAMC; joined Harold Gillies [q.v.] at Aldershot and Sidcup recording pastel drawings of facial reconstruction cases.

1918: official war artist on the Western Front.

1918–30: Professor of Fine Art at the Slade. 1930: retired. 1936: an exhibition of his work held at the Tate Gallery, a rare honour for a living artist.

Arnold Toynbee (1889–1975) Historian

Born: London into a well-known academic and reforming family.

Educ: Winchester College; Balliol College, Oxford (1907–11).

1912: Fellow of Balliol College and historian of ancient Greece and Rome; talented linguist.

1915: joined Wellington House and wrote about Turkish atrocities against Armenians. 1917–18: Foreign Office intelligence dept.

1919: attended Versailles Peace Conference.

Post-war: 1920–4: Koraes Prof of Greek at King's College, London. 1925–38: Dir of Royal Inst for International Affairs (Chatham House) and editor of *Survey of International Affairs*, annual survey of key contemporary events; author *A Study of History* (6 vols, 1934–9), a huge encyclopaedic survey of rise and fall of world civilisations.

1939–46: Dir of foreign research and press, Foreign Office (similar to WW1 role).

Post-WW2: famous in America but severely mauled by historians in Europe for his broad brush-stroke approach to history.

FBA 1937.

Charles Urban (1867–1942)
Cinema producer and pioneer of factual films

Born: Cincinatti, Ohio, son of Austrian and Prussian immigrants.

Educ: locally.

1889: opened a stationery store in Detroit specialising in new technology like the typewriter and then Edison's Kinetoscope. 1897: to Britain to sort out Maguire and Baucus (M&B) London office. 1899: started Warwick Trading Company as film distribution company, part of M&B. 1903: left M&B and formed his own film production and distribution company, Charles Urban Trading Company; developed all forms of factual production. 1911: filmed Delhi Durbar in Kinemacolor.

1915: member of Wellington House Cinema Committee; filmed and helped edit *Britain Prepared* propaganda film. 1916–17: struggled to distribute *Britain Prepared* and *The Battle of the Somme* in US.

Post-war: founded Kineto Company in US and tried to develop Kinekrom, new colour system. 1924: business bankrupt. 1929: retired to Britain and lived last years in Brighton.

Sir Robert Watson-Watt (1892–1973) Inventor of radar

Born: Brechin, Forfarshire, Scotland, the son of a carpenter.

Educ: Brechin High School; University College, Dundee.

1912: on graduation appointed assistant lecturer and developed an interest in radio waves.

1915: meteorologist at Royal Aircraft Factory, Farnborough; tried to develop system of using radio waves to predict the approach of thunderstorms.

Post-war: 1923: started using cathode ray oscilloscopes for work on the detection of radio waves. 1926: began to investigate the ionosphere. 1933: superintendent of Radio Research Station, Slough, part of the National Physical Laboratory. 1935: approached by the Tizard [q.v.] Committee looking for ways to detect the approach of enemy aircraft, wrote report *The Detection of Aircraft by Radio Methods* laying down the theoretical basis of radar; his team undertook preliminary work at Orford Ness.

1936: in charge of research station at Bawdsey Manor, Suffolk, where a practical radar system (the Chain Home system) was developed and built by 1940. Known as the 'father of radar'.

1940: scientific adviser on telecommunications.

1946: Royal Commission on Awards gave him £52,000 for his work on radar.

Post-WW2: attended many international conferences on radio aids to shipping and civil aviation. Lectured around the world. Spent many years living in Canada. 1966: returned to Scotland.

FRS 1941. Knighted 1942

Chaim Weizmann (1874–1952) Chemist and Zionist

Born: Pinsk, Belorussia, third of fifteen children.

Educ: Gymnasium, Pinsk; Polytechnic, Berlin; University of Fribourg.

1904: invited by William Perkin to come to England to work at Manchester University on synthetic dyes; became successful biochemical researcher; spent much of spare time devoted to Zionist cause of establishing a Jewish homeland in Palestine. 1911: elected Vice-Pres of English Zionist Federation.

1915–18: discovered and managed a fermentation process to produce acetone, an essential ingredient in the manufacture of high explosives; recognised as making vital contribution to Britain's war effort.

1917: helped persuade British government to issue Balfour Declaration recognising Zionist claims on Palestine on eve of Gen Allenby's conquest of the region.

1918: part of Zionist Commission to Palestine; began work on establishing a Hebrew University in Jerusalem (inaugurated 1925).

Post-war: 1920: elected Pres of World Zionist Organisation; became full time leader of Zionist movement based in London. 1934: founded Sieff (later Weizmann) Inst of Science in Rehovot, Palestine.

1948: appointed first Pres of the newly declared state of Israel.

Acknowledgements

There are many people to thank for their help in the writing of this book. Prof. Gary Sheffield provided invaluable advice in the early stages. Clive Coultass gave help on naval matters. Kevin Brownlow offered access to his unique papers on the silent cinema. Alan Russell advised on mathematics. Andrew Johnston provided inspiration on the life of an Edwardian. Roger Smither and Sarah Henning offered much information on the early history of the Imperial War Museum. Jane Fish provided help with the films at the Imperial War Museum. I am grateful to many librarians and archivists including those at the National Archives, Kew; the Imperial War Museum, London; the Churchill Archive Centre at Churchill College, Cambridge; the Royal Society Archives, London; the Liddell Hart Centre for Military Archives at King's College, London; the Wellcome Institute, London; the Institute of Historical Research and the London Library. It is archivists and librarians who keep the wheels of history rolling and all historians know how much we owe to them.

There is of course an abundance of excellent material on the First World War with a stream of new books coming out in the centenary year. I have drawn upon the work of many scholars who are credited in the endnotes and I am grateful to them all. Where possible I have tried to draw upon the material written soon after the war, in the 1920s and 1930s. In this period there was a publishing boom in books about all aspects of the war as people sought to understand and explain the enormity of what had happened in the war years. Of course, there can be several motives for writing up memoirs and autobiographies or publishing diaries, some not always worthy or benign. But a critical reading of this material written while memories were fresh and scars still deep is I think very helpful in a book like this. I have listed this material as Primary Sources in the Bibliography in that it is material written by those who had taken part or led the events described in the central narrative of the book. I have found these accounts, from soldiers, politicians, aviators, doctors, surgeons, psychiatrists, propagandists and scientists particularly helpful in writing this book. As someone who has spent many years interviewing participants in events, this published material from soon after the war offers the nearest experience to hearing the genuine voices of the Great War, now that sadly there is no one left alive from that conflict.

At Little, Brown I have had the great good fortune to work with a splendid team again. I should very much like to thank Iain Hunt for his help and support on the editorial side, Linda Silverman for tracking down the photographs and Steve Gove for his meticulous work on the manuscript. Once again, Tim Whiting has overseen the whole project and brought great encouragement and clear direction from start to finish.

Anne has helped and advised in so many ways. As always, my final thanks are to her.

Taylor Downing

Notes

Abbreviations

CAC	Churchill Archives Centre, Churchill College, Cambridge
IWM	Central Archive, Imperial War Museum, London
IWM ART	Art Department, Imperial War Museum
IWM DOCS	Documents Department, Imperial War Museum
IWM FILM	Film Archive, Imperial War Museum
NA	National Archives, Kew
NAM	National Army Museum, London
RS	The Archives of the Royal Society, London
WELLCOME	The Archive of the Wellcome Institute, London

Prologue

1 Jonathan Winkler, *Nexus: Strategic Communications and American Security in World War I*, pp. 5–7. In Barbara Tuchman, *The Zimmerman Telegram*, pp. 10–11, and David Kahn, *The Code Breakers*, p. 266, it is said that the CS *Teleconia* destroyed these cables. A study of the GPO records has now revealed that this is incorrect and that it was the CS *Alert* that did the job.

2 Sir Philip Joubert de la Ferté, *The Fated Sky*, pp. 30–43; Maurice Baring, *Flying Corps Headquarters 1914–1918*, pp. 14–18; Ralph Barker, *The Royal Flying Corps in France*, pp. 22–7.

3 RS: CMB/36: Minutes of the War Committee of the Royal Society, 12 November 1914.

4 Hew Strachan's mammoth three-volume global history of the war, *The*

First World War, will do much to correct this obsession with the Western Front by devoting a great deal of its space to the war in the Pacific, Asia, the Middle East, the Balkans and Africa.

5   Patricia Fara, *Science*, p. 309

6   See for instance the *Guardian* leader, 4 October 2013, in a debate about the links between GCHQ and the US National Security Agency.

7   J.A. Fleming, 'Science in the War and After the War', a public lecture at University College, London, 10 October 1915, reported in *Nature*, 14 October 1915, p. 184.

1 New Century, New World

1   Examples are Barbara Tuchman, *The Proud Tower: A Portrait of the World Before the War 1890–1914*; George Dangerfield, *The Strange Death of Liberal England*. In economic history Eric Hobsbawm wrote of the 'long' nineteenth century from 1789 to 1914 and the 'short' twentieth century from 1914 to 1989; see Hobsbawm, *Age of Extremes*, pp. 3ff.

2   Lord Brabazon, *The Brabazon Story*, p. 48.

3   Eric Hobsbawm, *The Age of Empire*, p. 243.

4   Winkler, *Nexus*, pp. 6ff.

5   Michael Freemantle, *Gas! Gas! Quick, Boys!*, p. 22.

6   J. Lee Thompson, *Northcliffe*, p. 206.

7   Trevor Williams, *A Short History of Twentieth Century Technology*, p. 5.

8   *The Neglect of Science*, report of proceedings of a conference held in the rooms of the Linnean Society, Burlington House, Piccadilly, 3 May 1916, presided over by Lord Rayleigh. The 'two cultures debate' was still raging in the 1960s with the dispute between C.P. Snow and F.R. Leavis.

9   H.G. Wells, *Mr Britling Sees It Through*, p. 237, cited in Michael Howard, *Lessons of War*, pp. 78–9.

10   John Terraine, *White Heat*, p. 11.

2 The Pioneers

1   Brabazon, *The Brabazon Story*, pp. 58–9. The first ever powered flight in Britain had been by an American, Samuel Cody, six months earlier on 16 October 1908.

2   Ibid., p. 55.

3   Ibid., p. 3.

4   Ibid., p. 40.

5   Alliott Verdon Roe, *The World of Wings and Things*, p. 23.

6   Ibid., pp. 36–7.

7   See Hugh Driver, *The Birth of Military Aviation*, Appendix I, 'The "first British flight" controversy', pp. 275–8.

8   Ibid., p. 40.

9   Brabazon, *The Brabazon Story*, pp. 60–2.

10 *Daily Express*, 27 July 1909; *Daily Mail*, 26 July 1909; H.G. Wells, 'Of a cross-channel passage' in the *Daily Mail*, 27 July 1909.

11 David Edgerton, *England and the Aeroplane*, pp. 2–3.

12 For instance, in Basil Liddell Hart, *History of the First World War*, p. 355, and Terraine, *White Heat*, pp. 30ff.

13 NAM: *Manual of Military Ballooning*, School of Ballooning, Aldershot, 1896, p. 14.

14 Driver, *The Birth of Military Aviation*, p. 182.

15 Edward Spiers, *Haldane: an Army Reformer*, pp. 11ff.

16 Peter Reese, *The Flying Cowboy*, pp. 110ff.

17 Driver, *The Birth of Military Aviation*, p. 192.

18 Thompson, *Northcliffe*, p. 168.

19 NA: CAB 16/7 quoted in Driver, *The Birth of Military Aviation*, p. 209.

20 NA: Cd 5282: *The First Report of the Advisory Committee for Aeronautics*, p. 4.

21 The Advisory Committee for Aeronautics in 1919 changed its name to the Aeronautical Research Committee and later still to the Aeronautical Research Council. Its members continued to advise the British government on aviation policy for much of the century until the body was finally dissolved in the Thatcher defence cuts in 1980.

3 The New Science

1   NA: Cd 5282: *The First Report of the Advisory Committee for Aeronautics*, pp. 5–13.

2   Edgerton, *England and the Aeroplane*, p. 6.

3   NA: Cd 5282: *The First Report of the Advisory Committee for Aeronautics*, p. 5.

4   Edgerton, *England and the Aeroplane*, p. 13.

5   Driver, *The Birth of Military Aviation*, p. 227.

6   Brabazon, *The Brabazon Story*, p. 66.

7   Geoffrey de Havilland, *Sky Fever*, p. 47.

8   Gary Sheffield, *The Chief*, p. 62; Joubert, *The Fated Sky*, p. 32.

9   Driver, *The Birth of Military Aviation*, p. 263.

10 Ibid., p. 71.

11 Winston Churchill, *The World Crisis Vol 1*, p. 49.

12 See Taylor Downing, *Churchill's War Lab*, pp. 25ff for a summary of these changes.

13 Randolph Churchill, *Winston S. Churchill Vol II: Young Statesman 1901–1914*, p. 694.

14 Reese, *The Flying Cowboy*, p. 203.

15 Peter Mead, *The Eye in the Air*, p. 46.

16 Gary Sheffield argues that far from being opposed to the use of aircraft for reconnaissance, Haig was keen on it and his failure in the 1912 exercises was down to the fact that he put too much reliance on his aviators, who failed to spot the 'Blue' army approaching. See Sheffield, *The Chief*, pp. 61–2.

17 Churchill, *Winston S. Churchill Vol II: Young Statesman 1901–1914*, pp. 690–1.

18 See for instance Driver, *The Birth of Military Aviation*, pp. 241ff; Edgerton, *England and the Aeroplane*, pp. 4ff.

4 Observing the War

1 Hew Strachan, *The First World War: Vol 1*, pp. 104, 137. Strachan also points out that enthusiasm for the war was much stronger in the cities of Europe than in the countryside; see Strachan, *The First World War: Vol 1*, pp. 142ff.

2 The great exponent of the notion of 'war by railway timetable' in August 1914 was A.J.P. Taylor; see Taylor, *The First World War*, pp. 16ff, *War By Timetable: How the First World War Began*, pp. 25ff. A more recent account talks of European diplomats and leaders 'sleep walking to war' in the summer of 1914; see Christopher Clark, *The Sleep Walkers*, passim.

3 Holger Afflerbach and David Stevenson (eds), *An Improbable War?*, pp. 1–17.

4 Norman Macmillan, *Sir Sefton Brancker*, p. 61.

5 Churchill, *Winston S. Churchill Vol II: Young Statesman 1901–1914*, p. 697.

6 Baring, *Royal Flying Corps Headquarters 1914–1918*, pp. 49–50.

7 Joubert de la Ferté, *The Fated Sky*, pp. 43–4.

8 Terence Finnegan, *Shooting the Front*, p. 32.

9 Joubert de la Ferté, *The Fated Sky*, p. 44.

10 Mead, *The Eye in the Air*, p. 57; Sir Walter Raleigh, *The War in the Air Vol 1*, p. 329.

11 Brabazon, *The Brabazon Story*, p. 87.

12   Ibid., pp. 92–3.

13   RS: CMB/36 War Committee of the Royal Society, p. 11; Letter War Office to War Committee, 6 Jan 1915.

14   John Charteris, *At G.H.Q.*, p. 77.

15   Raleigh, *War in the Air Vol 1*, p. 9.

16   Finnegan, *Shooting the Front*, p. 45.

17   Driver, *The Birth of Military Aviation*, pp. 24–44; Edgerton, *England and the Aeroplane*, pp. 15–16.

18   Pemberton Billing also founded an aircraft design and manufacturing company in Southampton specifically to produce seaplanes for the navy, calling them 'supermarines'. This company went on in the mid-1930s, long after Pemberton Billing had lost control of it, to design and build the Spitfire.

19   Col. Roy Stanley, *World War II Photo Intelligence*, p. 26.

20   Roy Conyers Nesbit, *Eyes of the RAF*, p. 37.

21   For an analysis of the techniques of photo interpretation in the Second World War, many of which had been developed in the First, see Downing, *Spies in the Sky*, pp. 82–97, 112–30.

22   Finnegan, *Shooting the Front*, p. 75.

23   Denis Winter, *The First of the Few*, p. 154.

24   Ibid, p. 139; Cecil Lewis, *Sagittarius Rising*, pp. 169ff.

25   Peter Hart, *Bloody April*, p. 249.

26   Winter, *The First of the Few*, p. 153.

27   Finnegan, *Shooting the Front*, pp. 97–8.

28   A.J.P. Taylor, *English History 1914–1945*, p. 102.

29   Douglas Haig, *War Diaries and Letters 1914–1918*, edited by Gary Sheffield and John Bourne, p. 403.

30   Nesbit, *Eyes of the RAF*, pp. 43–4.

31   H.A. Jones, *War in the Air Vol V*, p. 228.

5 Room 40

1   David Kahn, *The Code Breakers*, p. 266.

2   The description of his appearance is by his son in A.W. Ewing, *The Man of Room 40 – The Life of Sir Alfred Ewing*, p. 178.

3   CAC: MISC 20, The Papers of Walter Horace Bruford: A short account entitled 'Room 40'.

4   Churchill, *The World Crisis Vol I*, p. 194.

5   CAC: GBR/0014/DENN/1, The Papers of Alexander Guthrie Denniston: 1/2 A short history of the setting up of Room 40, p. 4.

6   William James, *The Eyes of the Navy*, p. 18.

7   CAC: MISC 20, The Bruford Papers: A short account entitled 'Room 40'.

8   Patrick Beesly, *Room 40*, p. 19.

9   Churchill, *The World Crisis Vol I*, p. 417.

10 Ibid., p. 419.

11 Ibid., p. 429.

12 Ibid., p. 562.

13 Robert Massie, *Castles of Steel*, pp. 375ff.

14 CAC: GBR/0014/HALL/3/1–6, The Papers of Admiral Sir William Reginald Hall: Chapter 6 of a draft memoir written by Hall c. 1932.

15 Downing, *Churchill's War Lab*, pp. 43ff.

16 CAC: GBR/0014/CLKE/3, The Papers of William Francis Clarke: Draft of an unpublished history of Room 40 written by Clarke in 1948/9.

17 Andrew Gordon, *The Rules of the Game*, pp. 441ff.

18 Beesly, *Room 40*, p. 162.

19 CAC: GBR/0014/CLKE/3, The Clarke Papers: Draft of an unpublished history of Room 40 written by Clarke in 1948/9.

6 The Great Game

1   See Christopher Sykes, *Wassmuss 'The German Lawrence'*, passim.

2   There are several different versions of this strange story. Some say that it was not in the bags of the vice consul that the code book was discovered but in those of the consul himself, Dr Helmuth Listemann. Another version had Wassmuss giving the code book to a Persian to pass on to Listermann who then betrayed him, while yet another claimed that the mission was not against Persia but against the British in Afghanistan. Whatever actually happened, the code book ended up in the India Office in London where Hall's representative found it. See Beesly, *Room 40*, pp. 130–2; Strachan, *The First World War: Vol 1*, pp. 770–81.

3   Christopher Andrew, *The Defence of the Realm* pp. 3-4 & 21-28

4   CAC: MISC 20, The Bruford Papers: A short account entitled 'Room 40'.

5   CAC: GBR/0014/HALL/3/1–6, The Papers of Admiral Sir William Reginald Hall: Chapter 1 of a draft memoir written by Hall c. 1932.

6   James, *The Eyes of the Navy*, pp. 44–53.

7   Ibid., pp. 112–14.

8   CAC: GBR/0014/HALL/3/1–6, The Hall Papers: Chapter 3 of draft memoir by Hall.

9   James, *The Eyes of the Navy*, pp. 76–8.

10 CAC: GBR/0014/HALL/3/1–6, The Hall Papers: Chapter 6 of draft memoir by Hall.

11 James, *The Eyes of the Navy*, p. 136.

12 Beesly, *Room 40*, p. 206.

13 CAC: GBR/0014/HALL/3/1–6, The Hall Papers: Chapter 6 of draft memoir by Hall.

14 Beesly, *Room 40*, p. 218.

15 James, *The Eyes of the Navy*, p. 154.

16 Beesly, *Room 40*, p. 221.

17 Tuchman, *The Zimmermann Telegram*, p. 199; Beesly, *Room 40*, p. 224.

18 Beesly, *Room 40*, p. 37.

19 James, *The Eyes of the Navy*, p. 34.

7 The Gunners' War

1 William Van der Kloot, 'Lawrence Bragg's role in the development of sound ranging in World War One' in *Notes and Records of the Royal Society*, Vol. 59 No. 3, September 2005, p. 276.

2 Bertrand Russell, *Autobiography Vol 2*, p. 15.

3 David Phillips, 'William Lawrence Bragg' in *Biographical Memoirs of Fellows of the Royal Society*, Vol. 25, 1979, p. 93.

4 In the spring of 1916 GHQ moved further west to Montreuil where Haig lived in a chateau about two miles from the town.

5 Guy Hartcup, *The War of Invention*, p. 71.

6 The French and the Germans largely described an artillery piece by the diameter of the shell it fired, the calibre, while the British often used the weight of the shell. A German 150mm howitzer fired a shell of about 120 lb. A British 60lb shell was fired by a 5in or 127mm calibre gun.

7 Terraine, *White Heat*, pp. 51–2, 68, 79.

8 Strachan, *The First World War Vol 1*, p. 160.

9 Hartcup, *The War of Invention*, p. 24.

10 Ibid., p. 24.

11 Max Egremont, *Siegfried Sassoon*, p. 103.

12 Siegfried Sassoon, *Memoirs of an Infantry Officer*, pp. 90–2. In the account of this action by Robert Graves (who was in the same battalion as Sassoon), having occupied the German trench Sassoon is supposed to have pulled out a book of poetry and quietly read it before returning to his own lines: 'It was a pointless feat; instead of reporting or signalling for reinforcements he sat down in the German trench and began dozing

over a book of poems which he had brought with him.' Robert Graves, *Goodbye To All That*, p. 262. But there is no mention of the book of poems in Sassoon's own account. However, Sassoon was reprimanded by Colonel Clifton Stockwell, the battalion commander, that evening for not having reported the capture of the trench so it could have been consolidated.

13 Hew Strachan, 'Command, Strategy and Tactics 1914–18' in John Horne (ed.), *A Companion to World War I*, p. 43.

14 See Alan Clark, *The Donkeys*, although the origin of the phrase is unclear. There was a book published in 1927 by Capt. P.A. Thompson with the title *Lions Led by Donkeys*.

15 Michael Freemantle, *Gas! Gas! Quick, Boys!*, p. 118; Hartcup, *The War of Invention*, p. 44.

16 Freemantle, *Gas! Gas! Quick, Boys!*, p. 31.

17 The Lochnagar Crater near la Boiselle; it is now privately owned and run as a memorial to all those who fell in the Great War.

18 Freemantle, *Gas! Gas! Quick, Boys!*, p. 14.

19 Williams, *A Short History of Twentieth Century Technology*, p. 135.

20 Hartcup, *The War of Invention*, p. 45.

21 RS: CMB/36 Meeting of War Committee, 9 Dec 1915.

22 RS : CMB/36 Letter of Royal Society to universities, 10 Jan 1916; Meeting of War Committee, 30 March 1916.

23 Arthur Marwick, *Women at War*, p. 57.

24 Strachan, *The First World War Vol 1*, p. 167.

25 Freemantle, *Gas! Gas! Quick, Boys!*, pp. 61–3.

26 Sandra Gilbert, 'Soldier's Heart: Literary Men, Literary Women and the Great War' in Sandra Gilbert and Susan Gubar (eds), *No Man's Land*, p. 204.

27 Taylor, *English History 1914–1945*, p. 35.

8 The Yellow-Green Cloud

1 http://avalon.law.yale.edu/19th_century/dec99–02.asp

2 Robert Foley, *German Strategy and the Path to Verdun*, pp. 156ff.

3 Simon Jones, *World War I Gas Warfare*, p. 6.

4 *The Times*, 29 April 1915.

5 Ben Shephard, *A War of Nerves*, p. 63.

6 Charteris, *At G.H.Q.*, p. 89.

7 Hartcup, *The War of Invention*, p. 99.

8 RS: CMB/36 Meeting of War Committee, 15 July 1915.

9 Sheffield, *The Chief*, p. 127.

10 Jones, *World War I Gas Warfare*, p. 14.

11 Hartcup, *The War of Invention*, p. 108.

12 RS: CMB/36 Letter from War Office to RS, 3 July 1915.

13 Vera Brittain, *Testament of Youth*, p. 395.

14 Freemantle, *Gas! Gas! Quick, Boys!*, pp. 135–6. Despite the horror of mustard gas, the total number of British deaths from it was relatively low at 1,859 or 1.5 per cent of those infected.

15 Freemantle, *Gas! Gas! Quick, Boys!*, pp. 43–4.

16 Ibid., pp. 118–21. Some historians argue that these figures are too low and underestimate an even larger number of gas deaths in the Russian army.

17 Under the terms of the Treaty of Versailles in 1919 Germany was not allowed to use or develop poison gas. The 1925 Geneva Protocol prohibited the use of chemical weapons by any power. During the Second World War chemical weapons were not used on the battlefield. However, the Germans did use the chemical Zyklon B to murder millions of innocent civilians in their extermination camps at Auschwitz, Treblinka and elsewhere. Zyklon B came out of work done by Fritz Haber after the war and disguised as pest control. See Jones, *World War I Gas Warfare*, p. 59.

9 Breaking the Stalemate

1 Liddell Hart, *History of the First World War*, pp. 254–5.

2 Foley, *German Strategy and the Path to Verdun*, p. 103.

3 Brigadier-General Sir James E. Edmonds, *British Official History of the Great War, Military Operations France and Belgium 1915 Vol 1*, p. 51.

4 H.G. Wells, 'The Land Ironclads', originally published in the *Strand Magazine*, December 1903. The short story is available in many editions of the writings of Wells.

5 David Fletcher, *The British Tanks*, p. 14.

6 Churchill, *The World Crisis Vol II*, p. 512.

7 Fletcher, *The British Tanks*, pp. 42–3.

8 Ibid., p. 45.

9 Sheffield, *The Chief*, p. 189.

10 Liddell Hart, *History of the First World War*, p. 251.

11 Churchill for instance was very critical and wrote, 'The immense advantage of novelty and surprise was thus squandered while the number of tanks was small, while their condition was experimental and their crews

almost untrained.' *The World Crisis Vol II*, Ch. IV, p. 525. A.J.P. Taylor wrote, 'The surprise of a really heavy attack by tanks was lost.' *The First World War*, p. 140.

12   Captain Wilfrid Miles, British Official History of the Great War, *Military Operations France and Belgium 1916 Vol 2*, p. 367.

13   Bryn Hammond, *Cambrai 1917*. The subtitle of this book is *The Myth of the First Great Tank Battle*.

14   Sheffield, *The Chief*, pp. 272ff; Taylor, *English History 1914–1945*, p. 102.

15   NA: MUN 5/210/1940.

16   Terraine, *White Heat*, p. 246.

17   Liddell Hart, *History of the First World War*, p. 262.

10 The Body

1   Philip Gibbs, *Realities of War*, p. 304 .

2   Mark Harrison, *The Medical War*, p. 71.

3   Guy Chapman (ed.), *Vain Glory*, pp. 324–5; R.H. Tawney's account is of being wounded on the first day of the Battle of the Somme.

4   Geoffrey Noon, 'The Treatment of Casualties in the Great War' in Paddy Griffith (ed.), *British Fighting Methods in the Great War*, p. 87.

5   Quoted in Harrison, *The Medical War*, p. 125.

6   Hartcup, *The War of Invention*, p. 171.

7   Harrison, *The Medical War*, p. 23.

8   Charteris, *At G.H.Q.*, p. 129.

9   Harrison, *The Medical War*, p. 37.

10   RS: CMB/36 Meetings of the Chemical Committee on 26 Nov. and 16 Dec. 1914.

11   Noon, 'The Treatment of Casualties' in Griffith (ed.), *British Fighting Methods in the Great War*, p. 91.

12   Hartcup, *The War of Invention*, p. 166.

13   Ian Whitehead, *Doctors in the Great War*, p. 182.

14   Brittain, *Testament of Youth*, pp. 152, 213–14, 215.

15   Harrison, *The Medical War*, p. 96.

16   Ibid., p. 97.

17   Sheffield, *The Chief*, p. 143; Harrison, *The Medical War*, pp. 34–5.

18   Harrison, *The Medical War*, p. 73.

19   Martin Middlebrook, *The First Day on the Somme*, pp. 246–7.

20   T. Howard Somervell, *After Everest*, pp. 25–7.

21   Harrison, *The Medical War*, p. 75.

22   Ibid., p. 90.

23 The short film can be seen online at: www.britishpathe.com/video/ amputees-learn-to-use-artificial-limbs/query.

24 *Help for Wounded Heroes* was the title of a book by H.H. Thomas published in 1920.

25 Harold Gillies and Ralph Millard, *The Principles and Art of Plastic Surgery*, p. 7.

26 Reginald Pound, *Gillies*, p. 24.

27 Gillies and Millard, *The Principles and Art of Plastic Surgery*, pp. 8–9.

28 Ibid., p. 10.

29 Harold Gillies, *Plastic Surgery of the Face*, pp. 24ff.

30 Tonks's before-and-after drawings can be seen at: http://www.gillies archives.org.uk/Tonks%20pastels/index.html. The association between Tonks and Gillies provides the backdrop to Pat Barker's novels *Life Class* (2008) and *Toby's Room* (2012).

31 Gillies and Millard, *The Principles and Art of Plastic Surgery*, p. 31.

32 Pound, *Gillies*, p. 42.

33 Ibid., p. 44. The sailor was Able Seaman Vicarage from HMS *Malaya*. See also Gillies, *Plastic Surgery of the Face*, pp. 356–8; Gillies and Millard, *The Principles and Art of Plastic Surgery*, pp. 33–4. The last operation on the poor man's face took place on 6 March 1919.

34 Pound, *Gillies*, p. 52.

35 In January 1918 there were 12,720 medical practitioners in the armed services for 6 million servicemen and 11,482 for the rest of the population of around 46 million at home, that is approximately 1 per 480 in the military and 1 per 4000 at home; see Noon, 'The Treatment of Casualties' in Griffith (ed.), *British Fighting Methods in the Great War*, p. 89.

36 The precise numbers were: 1,988,669 admissions to hospital as battle casualties, of whom 151,356 died, 584,959 returned to duty and 1,245,535 were evacuated home. See Mitchell and Smith, *History of the Great War: Medical Services*, p. 110.

37 The Commonwealth War Graves Commission currently lists the number of deaths as 886,939 including the dead from the then colonies of Ireland and Newfoundland. In 1922 the War Office listed a smaller total number of 'killed in action', died as prisoners, died of wounds and 'missing' as 702,410, including the Royal Navy and the RAF but not including the merchant navy or civilian and military deaths from enemy bombardment or bombing of the UK mainland; see The War Office, *Statistics of the Military Effort of the British Empire During the Great War 1914–1920*, London: HMSO, 1922.

11 The Mind

1 Shephard, *A War of Nerves*, pp. 1, 21.

2 C.S. Myers, *Shell Shock in France*, pp. 13–14.

3 *The Lancet*, 13 February 1915, pp. 316–20.

4 Fiona Reid, *Broken Men*, p. 11.

5 Myers, *Shell Shock in France*, pp. 95–6; Shephard, *A War of Nerves*, p. 29.

6 Peter Leese, *Shell Shock*, pp. 38–9.

7 Shephard, *A War of Nerves*, p. 2.

8 Myers, *Shell Shock in France*, pp. 97–8.

9 Ibid., p. 39.

10 Ibid., p. 90; Shephard, *A War of Nerves*, p. 27.

11 Frederick W. Mott, *War Neuroses and Shell Shock*, pp. 1–5, 130ff.

12 Myers, *Shell Shock in France*, p. 26.

13 Shephard, *A War of Nerves*, p. 41.

14 WELLCOME: RAMC/446/18.

15 Middlebrook, *The First Day on the Somme*, Appendix 5, p. 330.

16 WELLCOME: RAMC/446/18; Leese, *Shell Shock*, pp. 41–2; Shephard, *A War of Nerves*, pp. 42–3.

17 Shephard, *A War of Nerves*, p. 49.

18 IWM DOC 42/TIS for Claire Tisdall's unpublished memoir; Leese, *Shell Shock*, p. 37

19 Harrison, *The Medical War*, p. 117.

20 The films are available to view on: www.britishpathe.com/workspaces/BritishPathe/shell-shock. Sadly, being silent films, it is impossible to hear Hurst's voice as he persuades, cajoles and instructs his patients.

21 Holden, *Shell Shock*, p. 55, based on an interview with Pear's daughter.

22 Siegfried Sassoon, *Sherston's Progress*, pp. 13–89; Pat Barker's trilogy *Regeneration* (1992), *The Eye in the Door* (1994) and *The Ghost Road* (1996). The film *Regeneration* (1997), directed by Gillies MacKinnon, starred Jonathan Pryce as William Rivers, James Wilby as Siegfried Sassoon and Stuart Bunce as Wilfred Owen.

23 The full text of Sassoon's declaration is to be found in *Memoirs of an Infantry Officer*, p. 218.

24 Winter, *The First of the Few*, p. 146, 191.

25 Lewis Yealland, *Hysterical Disorders of Warfare*, pp. 3–4.

26 Ibid., pp. 7–15.

27 Shephard, *A War of Nerves*, p. 97.

28 Sophie Delaporte, 'Military Medicine' in Horne (ed.), *A Companion to World War I*, p. 303.

29 Sheffield, *The Chief*, p. 145.

30 Haig, *War Diaries and Letters*, p. 259, entry for 6 December 1916.

31 There are several websites devoted to the cases of those 'shot at dawn', for instance: www.ww1cemeteries.com/othercemeteries/shotatdawnlist.htm.

32 Shepard, *A War of Nerves*, p. 70.

12 The War of Words

1 IWM ART: PST 2734.

2 Douglas Sutherland, *Tried and Valiant*, p. 130.

3 Michael Sanders and Philip Taylor, *British Propaganda during the First World War*, p. 104.

4 In a letter to Venetia Stanley dated 5 September 1914, cited in Martin Gilbert, *Winston S. Churchill Vol III*, p. 71.

5 Cate Haste, *Keep the Home Fires Burning*, p. 33.

6 Sanders and Taylor, *British Propaganda during the First World War*, p. 9.

7 *The Times*, 30 August 1914 (in a special Sunday edition).

8 Gilbert, *Winston S. Churchill Vol III*, p. 70.

9 Thompson, *Northcliffe*, p. 227.

10 Hansard, House of Commons, 6 August 1914.

11 *Daily Telegraph*, 4 September 1914, cited in Haste, *Keep the Home Fires Burning*, p. 57.

12 Haste, *Keep the Home Fires Burning*, p. 83.

13 In May 1915 the British government issued the Bryce Report, an investigation into the evidence on which the atrocity stories were based. It was a complete whitewash. None of the witnesses were identified and much of their evidence was second hand. Not surprisingly the report concluded that there was evidence for many of the stories that had been reported.

14 Haste, *Keep the Home Fires Burning*, p. 122.

15 Spiers, *Haldane: an Army Reformer*, pp. 11ff.

16 Haste, *Keep the Home Fires Burning*, p. 81.

17 Christophe Prochasson, 'Intellectuals and Writers' in Horne (ed.), *A Companion to World War I*, pp. 326–33.

18 Sanders and Taylor, *British Propaganda during the First World War*, p. 3.

19 Ibid., p. 40.

20 A.J.P. Taylor, *Beaverbrook*, p. 125.

21 Ibid., p. 127.

22 Lucy Masterman, *C.F.G. Masterman*, p. 274.

23 Ibid., pp. 281–2, quoting from the first official report of Wellington House, published in June 1915.

24 In 1916 Reuters went through a complete share restructuring, financially supported by the British government, in order to make it less reliant upon potentially troublesome overseas shareholders. See Donald Read, *The Power of News*, pp. 122ff.

25 Read, *The Power of News*, pp. 127–8.

26 Graham Storey, *Reuters' Century 1851–1951*, pp. 148ff; Read, *The Power of News*, pp. 127ff.

27 *The Thirty Nine Steps* was turned into a film directed by Alfred Hitchcock in 1935, starring Robert Donat as Richard Hannay and Madeleine Carroll as the woman he meets on the train. Hitchcock made various changes to the storyline to make it work as a film and it rapidly became a classic. Later film versions, with Kenneth More as Hannay (1959) and Robert Powell as Hannay (1978), were truer to the original storyline but less successful on the screen.

28 Andrew Lownie, *John Buchan*, p. 123.

29 Thompson, *Northcliffe*, pp. 244, 240.

30 Sanders and Taylor, *British Propaganda during the First World War*, p. 30.

31 Lownie, *John Buchan*, p. 124.

32 Ibid., p. 125.

13 The War in Pictures

1 Rachael Low, *The History of British Film 1914–1918*, pp. 16–23, 49–55; some of these figures come from 'Enquiry into the Cinema' by the National Council of Public Morals in 1917.

2 See for instance BFI Screenonline on the Topical Budget: http://www.screenonline.org.uk/film/id/583128/index.html.

3 Low, *The History of British Film*, p. 32.

4 Nicholas Reeves, *Official British Film Propaganda During the First World War*, p. 46.

5 Masterman, *C.F.G. Masterman*, pp. 282–3.

6 Reeves, *Official British Film Propaganda During the First World War*, p. 54.

7 Luke McKernan, *Charles Urban*, pp. 31ff.

8 Kevin Brownlow, *The War, The West and the Wilderness*, p. 51.

9 McKernan, *Charles Urban*, p. 140.

10 A later version of the film is held by the Imperial War Museum Film Archive, ref: IWM 580, but in this version the colour scenes have only survived in black and white.

11 Brownlow, *The War, The West and the Wilderness*, p. 52.

12 Masterman, *C.F.G. Masterman*, pp. 284–5.

13   Brownlow, *The War, the West and the Wilderness*, p. 52.

14   McKernan, *Charles Urban*, p. 145.

15   Jane Carmichael, *First World War Photographers*, pp. 46ff.

16   When sound film was introduced, in the late 1920s, the speed at which film was pulled through the camera needed to be increased from 16 to 24 frames per second (fps) to ensure the audio signal played smoothly. By this time both cameras and projectors were motorised and set at precisely 24 fps. As a consequence much film from the silent era looks speeded up and comic when played at modern film or television speeds. This can easily be corrected using variable speed playback devices, but it is still far too common in television documentaries to see footage from the silent era run at the incorrect speed, giving a totally false impression of how it would have looked at the time.

17   Geoffrey Malins, *How I Filmed the War*, passim; Brownlow Papers: Interview with Bertram Brookes-Carrington, the last surviving First World War cameraman on 15 and 20 October 1972 (Brookes-Carrington filmed the battle of Arras in 1917). See also Brownlow, *The War, the West and the Wilderness*, pp. 66–8.

18   Nick Hiley, Introduction to 1993 edition of Malins, *How I Filmed the War*, pp. xv–xxi.

19   Malins, *How I Filmed the War*, pp. 162–3.

20   John Buchan, *Nelson's History of the War Vol XVI: The Battle of the Somme*, p. 32.

21   Malins, *How I Filmed the War*, p. 197. In all wars, those far away from the combat request cameramen to get ever more dramatic action on film. In the Second World War, for instance, combat cameraman Peter Hopkinson was constantly being asked to shoot more action when he found himself covering events at the Battle of El Alamein, in which most of the action happened at night when he could not film or at such a great distance that he could not see what was going on. So, once again, he had to resort to staging sequences to create the required effect. See Peter Hopkinson, *Split Focus*, pp. 37ff.

22   Brownlow Papers: Interview with Brookes-Carrington, 15 Oct 1972. See also Brownlow, *The War, the West and the Wilderness* pp. 65ff.

23   Roger Smither, '"A Wonderful Idea of the Fighting": the question of fakes in the Battle of the Somme' in *Historical Journal of Film, Radio and Television*, Vol. 13 No. 2, 1993, pp. 149ff.

24   J. Morton Hutcheson, 'Music in the Cinema' in *The Bioscope*, 17 August 1916.

25   Lloyd George's full address was as follows:
      You are invited here to witness by far the most important and

imposing picture of the war that our staff has yet procured. The Battle of the Somme, furious and desperate as it has been, is a first and most important phase in what is an historical struggle, unique in its scope and world-wide significance. I am convinced that when you have seen this wonderful picture, every heart will beat in sympathy with its purpose, which is no other than that every one of us at home and abroad shall see what our men at the Front are doing and suffering for us, and how their achievements have been made possible by the sacrifices made at home. Now gentlemen, be up and doing also! See that this picture which is an epic of self-sacrifice and gallantry, reaches everyone. Herald the deeds of our brave men to the ends of the earth. This is your duty. Ladies, I feel that no word is necessary to urge upon you the importance of throwing in the whole ardour and strength of your invaluable aid. Mothers, wives, sisters and affianced ones, your hearts will beat, your voices speak in honour and glory of the living and the dead. You are great and powerful. This is your mission.

26   The film (ref: IWM 191) is the jewel in the crown of the Imperial War Museum Film Archive's First World War collection. In 2005 *The Battle of the Somme* was recognised internationally and became the first ever documentary film to be recorded in UNESCO's Memory of the World Register. In the same year the film was given a complete digital restoration by the Imperial War Museum. This new version with background features about the making of the film is available on DVD, published by the IWM.

27   *Evening News*, 11 August 1916, cited in Hiley, Introduction to 1993 edition of Malins, *How I Filmed the War*, p. xxvii.

28   *The Star*, 25 August 1916, cited in Brownlow, *The War, the West and the Wilderness*, p. 61.

29   *Manchester Guardian*, 11 August 1916; Kine Weekly, 10 August 1916; The Cinema, 10 August 1916. Cited in Reeves, *Official British Film Propaganda During the First World War*, p. 238.

30   Stephen Badsey, 'Battle of the Somme: British war propaganda' in *Historical Journal of Film, Radio and Television*, Vol. 3 No. 2, October 1983, p. 110.

31   From Frances Stevenson, *Lloyd George: A Diary by Frances Stevenson*, p. 112, cited in Smither, '"A Wonderful Idea of the Fighting": the question of fakes in the Battle of the Somme' in *Historical Journal of Film, Radio and Television* Vol 13 No. 2, 1993, p. 149.

32   Letters to the Editor, *The Times*, 5 September 1916.

33 'The Film Coming Into Its Own', *The Times*, 6 September 1916.

34 Low, *The History of the British Film*, p. 29.

35 Pierre Sorlin, 'Film and the War' in Horne (ed.), *A Companion to World War I*, p. 358.

36 Reeves, *Official British Film Propaganda During the First World War*, p. 239.

37 McKernan, *Charles Urban*, pp. 147ff; Brownlow, *The War, the West and the Wilderness*, p. 54.

38 Masterman, *C.F.G. Masterman*, p. 285; M.I. Sanders, 'British Film Propaganda in Russia, 1916–1918' in *Historical Journal of Film, Radio and Television*, Vol. 3 No. 2, October 1983, pp. 119–24.

39 The film is also held at IWM FILM, ref: MGH 3556.

40 *Daily Mail*, 23 December 1916, and the *Evening News*, 9 June 1917, cited in Reeves, *Official British Film Propaganda During the First World War*, p. 239.

41 All these films are preserved at IWM FILM: *A Day in the Life of a Munitions Worker*, ref: IWM 510; *Mrs John Bull Prepared*, ref: IWM 521.

42 Masterman, *C.F.G. Masterman*, p. 287.

43 Most of these works are now held at IWM ART, for instance C.R.W. Nevinson, *Paths of Glory*, ref: IWM ART 518; Paul Nash, *The Ypres Salient at Night*, ref: IWM ART 1145 and *The Menin Road*, ref: IWM ART 2242; John Sargent, *Gassed*, ref: IWM ART 1460.

44 William Philpott, *Bloody Victory*, p. 291.

14 Masters of Information

1 NA: CAB 23/1/1, Minutes of the War Cabinet, 9 December 1916.

2 Sanders and Taylor, *British Propaganda During the First World War*, p. 61.

3 Roderick Jones was knighted for his efforts in 1918; see Read, *The Power of News*, pp. 129–30.

4 Haste, *Keep the Home Fires Burning*, p. 41.

5 Sanders and Taylor, *British Propaganda During the First World War*, pp. 74–5.

6 Ibid., p. 75.

7 The appointment was initially held up by the King, who did not like Beaverbrook's brashness and refused to approve the choice. After Beaverbrook threatened to turn the offer down unless he was given Cabinet rank, Lloyd George overruled the King's objections and the appointment was confirmed.

8 Taylor, *Beaverbrook*, pp. 190–2.

9 Stephen Roskill, *Hankey: Man of Secrets Vol 1*, pp. 502–3.

10 Taylor, *Beaverbrook*, p. 204.

11 During the Second World War Churchill called Beaverbrook into government as Minister of Aircraft Production in the crucial period of the 'Spitfire Summer' of 1940. He made a brilliant success of this and increased aircraft output considerably, introducing round-the-clock shifts and a far more efficient supply chain to the war factories. The key difference between 1918 and 1940 was that Beaverbrook had the total backing of Churchill in the Second World War to do things his way, which he did not have from Lloyd George in the First.

12 Taylor, *Beaverbrook*, p. 210.

13 Sir Campbell Stuart, *Secrets of Crewe House*, p. 49. Stuart is not a totally independent witness as he went on to work for Northcliffe after the end of the war and in 1920 was appointed managing director of *The Times*, then still owned by Northcliffe.

14 Sanders and Taylor, *British Propaganda During the First World War*, p. 226.

15 Stuart, *Secrets of Crewe House*, p. 47.

16 Ibid., p. 62.

17 Sanders and Taylor, *British Propaganda During the First World War*, p. 237.

18 Stuart, *Secrets of Crewe House*, p. 93.

19 Ibid., pp. 94, 107–8.

20 Sanders and Taylor, *British Propaganda During the First World War*, p. 211.

21 The calendar included four crimes against Belgium and five U-boat crimes in which innocent civilians had died; the remaining three crimes were the bombardment of Scarborough, the Zeppelin raids and the Turkish massacre of Armenians. See Sanders and Taylor, *British Propaganda During the First World War*, pp. 141–2.

22 Masterman, *C.F.G. Masterman*, p. 306.

23 Lownie, *John Buchan*, p. 135.

24 The other novel was *Mr Standfast*, written between July 1917 and July 1918 and published in 1919. Again it features Richard Hannay and includes several tricks of the trade Buchan picked up in the propaganda department.

25 Charles Ffoulkes, *Arms and the Tower*, p. 110; Brownlow Papers: Interview with Brookes-Carrington, 20 October 1972.

26 IWM: Second Annual Report of the Committee of the Imperial War Museum 1918–1919 (Cmd 138).

27 Ffoulkes, *Arms and the Tower*, p. 116.

28 'The Greatest War Memorial' in *The Times*, 10 June 1920.

29 Ffoulkes, *Arms and the Tower*, p. 142.

30 IWM: Third Annual Report of the Imperial War Museum 1920–1921 (Cmd 1353).

31 A good example of this was Charles Edward Montague in *Disenchantment*.

32 Stuart, *Secrets of Crewe House*, p. 120, 130–1; Ludendorff's memoirs were published in Germany in 1919.

Epilogue: The First Boffins

1   Cited in John Stevenson, *British Society, 1914–1945*, p. 90.

2   Pound, *Gillies: Surgeon Extraordinary*, pp. 71, 86, 122ff.

3   NA: FD2/4, Annual Report of the Medical Research Committee 1917–18, Cd 8825.

4   Lord Birkenhead, *The Prof in Two Worlds*, p. 66.

5   Adrian Fort, *Prof*, pp. 47–51.

6   Birkenhead, *The Prof in Two Worlds*, pp. 70–8; Birkenhead notes there was later some dispute about the date and the nature of these legendary test flights but concludes they did take place in June–July 1917.

7   Downing, *Churchill's War Lab*, pp. 153ff.

8   Birkenhead, *The Prof in Two Worlds*, p. 211.

9   Ronald Clark, *Tizard*, p. 33.

10 IWM DOCS: Tizard Papers; HTT 713; unpublished autobiography, p. 103.

11 Ibid., p. 113.

12 Ibid., p. 124.

13 Robert Watson-Watt, *Three Steps to Victory*, p. 43.

14 J.A. Ratcliffe, 'Robert Alexander Watson-Watt' in *Biographical Memoirs of Fellows of the Royal Society*, 1975, pp. 549–68.

15 For a fuller account of the development of radar see Downing, *Night Raid*, pp. 7–34; Robert Buderi, *The Invention that Changed the World*, pp. 52–76.

16 Downing, *Churchill's War Lab*, pp. 173ff; Ronald Clark, *The Rise of the Boffins*, pp. 209–24. The author's father, Peter Downing, a mathematics graduate in 1942, was recruited straight from university into operational research and served in the RAF in Egypt and the Mediterranean throughout the war. The various projects he worked on included assessing the strike rate of bombs versus torpedoes at ships at sea (torpedoes had a higher success rate), the best way to spot U-boats, and the ideal colour for life vests and rafts to be spotted at sea (orange rather than yellow, the colour used before the war).

17 Michael M. Postan, *British War Production*, p. 5; other examples of this 'declinist' view include Taylor, *English History 1914–1945* and Barnett, *The Collapse of British Power*.

18 David Edgerton, *Warfare State Britain 1920–1970*, pp. 15–58. Edgerton

shows that the Royal Navy in 1935 still had 15 battleships (cf. America 15; Japan 10; Germany 0); that the army did much to develop new and modern tanks; that the RAF had by comparison to France and Germany a modern air force until overtaken by Germany in the late 1930s; and that 80,000 people worked in the British armaments industry, which was sustained by a powerful export market particularly of military aircraft.

19 For a listing of all the military research and development establishments and the numbers of scientists working in each, see Edgerton, *Warfare State*, p. 120.

20 These included 1270 engineers (of whom half were GPO engineers running the telephone service), 740 chemists and scientific research officers, 460 medical research doctors and 200 technical officers engaged in other sorts of scientific research, see Edgerton, *Warfare State*, p. 111; the number of 6500 in the technical and scientific grades compares with 1150 in the administrative grades (the ruling group of permanent and assistant secretaries in the senior civil service) and 4350 in executive grades (the middle management of the civil service).

21 Edgerton, *Warfare State*, p. 118; equivalent sum calculated on www.measuringworth.com according to the retail price index.

22 It was more than a decade before Frank Whittle's theoretical design for a gas turbine jet engine could become a practical reality as new metals needed to be developed to sustain the stresses of jet propulsion before an engine could be built; although Whittle later complained of lack of support from the Air Ministry, the record shows that several senior RAF figures continued to provide him and his private company, Power Jets Ltd, with considerable support. See Andrew Nahum, *Frank Whittle*, pp. 52ff.

23 Clark, *The Rise of the Boffins*, p. 12.

24 Speaking to the Parliamentary and Scientific Committee in February 1942. See Downing, *Churchill's War Lab*, p. 178.

Bibliography

Primary Sources

Official histories:

Brigadier-General Sir James E. Edmonds (General Editor), *History of the Great War Based on Official Documents*. London: HMSO, 29 volumes, 1923–1949 [also known as the *British Official History of the Great War* and as *Military Operations*]

Major T.J. Mitchell and Miss G.M. Smith, *History of the Great War: Medical Services – Casualties and Medical Statistics of the Great War*. London: HMSO, 1931

Sir Walter A. Raleigh and H.A. Jones, *The War in the Air*. London: HMSO, 6 volumes, 1922–8

Memoirs, autobiographies, diaries, etc.:

Maurice Baring, *Royal Flying Corps Headquarters 1914–1918*. London: William Heinemann, 1930

Lord Brabazon of Tara, *The Brabazon Story*. London: William Heinemann, 1956

Vera Brittain, *Testament of Youth: An Autobiographical Study of the Years 1900–1925*. London: Virago, 1978 [originally London: Victor Gollancz, 1933]

John Buchan, *Nelson's History of the War*. London: Thomas Nelson & Sons, 24 volumes, 1915–20

John Charteris, *At G.H.Q.* London: Cassell & Co., 1931

Winston Churchill, *The World Crisis 1911–1918*. London: Odhams, 5 volumes, 1923–31; republished in 2 volumes London: Odhams, 1939 (page references in endnotes from the 1939 edition)

Charles Ffoulkes, *Arms and the Tower*. London: John Murray, 1939

Philip Gibbs, *Realities of War*. London: William Heinemann, 1920

Harold Gillies, *Plastic Surgery of the Face: Based on Selected Cases of War Injuries of the Face Including Burns*. London: Hodder and Stoughton, 1920

Sir Harold Gillies and D. Ralph Millard Jnr, *The Principles and Art of Plastic Surgery*. London: Butterworth & Co., 1957

Robert Graves, *Goodbye To All That: An Autobiography*. London: Jonathan Cape, 1929

Douglas Haig (eds Gary Sheffield and John Bourne), *War Diaries and Letters 1914–1918*. London: Phoenix, 2006

Geoffrey de Havilland, *Sky Fever: The Autobiography of Sir Geoffrey de Havilland*. London: Hamish Hamilton, 1961

Sir Philip Joubert de la Ferté, *The Fated Sky: An Autobiography*. London: Hutchinson, 1952

Cecil Lewis, *Sagittarius Rising*. London: Peter Davies, 1936. Republished by Greenhill Books, 1993, with a new introduction by the author

Geoffrey Malins, *How I Filmed the War: A Record of the Extraordinary Experiences of the Man who Filmed the Great Somme Battles, etc*. London: Herbert Jenkins, 1920. Republished by the Imperial War Museum, London, and The Battery Press, Nashville, 1993, with an Introduction by Nick Hiley

Lucy Masterman, *C.F.G. Masterman: A Biography*. London: Nicholson & Watson, 1939

Charles Edward Montague, *Disenchantment*. London: Chatto & Windus, 1929

Frederick W. Mott, *War Neuroses and Shell Shock*. London: Hodder and Stoughton, 1919

Charles S. Myers, *Shell Shock in France 1914–18: Based on a War Diary*. Cambridge: Cambridge University Press, 1940

Arthur Ponsonby, *Falsehood in Wartime: Containing an Assortment of Lies Circulated Throughout the Nations during the Great War*. London: Dutton, 1924 [republished by Kessinger Publishing, 2010]

Alliott Verdon Roe, *The World of Wings and Things*. London: Hurst & Blackett, 1939

Bertrand Russell, *Autobiography Vol. 2 1914–1944*. London: George Allen and Unwin, 1968

Siegfried Sassoon, *Memoirs of an Infantry Officer*. London: Faber and Faber, 1930

Siegfried Sassoon, *Sherston's Progress*. London: Faber and Faber, 1936

T. Howard Somervell, *After Everest: The Experiences of a Mountaineer and Medical Missionary*. London: Hodder and Stoughton, 1936.

Sir Campbell Stuart, *Secrets of Crewe House; The Story of a Famous Campaign*.

London: Hodder and Stoughton, 1920

Robert Watson-Watt, *Three Steps to Victory: A Personal Account by Radar's Greatest Pioneer*. London: Odhams, 1957

H.G. Wells, *Mr Britling Sees it Through*. London: Cassell, 1916

Lewis R. Yealland, *Hysterical Disorders of Warfare*. London: Macmillan & Co., 1918

Secondary Sources

Holger Afflerbach and David Stevenson (eds), *An Improbable War? The Outbreak of World War I and European Political Culture Before 1914*. Oxford: Berghahn Books, 2007

Christopher Andrew, *Defence of the Realm: The Authorized History of MI5*. London, Penguin, 2009

Ralph Barker, *The Royal Flying Corps in France: From Mons to the Somme*. London: Constable, 1994

Correlli Barnett, *The Collapse of British Power*. London: Eyre, Methuen, 1972

Patrick Beesly, *Room 40: British Naval Intelligence 1914–18*. London: Hamish Hamilton, 1982

Earl of Birkenhead, *The Prof in Two Worlds: The Official Life of Prof F.A. Lindemann, Viscount Cherwell*. London: Collins, 1961

Kevin Brownlow, *The War, the West and the Wilderness*. London: Secker & Warburg, 1979

Robert Buderi, *The Invention that Changed the World: The Story of Radar from War to Peace*. London: Little, Brown, 1997 [originally New York: Simon & Schuster, 1996]

Jane Carmichael, *First World War Photographers*. London: Routledge, 1989

Guy Chapman (ed.), *Vain Glory: A Miscellany of the Great War 1914–18 Written by Those Who Fought in it on Each Side and on All Fronts*. London: Cassell, 1968

Randolph Churchill, *Winston S. Churchill Vol. II: Young Statesman 1901–1914*. London: William Heinemann, 1967

Alan Clark, *The Donkeys*. London: Hutchinson, 1961

Christopher Clark, *The Sleep Walkers: How Europe Went to War in 1914*. London: Allen Lane, 2012

Ronald Clark, *The Rise of the Boffins*. London: Phoenix House, 1962

Ronald Clark, *Tizard*. London: Methuen, 1965

George Dangerfield, *The Strange Death of Liberal England 1910–1914*. London: Perigree, 1961

Taylor Downing, *Churchill's War Lab: Code-Breakers, Boffins and Innovators: The Mavericks Churchill Led to Victory*. London: Little, Brown, 2010

Taylor Downing, *Spies in the Sky: The Secret Battle for Aerial Intelligence during World War Two*. London: Little, Brown, 2011

Taylor Downing, *Night Raid: The True Story of the First Victorious Para Raid of WWII*. London: Little, Brown, 2013

Hugh Driver, *The Birth of Military Aviation: Britain, 1903–1914*. Woodbridge: The Royal Historical Society & the Boydell Press, 1997

David Edgerton, *England and the Aeroplane: An Essay on a Militant and Technological Nation*. Basingstoke: Macmillan Academic, 1991

David Edgerton, *Warfare State: Britain, 1920–1970*. Cambridge: Cambridge University Press, 2006

Max Egremont, *Siegfried Sassoon: A Biography*. London: Picador, 2005

A.W. Ewing, *The Man of Room 40 – The Life of Sir Alfred Ewing*. London: Hutchinson, 1939

Patricia Fara, *Science: A Four Thousand Year History*. Oxford: Oxford University Press, 2009

Terence J. Finnegan, *Shooting the Front: Allied Aerial Reconnaissance in the First World War*. Stroud: Spellmount/The History Press, 2011

David Fetcher, *The British Tanks 1915-19*. Marlborough: Crowood Press, 2001

Robert Foley, *German Strategy and the Path to Verdun: Erich von Falkenhayn and the Development of Attrition, 1870–1916*. Cambridge: Cambridge University Press, 2005

Adrian Fort, *Prof: The Life of Frederick Lindemann*. London: Jonathan Cape, 2003

Michael Freemantle, *Gas! Gas! Quick, Boys! How Chemistry Changed the First World War*. Stroud: Spellmount/The History Press, 2012

Martin Gilbert, *Winston S. Churchill, Vol. III 1914–1916*. London: William Heinemann, 1971

Sandra Gilbert and Susan Gubar (eds), *No Man's Land: The Place of the Woman Writer in the Twentieth Century*. Yale: Yale University Press, 1988

Andrew Gordon, *The Rules of the Game: Jutland and British Naval Command*. London: John Murray, 1996

Paddy Griffith (ed.), *British Fighting Methods in the Great War*. London: Frank Cass, 1996

Bryn Hammond, *Cambrai 1917: The Myth of the First Great Tank Battle*. London: Weidenfeld & Nicolson, 2008

Mark Harrison, *The Medical War: British Military Medicine in the First World War*. Oxford: Oxford University Press, 2010

Peter Hart, *Bloody April: Slaughter in the Skies over Arras, 1917*. London: Cassell, 2006

Guy Hartcup, *The War of Invention: Scientific Developments, 1914–18*. London: Brassey's Defence Publishers, 1988

Cate Haste, *Keep the Home Fires Burning: Propaganda in the First World War*. London: Allen Lane, 1977

Eric Hobsbawm, *The Age of Empire: 1875–1914*. London: Weidenfeld & Nicolson, 1987

Eric Hobsbawm, *Age of Extremes: The Short Twentieth Century 1914–1991*. London: Michael Joseph, 1994

Wendy Holden, *Shell Shock: The Psychological Impact of War*. London: Macmillan, 1998

Peter Hopkinson, *Split Focus*. London: Rupert Hart-Davis, 1969

John Horne (ed.), *A Companion to World War I*. Oxford: Wiley-Blackwell, 2010

Michael Howard, *The Lessons of History*. Oxford: Oxford University Press, 1991

Admiral Sir William James, *The Eyes of the Navy: A Biographical Study of Admiral Sir Reginald Hall*. London: Methuen, 1955

Simon Jones, *World War I Gas Warfare: Tactics and Equipment*. Oxford: Osprey, 1994

David Kahn, *The Code Breakers: The Story of Secret Writing*. New York: Scribner, 1967

Peter Leese, *Shell Shock: Traumatic Neurosis and the British Soldiers of the First World War*. Basingstoke: Palgrave Macmillan, 2002

Basil Liddell Hart, *History of the First World War*. London: Cassell, 1970 [originally published as *The Real War 1914–18* London: Faber and Faber, 1930]

Rachael Low, *The History of the British Film 1914–1918*. London: George Allen & Unwin, 1950

Andrew Lownie, *John Buchan: The Presbyterian Cavalier*. London: Constable, 1995

Norman Macmillan, *Sir Sefton Brancker*. London: William Heinemann, 1935

Arthur Marwick, *Women at War, 1914–1918*. London: Croom Helm, 1977

Robert K. Massie, *Castles of Steel: Britain, Germany and the Winning of the Great War at Sea*. London: Jonathan Cape, 2004

Luke McKernan, *Charles Urban: Pioneering the Non-Fiction Film in Britain and America 1897–1925*. Exeter: Exeter University Press, 2013

Peter W. Mead, *The Eye in the Air: A History of Air Observation and Reconnaissance in the Army 1785–1945*. London: HMSO, 1983

Martin Middlebrook, *The First Day on the Somme: 1 July 1916*. London: Allen Lane, 1971

Andrew Nahum, *Frank Whittle: Invention of the Jet*. Cambridge: Icon Books, 2004

Roy Conyers Nesbit, *Eyes of the RAF: A History of Photo-Reconnaissance*. Stroud: Sutton, revised edition 2003

William Philpott, *Bloody Victory: The Sacrifice on the Somme*. London: Little, Brown, 2009

Michael M. Postan, *British War Production: Official History of the Second World War*. London: HMSO, 1952

Reginald Pound, *Gillies: Surgeon Extraordinary*. London: Michael Joseph, 1964

Douglas Read, *The Power of News: The History of Reuters*. Oxford: Oxford University Press, 1992

Peter Reese, *The Flying Cowboy: Samuel Cody, Britain's First Airman*. Stroud: Tempus, 2006

Nicholas Reeves, *Official British Film Propaganda During the First World War*. London: Croom Helm, 1986

Fiona Reid, *Broken Men: Shell Shock, Treatment and Recovery in Britain 1914–30*. London: Continuum, 2010

Stephen Roskill, *Hankey: Man of Secrets, Vol 1 1877–1918*. London: Collins, 1970

Michael Sanders and Philip M. Taylor, *British Propaganda during the First World War 1914–18*. London: Macmillan, 1982

Gary Sheffield, *The Chief: Douglas Haig and the British Army*. London: Aurum Press, 2011

Ben Shephard, *A War of Nerves: Soldiers and Psychiatrists 1914–1994*. London: Pimlico, 2002

Edward Spiers, *Haldane: An Army Reformer*. Edinburgh: Edinburgh University Press, 1980

Col. Roy Stanley, *World War II Photo Intelligence*. London: Sidgwick and Jackson, 1982

John Stevenson, *British Society 1914–1945*. Harmondsworth: Penguin, 1984

Dorothy Stimson, *Scientists and Amateurs: A History of the Royal Society*. New York: Greenwood Press, 1948

Graham Storey, *Reuters' Century 1851–1951*. London: Max Parrish, 1951

Hew Strachan, *The First World War: Vol 1 To Arms*. Oxford: Oxford University Press, 2001

Hew Strachan, *The First World War: A New History*. London: Free Press, 2006

Douglas Sutherland, *Tried and Valiant: The History of the Border Regiment 1702–1959*. London: Leo Cooper, 1972

Christopher Sykes, *Wassmuss 'The German Lawrence'*. New York: Longman, Green & Co., 1936

A.J.P. Taylor, *English History 1914-1945*. Oxford: Oxford University Press, 1965

A.J.P. Taylor, *The First World War: An Illustrated History*. Harmondsworth: Penguin, 1966

A.J.P. Taylor, *War by Timetable: How the First World War Began*. London: Macdonald, 1969

A.J.P. Taylor, *Beaverbrook*. London: Hamish Hamilton, 1972

John Terraine, *White Heat: The New Warfare 1914–18*. London: Leo Cooper, 1992

J. Lee Thompson, *Northcliffe: Press Baron in Politics, 1865–1922*. London: John Murray, 2000

Barbara Tuchman, *The Zimmermann Telegram*. London: Constable, 1959

Barbara Tuchman, *The Proud Tower: A Portrait of the World Before the War 1890–1914*. London: Hamish Hamilton, 1966

Ian R. Whitehead, *Doctors in the Great War*. Barnsley: Leo Cooper, 1999

Trevor I. Williams, *A Short History of Twentieth Century Technology c. 1900–c. 1950*. Oxford: Oxford University Press, 1982

Jonathan Reed Winkler, *Nexus: Strategic Communications and American Security in World War One*. Harvard: Harvard University Press, 2008

Denis Winter, *The First of the Few: Fighter Pilots of the First World War*. London: Allen Lane, 1982

Index

acetone, 169–70, 337

Admiralstab, Berlin, 104, 107, 116

Admiralty: acetone and, 169–70; allows reporting from the front, 288; Balfour as First Lord, 118, 201, 294, 296–7; Board of Invention and Research, 154–5; Churchill as First Lord of, 67–9, 72, 104, 117–18, 169, 196–7, 271–2; cinema and, 292, 294, 295–7, 332; code-breaking and, 11, 102–3, 106–14, 117, 118–19, 121–3, 125, 126–7, 128, 138, 339; communications interception, 102, 105–6, 109–14, 116–17, 118–19, 121–3, 138, 163; conservative values of, 36; development of the tank and, 197, 198–200; Director of Scientific Research at, 354; failures in processing of intelligence, 112–16, 118, 121–2, 123; Hall's recruitment of women, 127–8; Intelligence Division, 101, 109–11, 324; interception of diplomatic cables, 110, 111, 125, 126–7, 131–2, 138, 139–44, 145–6; Landships Committee, 198–200; propaganda and, 294, 295–7, 316; radio communications and, 104, 105–6, 155, 163; Research Laboratory, Teddington, 354; Room 40 code-breakers, 106, 108–9, 110–12, 117, 118–19, 121–3, 125, 128, 138, 339; Room 40 diplomatic mail readers (Room 45), 126–8, 131–2, 139–44, 145–6; Royal Society and, 7; sacking of Churchill, 117–18, 200, 277; Short brothers and, 62, 67, 72; suspicion of scientists, 155; *see also* Royal Navy

aerial combat: 'ace' fighter pilots, 93–4; 'dog fights', 91, 93; first examples, 89–90; new aircraft for, 91; tactics, 90

aerial reconnaissance: ballooning, 44, 52–3; Battle of Cambrai and, 206; Battle of Mons and, 81–2; Battle of the Somme and, 93; cameras used for, 86, 87, 89, 91, 92–3, 98; Capper on, 53; Dunne and, 57; fighter escorts for reconnaissance aircraft, 90; first ever flight (19 August 1914), 80–1; French as leaders in photographic field, 84, 97; German counter-measures, 95–6; German spring offensive (1918) and, 96; German techniques, 89; higher altitude flying, 91;

aerial reconnaissance – *continued*
 hobbyists as pioneers, 65; during
 Palestine campaign, 97–8; photo
 interpretation, 85, 92, 93, 97, 98;
 photo maps, 87–8, 98;
 photography and, 83–5, 86, 87–9,
 91–3, 95–7, 98, 347; poison gas
 and, 173; RE series aircraft, 73;
 RFC for Expeditionary Force, 72,
 80–3, 87–8; spotting for the
 artillery, 88, 98; summer
 manoeuvres (1912), 71; support
 staff for aerial observers, 91–2;
 trench warfare and, 83–4, 87–8,
 93
Aero Club, English, 42, 49, 57, 66
aeronautics: Advisory Committee on
 Aeronautics, 11, 59–60, 61–2, 63,
 71, 73, 74, 342, 349; Aeronautical
 Research Committee, 349, 355,
 397; 'H Department' at
 Farnborough ('Chudleigh
 Mess'), 342–3, 350; internal
 combustion engine and, 31–2, 42;
 lasting achievements in, 10,
 355–6; Lindemann and, 342–4;
 principles of aerodynamics, 31,
 61; research scholarships in, 61;
 universities and, 33; *see also*
 aviation
HMS *Africa*, 68
agriculture, 20, 196, 338
Air Ministry, 348, 351, 355; Director
 of Scientific Research at, 354–5;
 openness to scientific ideas,
 355–6
aircraft: Albatros DI, 94–5;
 Antoinette, 50; Avro, 48, 70, 73;
 BE (Blériot Experimental) series,
 3, 4, 65, 71, 73, 90–1, 301, 344;
 Blériots, 3, 4, 50–1, 65, 66, 70, 80;
 'Bristols', 56, 66, 70, 95; cockpit
 instruments, 79; criticisms of
 British design, 73, 90–1; DH2s,
 91; Farmans, 4, 42–3, 66, 70, 80,
 91; FE1 (Farman Experimental
 1), 64; FE2b (Farman

Experimental 2b), 64, 91;
 Fokkers, 90, 91, 94, 95; Gotha
 bombers, 348; Handley Page
 bombers, 335, 336; Hurricane
 fighter, 46; Lancaster bomber, 48;
 machine guns and, 90, 95;
 national markings on, 82;
 Nieuports, 91, 95; 'pusher' and
 'tractor' planes, 64–5, 70, 90; RE
 series, 73; RFC competition at
 Larkhill (1912), 70–1; Short
 Brothers, 45; Sopwiths, 46, 91, 95,
 348; Tizard and, 347–8; Vickers,
 90, 95, 335–6; Wright Flyers, 52,
 57, 62
aircraft engines, 3, 4, 8, 31–2, 41, 42,
 50; Antoinette, 47–8, 57; Anzani,
 51; Renault, 64, 73; Rolls-Royce,
 46
airfields, 79
airships, 56, 61, 68, 73
Aitken, Max, Lord Beaverbrook,
 281–2, 310–11, 313, 360, 412;
 dispute with Balfour, 325;
 Ministry of Information and,
 322–4, 325–6, 330
Alcock, John, 335–6
Aldershot, Cambridge Military
 Hospital, 233, 234–7
All Souls, Oxford, 127
Allenby, General, 97, 230, 311, 324
aluminium, 61, 159, 167
Amery, Leo, 288
Amiens, 4–5, 78
ammonia, 20, 166–7, 174
amputation, 213, 220, 223, 229, 231;
 prosthetics and, 232–3
Anderson, John, 219
Andrade, Edward, 153
Anglo-Persian Oil Company (later
 BP), 68, 124
Anti-Aircraft Experimental Station,
 352
antiseptics, 21, 157, 213–14, 219,
 226–7, 337, 338–9
Antoinette aircraft, 50
Antwerp, 197

appeasement policy, 1930s, 353
Appleton, Edward Victor, 150
argon, discovery of, 6
armaments and gunnery *see*
    munitions
Armentières, bombardment of, 188
armoured cars, 197
army, British: 1912 summer
    manoeuvres, 71–2, 73; Air
    Battalion, 67, 69; aviation and,
    53, 56, 57–8, 59, 60, 65–7; Balloon
    Factory, Farnborough, 52–3, 56,
    57, 58–9, 63, 64, 71; ballooning
    and, 52–3; communications at
    front, 163–4; as conservative-
    minded, 35, 36, 164; de
    Havilland as aircraft designer,
    64–5, 71; 'declinist' view of inter-
    war Britain, 353–4; development
    and use of poison gas, 179–80,
    181–3, 184, 185, 187–8, 203, 219,
    288, 337; development of gas
    masks, 178–9; Director of
    Scientific Research at, 354;
    executions for cowardice or
    desertion, 264–6; film
    cameramen with, 11, 293–4, 296,
    297, 298–309, 311, 314; first day
    of the Somme, 202–3, 228–9,
    250–1, 301–2, 306; GHQ at St
    Omer, 151; grenades and,
    158–60; Haig's *Field Service
    Regulations*, 54–5; Haldane's
    reforms, 54–5, 77, 215, 216; July
    1918 counterattack, 96–7, 208,
    231–2; khaki uniforms, 78;
    Kitchener's new army, 202–3;
    light mortars and, 161–3; 'Lions
    led by donkeys' phrase, 164;
    machine guns and, 154, 172, 192;
    mustard gas casualties, 187; new
    tactics at Cambrai, 206;
    obstructiveness to tank
    development, 198, 199, 204, 210;
    officers and 'other ranks', 36–7;
    'Pals' Battalions', 203, 250–1, 314;
    poison gas and, 177, 178–9;
    recruitment at start of war,
    269–71, 273; Territorial Force, 54,
    215, 216, 218, 221; as under-
    gunned in 1914, 153; *see also*
    British Expeditionary Force
artillery: accuracy of, 37, 78; aerial
    spotting for, 88, 98; Battle of
    Cambrai (1917) and, 206–7, 208;
    Battle of the Somme (1916) and,
    202, 302; the 'Big Bertha', 161,
    207–8; creeping barrage, 206,
    207, 208; explosions caused by,
    165–6; fifteen-inch guns on
    dreadnoughts, 38; German, 153,
    161, 207–8; heavy howitzers,
    153; heavy mortars, 37, 161;
    light mortars, 161–3; mobile
    field guns, 153–4; at
    Passchendaele, 205; shell
    shortages, 117, 156, 287; shells,
    9, 155, 165–7, 170, 172, 189;
    'sound ranging' and, 151–3, 206;
    static war and, 193, 194; against
    tanks, 206
artists, 11, 235, 272, 312–13, 318, 332,
    333; *Gassed* (Sargent), 313; *The
    Menin Road* (Nash), 313; *The
    Western Front* (Bone anthology),
    313; *The Ypres Salient at Night*
    (Nash), 313
Asdic, underwater sonar, 354
aspirin, invention of, 20
Asquith, Herbert, 54, 59, 67, 198, 271,
    273, 275, 278; collapse of Liberal
    Government, 117, 156, 287;
    ousted as Prime Minister, 315,
    322
atom bombs, 175, 346, 357
Austin, Herbert, 30, 44
Australia, 17, 106–7, 149, 311
Austro-Hungarian empire, 75, 76–7,
    161; disintegration of, 327; ethnic
    groups within, 326–7
aviation: 1912 summer manoeuvres,
    71–2, 73; Advisory Committee
    on Aeronautics, 11, 59–60, 61–2,
    63, 71, 73, 74, 342, 349; 'Aerial

aviation: – *continued*
    Navigation' sub-committee,
    58–9; Alcock and Brown cross
    the Atlantic, 335–6; Allied total
    air supremacy (1918), 96–7;
    arms manufacturers and, 55–6;
    'Aviators' Neurasthenia', 257–8;
    Balloon Factory, Farnborough,
    52–3, 56, 57, 58–9, 63, 64, 71;
    bomb-sights on bombers, 344,
    347; British army and, 53, 56,
    57–8, 59, 60, 65–7; British
    pioneers, 34, 41–2, 43–8, 49–50,
    57–8, 357; Capper and, 53, 56;
    Churchill and, 68, 69, 72; civil
    aviation industry, 336, 338; de
    Havilland and, 64–5, 71;
    'declinist' view of inter-war
    Britain, 353–4; fatal accidents,
    62–3; first Channel crossing
    (1909), 3, 50–1; first powered
    flight by Wright brothers, 32, 42;
    flying from ships at sea, 68, 72;
    France as centre of in Europe,
    42–3, 45, 49; French annual
    manoeuvres (1910), 66;
    Germany and, 56, 58; Haldane
    and, 57–8, 59, 60, 61, 62, 63;
    Handley Page and, 48, 335, 336;
    industry, 10, 55, 63, 73, 336, 338;
    inter-war science and
    investment, 355; Lindemann's
    work on fatal spins, 343–4;
    meteorology and, 350; military
    market, 51–2; Moore-Brabazon's
    historic flight (1909), 41–2, 43,
    59, 73–4, 336; moustaches and,
    70; naval, 62, 67, 68, 69, 72–3; as
    new science, 38, 44; Northcliffe
    and, 32, 49–50, 58, 62; rapid
    advances in, 8–9, 10, 55, 62, 74,
    335–6; RFC established (April
    1912), 69–70; seaplanes, 69,
    72–3; Short brothers and, 62, 67,
    72; Wright brothers and, 52, 53;
    see also aeronautics
Avro, 48, 70, 73

Babinski, Joseph, 23, 262
bacteriology, 14–15, 21, 214; shrapnel
    wounds and, 219–20, 221
Baker, Professor, 179–80
Baker-Wilbrahim, Philip, 127
Baldwin, Stanley, 325–6
Balfour, Arthur, 32, 53, 142, 143, 144,
    317, 326, 328; Balfour Declaration
    (1917), 170; Beaverbrook and,
    325; as First Lord of the
    Admiralty, 118, 201, 294, 296–7
Ball, Albert, 93–4
Balliol College, Oxford, 283
ballooning, 30–1, 44, 47, 52–3, 56, 57;
    drachen or 'sausage balloons', 89
barbed wire, 37, 83, 93, 193, 195, 198
Barclays Bank, 324
Baring, Maurice, 80
Barker, Pat, *Regeneration* trilogy, 257;
    Life Class and *Toby's Room*, 405
Barrie, J.M., 280
Barrow-in-Furness shipyards, 55
BASF (Badische Anilin- und Soda-
    Fabrik), 20, 167
Battenberg, Prince Louis of, 276
battleships: armour plating, 38, 115,
    120, 196; dreadnoughts, 38, 114,
    119; inadequacy of British
    armour, 115, 120; Queen
    Elizabeth class, 68
Bauer, Colonel Max, 174, 175, 176
Bawdsey Manor, Suffolk, 351
Bayer, 20, 167, 175, 185
Beatty, Vice Admiral, 112–13, 114, 115,
    116, 119, 120, 122
Beaverbrook, Lord *see* Aitken, Max
Beilby, George, 180
Belgium, 77, 161, 192–3, 197, 273, 274,
    275, 300
Bell, Alexander Graham, 17
Bell, Ed, 143–4
Belloc, Hilaire, 286
Bennett, Arnold, 280, 324
Benz, Karl, 27
Bergson, Henri, 277
Berlin University, 14, 341–2, 346
Bernstorff, Count, 131, 142

Bethlem Royal Hospital, 256, 332–3
Beveridge, William, 338
bicycles, 26–7, 29, 45
The Bioscope, 304
Birch, Frank, 127
Blackburne company, 63
Blackett, Patrick, 352–3, 361
Blackpool Pleasure Beach, 192
Blériot, Louis, 3, 50–1
Blériot company, 3, 4, 50–1, 65, 66, 70, 80
Bletchley Park, 123, 128, 339
Blücher (German warship), 114, 115
Board of Trade, 7, 338
Boelcke, Lieutenant Oswald, 90, 94
Boer War, 25, 53, 57, 69, 194, 214, 215, 218, 269, 273, 285
Bone, Muirhead, 313
Boots the chemists, Nottingham, 186
Bosch, Carl, 20
Bottomley, Horatio, 276
Boulogne Base Hospital, 224, 226, 244, 252–3, 338
Bovington, Dorset, 205
Bowlby, Sir Anthony, 218–19, 223, 227, 228, 229–30, 234, 362
Boyle, Robert, 6
Brabazon, John Moore-, *see* Moore-Brabazon, John
Brade, Sir Reginald, 310
Bragg, William Lawrence, 149–51, 362-3; 'sound ranging' and, 151–3
Bragg, William (senior), 149–50, 154–5
Brasenose College, Oxford, 285
'Bristol' aircraft (British & Colonial), 56, 66, 70, 95
Britain: anti-German hysteria, 274–7, 278, 282, 333, 342; Belgian neutrality and, 77, 273; Belgian refugees, 274; cinema industry, 26, 291; 'declinist' view of inter-war period, 353–4; development of the tank in, 196; developments in mass communications, 24; Edwardian era, 13–15, 16–17, 18–26, 28–30, 32–8, 150–1, 281, 284; fading industrial supremacy, 33–4; internment of Germans, 276; official war aims, 319–20; propaganda discredited in post-war era, 333, 334; public schools as anti-technology, 32–3, 34–5, 36; role of state post-War, 35, 338, 354; royal family changes name to Windsor, 277; spread of motoring, 28, 29–30, 35; Victorian era, 6, 13, 15–16, 17, 21–2, 24–8, 37, 167, 174, 215, 232
British & Colonial Aeroplane Company, 56, 66, 70
British and Colonial Film Company, 301
British Association for the Advancement of Science, 102
British Board of Film Censors, 293
British Broadcasting Corporation (BBC), 336–7
British Dyes, 170, 337, 338
British Empire, 17, 53, 55, 58, 75–6, 217, 269
British Expeditionary Force, 4, 5, 54, 72, 78–9, 192, 271; ambulance provision for, 216–17; Battle of Le Cateau (1914), 217; Battle of Mons and, 81–2, 217, 272–3; crosses channel, 77–8; shell shock and, 241–2
British government: coalition formed (1915), 117–18, 277, 287; collapse of Asquith government, 117, 156, 287; Department of Information, 317–19, 320, 321; Department of Scientific and Industrial Research, 157, 349, 350, 354; Foreign Office, 125, 142, 143, 279–80, 316, 317–18, 323, 325, 339; Home Office, 67, 125, 135–6; Lloyd George becomes Prime Minister (1916), 315–16; Ministry of Information, 322–4, 325–32; tradition of secrecy, 279–80; War Cabinet, 315–16, 317, 320, 324, 327, 331; *see also* Admiralty; Air Ministry; Ministry of Munitions; War Office

British Museum, 102
Brittain, Vera, 187, 225
broadcasting industry, 336–7, 338, 357
Broadway cinema, Hammersmith, 307
Brock, Arthur, 258
Bromhead, Captain, 309
Brooke, Rupert, 76
Brooke-Popham, Major Robert Henry, 79
Brooklands, Surrey, 46, 47
Brown, Arthur Whitten, 335–6
Brown, Des, 266
Brown, Dr William, 252, 258
Bryce Report, 407
Buchan, John, 284–7, 288–9, 303, 316, 324, 363; director of Department of Information, 317–19, 320, 321, 330–1; *Greenmantle*, 331; *The Thirty Nine Steps*, 286, 408
Bulgaria, 297, 329
Burlington House, 5–8

Cambridge Scientific Instrument Company, 59
Cambridge University, 5–6, 14, 35, 43, 44, 59, 97, 127, 154, 234, 279, 347; Cavendish Laboratory, 345, 352–3; department of psychology, 23, 242, 256; engineering at, 33, 102, 187; Trinity College, 33, 45, 149, 150, 352
Cammells, 196
Campbell, Charles, 84–5
Campbell-Bannerman, Henry, 53–4
Canadian troops, 173, 176, 177, 281, 282, 323
Canadian War Records Service, 282, 323
Capper, Lieutenant-Colonel John, 53, 56, 57, 58–9, 63
Carnarvon, Lord, 64
RMS *Carpathia*, 19
Carranza, President of Mexico, 140, 142, 143
Carrell, Dr Alexander, 227

Carson, Sir Edward, 129, 320, 321, 322
Casement, Sir Roger, 129–30, 131–2; trial and execution of, 132–3
Castner-Kellner, Wallsend, 180
cathode ray oscilloscope, 350, 351
cavalry, 5, 54, 78, 81, 84, 151, 206, 218, 272
Cavell, Edith, 275
Cavendish Laboratory, Cambridge, 345, 352–3
Cayley, Sir George, 31
celluloid, 26, 88, 178
cellulose, 19
cellulose nitrate (gun cotton), 78
censorship, 11, 12, 271–2; of overseas mail, 135–6
Chaplin, Charlie, 291, 312
Charles II, King, 6
Charteris, Brigadier John, 87, 179, 298
chemical industries, 15, 19, 20, 169; chemical warfare and, 174, 185; development of in Britain, 168, 169–72; foundations during War, 337–8; German domination of, 20, 21, 167–8, 172; Haber-Bosch process, 20, 167, 174; synthetic colours and materials, 19–20, 26, 168, 338; US ascendancy post-war, 172, 338
chemical warfare: history of, 173–4, 403; research station at Porton Down, 354; *see also* poison gas
chemists, 6, 7–8; committee of industrial chemists, 168; development of gas masks, 178–9; French, 185; German, 20, 21; poison gas and, 175–6, 179–80, 184; running of munitions factories by, 170; shortage of in Britain, 168; tear gas and, 174–5, 188
Cherwell, Lord *see* Lindemann, Frederick
Chesterton, G.K., 280
Childers, Erskine, *The Riddle of the Sands*, 105, 286

chlorine, 173, 175–8, 180, 181, 184, 189, 354

chloroform, 235

chloropicrin, 189

cholera, 21–2

Christian Socialist movement, 279

Christ's College, Cambridge, 279

Churchill, Winston: creates Chemical Warfare Committee, 188; development of the tank and, 196–7, 198–200, 210; as First Lord of the Admiralty, 67–9, 72, 104, 117–18, 169, 196–7, 271–2; as Home Secretary, 67, 125; learns to fly, 68, 79; Lindemann and, 341, 345–6; on naval battles, 114; naval intelligence and, 108, 110–11, 112–14, 115, 118, 122; on Northcliffe's war reporting, 273; Press Bureau and, 271, 273; on radio communications, 104; sacking of from Admiralty, 117–18, 200, 277; Second World War and, 123

cinema: *The Battle of the Ancre and the Advance of the Tanks*, 310; *The Battle of the Somme*, 301–10, 314, 410; *Bei unseren Helden an der Somme (With Our Heroes on the Somme)*, 310; *Britain Prepared*, 294–8, 309; colour film, 295, 296; *A Day in the Life of a Munitions Worker*, 311; Department of Information and, 318; developments in 1895-1914 period, 290–1; documentary films, 291, 294–8, 301–11, 314; 'electric picture palaces', 290; fake footage of 'over the top' moment, 303–4, 306, 308; film cameramen at the front, 11, 293–4, 298–309, 311, 314; film cameramen with army, 11, 293–4, 296, 297, 298–309, 311, 314; film cameramen with Royal Navy, 294, 295–7; German propaganda films, 292–3, 310; *The German*

Retreat and the Battle of Arras, 310; Hollywood, 291, 312, 357; images of the dead and the dying, 304, 307, 308; initial banning of film-makers from front, 292, 293, 295; international nature of, 291, 292; invention of, 26; limitations of film cameras, 299; *Mrs John Bull Prepared*, 311; newsreel companies in Britain (Topical companies), 291–2, 293–4, 300–1, 311; NWAC and, 330; official films at Imperial War Museum, 332, 333; *Pictorial News (Official)*, 311; popularity of, 35, 290, 305–6; recruitment pictures, 292; in Second World War, 334; severe restrictions at front, 299–300; standard 50mm lens, 299, 306; US Army Signal Corps film unit, 312; War Office Cinematographic Committee, 310–11; War Office Topical Film Committee, 293, 301, 304; Wellington House Cinema Committee, 294

Circuit des Ardennes, 28, 44

Clarendon Films, Croydon, 300, 303

Clarendon Laboratory, Oxford, 345

Clark, Alan, 164

Clarke, Russell, 105–6

Clarke, William, 118–19

coal mines, industrial health in, 10

coal tar, 20

code-breaking, 8, 11, 12; Admiralty and, 11, 102–3, 106–14, 117, 118–19, 121–3, 125, 126–7, 128, 138, 339; capture of code books, 106–8, 116–17, 125, 126, 131; diplomatic service code books, 125, 126, 131, 144, 145; Ewing and, 101, 102, 103, 106, 108, 126; Government Code and Cypher School, 339; Government Communications Headquarters (GCHQ), 339–40; legacy of, 339–40; at War Office, 126

Cody, Samuel F., 56–7, 58, 59, 70
Coke, Sir Charles, 130–1
Cold War, 339
Columbia Broadcasting System
    (CBS), 338
Committee of Imperial Defence, 58,
    69, 194
communications and signals, 15, 18,
    19, 24, 52, 163, 354; on Eastern
    Front, 103–4; failures of the
    command-and-control system,
    164; interception of, 2, 8, 102,
    103–6, 109–14, 116–17, 118–19,
    121–3, 138, 163; interception of
    diplomatic cables, 110, 111, 125,
    126–7, 131–2, 138, 139–44, 145–6;
    lack of improvements in, 163–4;
    see also intelligence; radio
    (wireless telegraphy)
conscription, 168, 357
Conservative Party, 53, 117–18, 277, 315
Constantinople (Istanbul), 124
Cook, Sir Edward, 287–8
cordite, 166, 169–70
HMS Cornwallis, 131
Courtney, Captain Ivon, 79
courts martial for cowardice or
    desertion, 263–5
Coventry Sewing Machine Company,
    27
Craiglockhart Hospital for Officers,
    256–9
cricket, 36, 273, 318
Crimean War, 37, 215
Crippen, Dr, arrest of (1910), 19
Crompton, Colonel, 199
Crookes, Sir William, 7
Crystal Palace Technical College,
    London, 34, 63
CS Alert, 1–2, 101
Cunliffe-Owen, Sir Hugo, 324
Curie, Marie, 21, 341

Daily Chronicle, 316
Daily Express, 51, 281, 322
Daily Mail, 25, 49, 50, 51, 62, 276, 281,
    287, 310

Daily Mirror, 25, 275–6
Daily News, 279, 288
Daily Telegraph, 127, 274
Daimler, Gottlieb, 27
Daimler-Knight petrol engine, 201
Dakin, Henry, 227
Darracq Motor Company, 43
Dartmouth Naval College, 102
Darwin, Charles (grandson of
    evolutionist), 153
Darwin, Horace, 59, 156
de Grey, Nigel, 127, 139–40, 142, 144,
    366
de Havilland, Geoffrey, 34, 63–5, 71,
    91
de Valera, Eamon, 132
Defence of the Realm Act (DORA),
    105, 271–2
Dejerine, Jules, 23, 242, 255, 262
DeMille, Cecil B., 312
Denniston, Alexander, 102–3, 109, 364
Denniston, Mrs Alexander, 128
Deperdussin aircraft, 70
Deptford generating station, 16
Derby, Lord, 227
Derflinger (German warship), 114, 115
d'Eyncourt, Eustace Tennyson, 198,
    200, 210
Dickens, Charles, Our Mutual Friend,
    340
Dickson, Bertram, 66
diphosgene, 185, 189
Diplock, Bramah, 196
dirigibles, 58
disease and fever, 15, 21–2, 214–15,
    216; bacteria in shrapnel and,
    219–20, 221; gas gangrene, 220,
    231; inoculations, 216
DNA, discovery of, 153
Dodd, Francis, 313
Donald, Robert, 316–17, 320
Doncaster international air show
    (1909), 62
Dover, 1, 2, 3, 51, 156
Downing College, Cambridge, 97
Doyle, Sir Arthur Conan, 133, 195,
    280, 286

Driver, Hugh, 57
Duchess of Westminster's War
    Hospital, Le Touquet, 242–3
Dulux paint, 338
Dunne, John William, 57, 59
DuPont, 338
dyes, synthetic, 19, 168, 170
dynamite, 166
dysentery, 214

Earp, Private, 264
Eastchurch, Isle of Sheppey, 67, 68, 72
Eastern Telegraph Company, 17
Eastman, George, 26, 83–4
Edgerton, David, 354
Edison, Thomas, 16, 26, 191
education system, British, 24, 51, 349;
    public schools as anti-
    technology, 32–3, 34–5, 36;
    technical schools, 34, 48, 63
Edward VII, King, 32
Egypt, 97, 98, 269, 311
Ehrlich, Paul, 21
Einstein, Albert, 14, 149, 341
Electric Lighting Company, US, 191
electricity, 10, 15–16, 356; domestic
    consumption, 16–17, 357;
    electrical industries, 16–17, 21;
    generating stations, 16; as key
    technology, 35; motor cars and,
    29; power stations, 16;
    replacement of gas street
    lighting by, 16, 356
electrocardiograph devices, 21
endocrinology, 261
engineering, 9, 10, 12; at Cambridge,
    33, 102, 187; precision, 29;
    vocational training in, 33–4
engines, internal combustion, 15,
    27–8, 29, 30, 31–2, 33, 43, 60, 61,
    123, 201, 356; see also aircraft
    engines
epilepsy, 23
Esher, Lord, 58
ethnographic studies, 242, 256
Evelyn, John, 6
Evening News, 305, 310

Evening Standard, 322
Ewing, Sir Alfred, 101–3, 105, 106,
    108, 126, 364–5
HMS *Excellent*, 354
explosives, 9, 20, 33, 78, 155, 165–8,
    169, 170–2, 337, 338; amatol, 167,
    170; ammonal, 167, 171; Ballisite
    (propellant), 166

Fairbanks, Douglas, 312
Fairey, Richard, 34
Falkenhayn, General Erich von, 175,
    176, 193, 207
Faraday Laboratory, Royal
    Institution, 346
Farman, Henri, 42–3
Farman aircraft, 4, 42–3, 66, 70, 80, 91
Farnborough: Airship Company
    based at, 67; Balloon Factory at,
    52–3, 56, 57, 58–9, 63, 64, 71; RFC
    School of Photography at, 91–2;
    Royal Aircraft Establishment at,
    355; Royal Aircraft Factory at, 71,
    73–4, 90–1, 342–3, 350
Farr, Gertrude, 264, 265
Farr, Harry, 264, 265
Faunthorpe, Captain, 298, 301
Faversham, 'the Great Explosion' at,
    171
Ferranti company, 16
fertilisers, agricultural, 20, 338
Ffoulkes, Charles, 331, 332
film-makers see cinema
Finsbury Technical College, London,
    34, 48
First World War: American
    Expeditionary Force, 146;
    Armistice (November 1918), 97,
    208–9, 259, 329–30; British
    declaration of war (4 August
    1914), 1, 6, 77, 105, 129; British
    public support for, 273–4, 284,
    314; casualty figures, 93, 176,
    177, 179, 187, 189, 202, 217, 228,
    231, 240, 251; causes of, 76–7;
    Eastern Front, 103–4, 174–5, 189,
    206–7; first aerial combat, 89–90;

First World War: – *continued*
Hindenburg Line, 206; initial
patriotic fervour, 75, 76; outbreak
of (August 1914), 1, 6, 75–8, 105,
129, 150–1, 269–71; Russia's
departure from (1918), 95, 207,
321–2; scale of conflict, 38;
trenches on Western Front, 9,
83–4, 87–8, 93, 154, 161, 174, 193,
194–5, 222; US entry (6 April
1917), 97, 145, 207, 322, 328; vast
scale of casualties, 213, 230;
Ypres salient, 175, 179
First World War (battles and
campaigns): Allied counter-
offensive (July 1918), 96–7, 208,
231–2; Arras (1917), 95, 188, 230;
Britain's first offensive act (1914),
1–2; Cambrai (1917), 153, 205–7,
208, 209, 210, 321; Caporetto
(1917), 321; Dogger Bank (1915)
(naval), 113–16, 118; Gallipoli
(1915-16), 116, 117; German
advance through Belgium
(August 1914), 77, 161, 192–3;
German retreat in West (1918),
97, 208–9, 329; German spring
offensive (1918), 95–6, 188,
207–8, 231; Jutland (1916)
(naval), 118–22, 123, 138, 238,
352; Le Cateau (1914), 217; Loos
(1915), 181–3, 203, 227–8, 288; the
Marne (1914), 193; the Menin
Road Bridge (1917), 172; Mons
(August 1914), 81–2, 217, 272–3;
Neuve-Chapelle (March 1915),
87, 89, 117, 156, 287; Palestine
campaign (1917-18), 97–8, 311,
324–5; the Somme (1916), 93, 138,
160, 165, 167, 185, 202–4, 228–30,
236, 250–3, 264–5, 289, 301–10,
314, 315; Tannenberg (1914),
103–4; Verdun (1916), 138, 163,
185; Vimy Ridge (1917), 153; First
Battle of Ypres (October 1914),
242; Second Battle of Ypres
(1915), 175, 176–9, 282, 288; Third

Battle of Ypres (Passchendaele)
(1917), 153, 186–7, 205, 207,
237–8, 253, 311, 321
Fisher, Lord, 102, 111, 112, 113, 115,
118
flags, 163
Fleming, Alexander, 226, 338–9
Fleming, Professor John Ambrose, 10,
11, 365
Flint, Charles R., 52
Foch, Marshall, 207, 208
Fokker aircraft, 90, 91, 94, 95
food and drink, 35, 78, 157
Food Investigation Board, 157
Ford, Henry, 30
Foster & Co, Lincoln, 199, 200
Foulkes, Colonel Charles, 180, 181,
182
Fowler, R.H., 352
France: aircraft engines, 47–8; annual
manoeuvres (1910), 66;
ballooning in, 30–1, 44; as centre
of aviation in Europe, 42–3, 45,
49; cinema in, 26, 291; defeated
by Prussia (1870-1), 38;
development of motor car, 27,
43, 45; intellectuals and writers
support war, 277; motor races in,
28; patriotic fervour in, 75, 277;
population growth, 38;
Schlieffen Plan and, 77;
treatment of mental conditions
in, 23, 261–2; war correspondents
at front, 288; Wilbur Wright's
aeroplane displays in, 32, 64; *see
also* French army
Franz Ferdinand, Archduke,
assassination of, 77
Fraser, Lionel, 127
Freemantle, Michael, 165
French, General Sir John, 5, 81–2, 83,
193, 287
French army: aerial photography
and, 84, 97; Algerian divisions,
173, 176; annual manoeuvres
(1910), 66; artillery of, 153–4;
Battle of Mons and, 81–2, 272;

BEF on left flank of, 77, 79, 192; development of the tank and, 209; infantry uniforms, 78; July 1918 counterattack, 208; machine guns and, 154; poison gas and, 173, 176, 180, 185; protection against poison gas, 177–8; shell shock and, 261–2, 266; size of, 38; 'sound ranging' and, 151
Freud, Sigmund, 14, 23–4, 255
friendly fire incidents, 82
Fuel Research Board, 157
fuel supply industry, 29, 349
Fuller, Colonel J., 205
Fulton, John, 66

Gallipoli, 116, 117
Galsworthy, John, 280
gas engines, 27
gas masks, 177, 181, 184, 185–6
Gaumont, 291–2, 293–4, 300
gelignite, 166
gender reassignment surgery, 339
General Electric Company (GEC), 16–17, 336–7
George V, King, 277, 295, 298, 332
German army: advance through Belgium (August 1914), 77, 161, 192–3; aerial observation techniques, 89; anti-tank tactics, 206; atrocity and barbarism stories, 274–5, 278, 330, 333; British propaganda leaflets and, 329, 333–4; field grey uniforms, 78; film cameramen with, 310; grenades and, 158; heavy artillery of, 153, 161, 207–8; impact of of British tanks, 203; light mortars, 161; machine guns and, 154, 192; Pioneer Regiment, 175–7, 184; shell shock and, 261, 262–3, 266; size of, 38; 'stabbed in the back' argument, 333–4; use of mustard gas, 186–7, 188; use of poison gas, 173, 175–7, 179, 288; as victims of poison gas, 183
German Navy: challenge to British

supremacy, 25, 75; code books, 106–8, 116–17; radio communications, 104–6, 108–9, 110–17; shelling of Scarborough and Hartlepool, 112, 113; use of code, 101, 103, 105–8
Germany: armament production, 154; aviation and, 56, 58; British naval blockade of, 118, 122, 135, 136, 167; communications interception, 103–5; diplomatic mission in USA, 125, 129, 131–2, 138, 139–42; diplomatic service messages intercepted, 110, 111, 125, 126–7, 138, 139–44, 145–6; domination of chemical industry, 20; at forefront of technological changes, 38; Harmsworth's warnings on, 25; intellectuals and academics support war, 277–8; invention of motor car, 27; Irish nationalists and, 129–30; Mexico and, 139, 140–1, 142, 143, 145; patriotic fervour in, 75; population growth, 38; propaganda in Nazi era, 334; Schlieffen Plan, 77; 'stabbed in the back' argument, 333–4; tank development in, 209; *Technische Hochschulen* (technical schools), 34, 341; telegraphic cables cut, 1–2, 8, 101, 104, 138; unrestricted submarine warfare, 122, 138–40, 142, 146; use of code, 8, 101, 102, 103; vocational training in, 34; Zimmermann telegram, 139–42, 143–6
Gilbert, Sandra, 171
Gillies, Harold, 233–9, 339, 365–6
Glasgow University, 285
Glazebrook, Dr Richard, 59, 156
gliders, man-carrying, 31, 32
Goebbels, Joseph, 334
Gonville and Caius College, Cambridge, 234, 242
Gordon Bennett races, 28–9, 43
Gough, General Sir Hubert, 251

Government Communications
   Headquarters (GCHQ), 339–40
Graves, Robert, 401-2
Greece, 74
Green, Frederick, 63, 64
Gretna munitions factory, 170
Grey, Edward, 54
Grierson, General Sir James, 71–2
Griffith, D.W., 312; *The Birth of a
   Nation*, 291, 305
gunpowder, 166

Haber, Fritz, 20, 174–5, 176, 184, 187
Haggard, Lieutenant, 127
Haggard, Henry Rider, 306
Hague Conventions (1899/1907), 174,
   175
Hahn, Otto, 175
Haig, Douglas, 54, 65, 72, 87, 172, 301;
   Allied counter-offensive (July
   1918), 208; 'backs to the wall'
   Order (1918), 96, 207; becomes
   British commander-in-chief on
   Western Front, 202, 228, 288–9;
   executions for cowardice or
   desertion and, 264–5; *Field
   Service Regulations*, 54–5; poison
   gas and, 181–2, 203, 288; tanks
   and, 203, 204; treatment of the
   wounded and, 218, 228, 230
Haldane, Professor John, 10, 168, 178,
   181, 367
Haldane, Richard Burton, 53–5, 57–8,
   59, 60, 61, 62, 63, 77, 349; accused
   of pro-German sentiment, 276–7;
   on Keogh, 221; medical reforms
   of, 215, 216; Royal Aircraft
   Factory and, 73
Halifax bomber, 48
Hall, William, 109–10, 111, 115, 116,
   123; Casement's 'Black Diaries'
   and, 133; diplomatic service
   intelligence and, 125, 126–7, 138,
   139–44, 145–6; eastern
   Mediterranean network of, 138;
   Ireland and, 129–32; Mata Hari
   and, 134–5; monitoring of

overseas mail, 135–6;
   recruitment of women, 127–8;
   Sayonara escapade and, 131–2;
   use of misinformation, 110, 116,
   136–8; *Vergemere* cruise to Spain
   and, 133–4; Zimmermann
   telegram and, 139–42, 143–4,
   145–6
Hambro, Lady, 128, 324
Hambro, Sir Eric, 324
Hamshaw-Thomas, Hugh, 97–8, 367-
   8
Handley Page, Frederick, 34, 48, 55,
   335, 336, 368
Hands, Charlie, 50
Hankey, Maurice, 194–5, 198, 316, 324
Hanriot aircraft, 70
Hardinge, Lord, 142, 325
Hardy, Thomas, 280
Harmsworth, Alfred, *see* Northcliffe,
   Lord
Harmsworth Cup, 49
Harrison, Edward, 186, 368-9
Hart, Basil Liddell, 192, 210
Hart, William S., 312
Hartlepool, 112, 113
Harvey, Joan, 128
Hawker, Harry, 46, 55
Head, Henry, 242
Headlam-Morley, John, 327, 369
Hearst, William Randolph, 309
Heath, Frank, 157–8, 349, 354, 370
Hegel, philosophical school of, 54
HMS *Helga*, 132
Henderson, Miss, 128
Henderson, General Sir David, 69–70
Henry, Sir Edward, 180–1
Hepworth, Cecil, 292
Hertz, Heinrich, 18
HMS *Hibernia*, 68
Highclere Park estate, Hampshire, 64
Hill, Archibald Vivian, 150, 352, 353,
   370-1
Hindenburg, Marshal von, 328
Hipper, Franz von, 112, 113, 116, 118,
   119, 120, 121
Hippisley, Colonel, 105

Hitler, Adolf, 190, 334
Hobart (German merchant
  steamship), 106–7
Hobsbawm, Eric, 15
Hoechst, 20, 167, 185
Holmes, Gordon, 252–3, 371
the Holocaust, 334
Holt Company, California, 196, 197
Home Front, 11–12, 288, 311, 314, 316,
  330
Home Rule movement, Irish, 129
homosexuality, 257
Hong Kong, 17
Hope, Herbert, 111, 115, 121, 123
Hopkinson, Bertram, 33, 347, 348, 372
horses, 78, 197–8, 216–17
Howell, Major Wilfred, 130
Hunstanton coastguard station,
  105–6
Hurley, Frank, 311
Hurst, Dr Arthur, 254–5, 372-3
hydrotherapy, 262
hypnosis, 244, 254, 262

Imperial Airways, 336
Imperial Chemical Industries (ICI),
  337–8, 355
Imperial College, London, 54, 61, 156,
  174, 179–80, 215–16, 345
Imperial General Staff, 55
Imperial War Museum, 331–3
HMS *Indefatigable*, 120
India, 53, 138, 217, 256, 269; Delhi
  Durbar (1911), 295
industrial production, 16–17, 19–20,
  27, 28–30, 33–4, 38, 74, 157,
  167–9, 170–2, 332, 337–8, 353–4;
  health issues and, 10, 171; strikes
  and unrest, 14, 319
Industrial Revolution, 16, 33, 37–8
Inskip, Thomas, 127
intelligence, 10, 11; Churchill and,
  110–11, 112–14, 115, 118, 122;
  failures in processing of, 112–16,
  118, 121–2, 123; Henderson's
  principles, 69–70; interception of
  diplomatic cables, 110, 111, 125,

126–7, 138, 139–44, 145–6;
  propaganda and, 318; Secret
  Service Bureau, 125; 'the
  surveillance state', 10, 339–40;
  use of misinformation, 110, 116,
  136–8
Intelligence Corps, 84–5, 289
International Film Corporation, 309
HMS *Invincible*, 120
Ireland, 128–32, 281; Easter Rising
  (April 1916), 132, 298; Irish
  nationalists, 274; Irish
  Volunteers, 129
Irvine, James, 219
Italy, 77, 311, 321, 327

Jackson, Captain Sir Thomas, 118–19
Jackson, Colonel Louis, 155, 179, 200
Japan, 139, 140; Russo-Japanese War
  (1904-5), 38, 116, 161, 194, 295
Jardine, Brigadier J.B., 251
Jellicoe, Admiral Sir John, 111–12,
  114, 115, 119–20, 121, 122, 123,
  294, 296
jet engine, 10, 355, 357
Jews, 169; Balfour Declaration and,
  170
John, Augustus, 235
John Bull, 276
Johnson, William, 254
Jones, Roderick, 283–4, 318, 320, 373
Joubert de la Ferté, Captain Philip,
  2–5, 80–1, 83
Jung, Carl, 255
Jury, William, 294, 304, 310, 312, 314,
  373-4
Jury's Imperial Pictures, 293–4, 304

Kaiser Wilhelm Institute for Physical
  Chemistry, 174, 184
Kaufman, Dr Fritz, 263
Kaye, Captain H., 246–7
Kennedy, Sir Alexander, 156
Keogh, Sir Alfred, 215–16, 218, 221,
  227, 234, 237
Keynes, John Maynard, 150
Keystone Company, 291

Kiel, 112

Kine Weekly, 305

Kinemacolor, 295, 296

Kinematograph Manufacturers' Association, 293

Kings College, Cambridge, 127

Kipling, Rudyard, 277, 280

Kirkwood, Lieutenant, 251–2

Kitchener, Lord, 76, 89, 155, 172, 178, 181, 193, 201, 202; appeals to the public for funds, 217; death of (1916), 315; Press Bureau and, 271–2; recruitment at start of war, 269–71; views on the press, 271, 272; 'Your Country Needs YOU' poster, 269–71

Kling, Dr André, 178

Knox, Dilwyn 'Dilly', 127, 128, 374

Krupps, 154, 161, 196

La Pitié Hospital, Paris, 23

Labour Party, 315

Lambert, Bertram, 186

Lancashire Fusiliers, 301–2, 308

Lancaster bomber, 48

The Lancet, 243

Lanchester, Frederick, 27–8, 29, 30, 59–60, 156, 374–5

Lanchester Engine Company, 28

Lane, Sir William Arbuthnot, 236, 237–8

Langrishe, Sir Hercules, 133–4

Lansing, Robert, 144

lantern shows, 26

Larkhill field, Salisbury Plain, 66, 67, 70–1

Latham, Hubert, 50

Law, Andrew Bonar, 315

Lawrence, Colonel T.E. ('of Arabia'), 97, 124, 324–5

Laws, Victor, 85, 91–2

Le Queux, William, *Spies of the Kaiser*, 105

Lee, Arthur, 62

Lee Enfield rifles, 37, 81

Leeds University, 33, 149, 154–5

Leete, Alfred, 269–70, 375-6

Leishman, Sir William, 216

Lewis, Cecil, 94

Lewis machine guns, 90, 205

Leyden University, 21

Leysdown, Isle of Sheppey, 41–2, 43, 45, 62

Libau (German vessel), 132

Liberal Party, 53, 67, 117, 156, 278, 279, 287, 315

Liège, Belgium, 161

light bulb, electric, 16, 191, 356

Light Locomotives on Highways Act, 27

Lindemann, Frederick, 340–6, 348, 350, 351, 376-7

linguistics, 11, 127–8

HMS *Lion*, 114, 116, 120

literacy, 24, 285

literary figures, 11, 280, 284–7, 316, 324

Livens, William, 187–8

Liverpool University, 33

Lloyd George, David, 22, 156, 162, 172, 201, 305, 306; becomes Prime Minister, 315–16; Britain's official war aims, 319–20; 'Garden Suburb' of, 316; the press and, 322–4; propaganda and, 316–17, 320, 322–4, 326, 330

Lloyds the insurers, 102

Loch, Captain, 127

Lockhart, Robert Bruce, 310

Locock, Sir Guy, 136–7

Lodge, Sir Oliver, 155

London, 29, 34, 48, 135, 171, 180–1, 197, 225, 237, 279, 291, 356; 1908 Olympic Games, 102; electric underground railway, 35, 269; first regular aviation services from, 336; inequality in, 14; size of, 35

London School of Economics, 54, 283

London University, 10, 54, 61, 127, 156, 157, 174, 179–80, 215–16, 283, 345

Lonsdale, Earl of, 250–1, 270

lorries, articulated, 29

Lowry, Martin, 170
Ludendorff, General Erich, 95–6, 207–8, 333
Lumiere Brothers, 26
lunatic asylums, 23
Lutzow (German warship), 120
Lynch, J.P., 217

Macdonogh, General Sir George, 103
Machell, Lieutenant-Colonel Percy, 250–1
machine guns, 9, 37, 64, 90, 95, 154, 172; Battle of the Somme (1916) and, 202–3; Gatling gun, 191; Lewis, 90, 205; static war and, 192, 193, 194, 195, 198; Vickers, 191–2, 202, 205
Madge, H.A., 104
Madrid, 125
Magdalene College, Oxford, 346
Magdeburg (light cruiser), 107, 116
Maghull (later Moss Side) Hospital, 255–6
Majestic film theatre, Clapham, 290
Major, John, 265
Malins, Geoffrey, 293–4, 300–4, 306, 311, 314, 377
Manchester, 35
Manchester Guardian, 24, 305
Manchester University, 14, 33, 149, 151, 155, 169, 255
Mannock, Edward, 94
Marconi, Guglielmo, 18
Marconi Wireless Telegraph Company, 18, 104, 336–7
Martlesham Heath, Suffolk, 347
Mary, Queen, 295
Masefield, John, 280
Massachusetts Institute of Technology, 34
Masterman, Charles, 278–9, 280, 282–3, 284, 286–7, 318, 320–1, 377–8; cinema and, 293, 294, 297; end of war and, 330; war artists and, 312–13
Mata Hari (Margaretha Zelle), 134–5

mathematics, 8, 33, 59, 61, 152, 153, 344, 347, 352, 353; *Principia Mathematica* (Russell and Whitehead), 14, 150
Matthews, Private, 228
Maudsley Hospital, London, 249
Maxim, Hiram Percy, 191–2
McCudden, James, 94
McDougall, William, 242
McDowell, John Benjamin, 301, 303, 304, 311, 314
McKenna, Reginald, 135–6
Medical Research Committee, 11, 22, 226–7, 339
medicine: advances in Edwardian era, 15, 20–2; anaesthetics, 20–1, 219, 235; antiseptics, 21, 157, 213–14, 219, 226–7, 337, 338–9; cosmetic surgery, 339; discovery of radium, 21; drugs, 8, 11, 20, 21, 157, 219, 338–9, 357; foundations of progress laid during War, 10, 337, 338–9; preventative, 21–2, 216; research, 10–11, 22; shortage of doctors on Home Front, 239; surgery, 12, 20–1, 213–14, 218, 220–1, 224, 226–7, 228, 229, 230, 233–9, 339; X-rays, 21, 149–50; *see also* bacteriology; military medicine
mental disorders: electric shock treatment, 254, 259–60; French treatments, 23, 261–2; insanity, 23; post-traumatic stress disorder, 245; social class and, 245, 248; as un-masculine, 245, 251; *see also* shell shock
merchant shipping, 138–9, 146
metal fatigue, 33
Meteorological Office, 59
meteorology, 350
Metropolitan Carriage, Wagon and Finance Company, Birmingham, 199
Metropolitan Police, 181
Metropolitan Vickers, 336–7
Mexico, 139, 140–1, 142, 143, 145

Middle Eastern region, 68, 97–8, 124–5, 311, 324–5; Balfour Declaration (1917), 170

Middlebrook, Martin, 251

Miles, Lilian, 171

Military Intelligence 5 (MI5), 125

military medicine: ambulance trains or barges, 224, 226, 228, 230, 231, 232, 253; anaesthetics, 219, 235; antiseptics, 213–14, 219, 226–7, 337, 338–9; appeals to the public for funds, 217; army morale and, 214, 240; base camp General Hospitals, 224–5, 226, 232, 244, 249, 338; basic field dressing parcel, 221; Battle of the Somme (1916) and, 228–30, 236, 250; blood transfusion techniques, 220–1, 338; care and rehabilitation of wounded soldiers, 232–3, 234–9; Carrell-Dakin solution, 226–7; casualties through disease and fever, 214–15; the Casualty Clearing Station (CCS), 222–4, 226, 228, 229, 230–2, 241, 249, 252, 254, 303, 357; consulting surgeons to army, 218–19; convalescence and treatment in Britain, 225, 232–3, 234–9; doctors in the army, 213, 215, 239, 240; dramatic improvements during war, 239; drugs, 219, 338–9, 357; evacuation of the wounded, 222–6, 227–8, 229–30, 231–2; face reconstruction work, 233–9; Field Ambulance units, 222, 226, 228, 231; flexibility in final phase of the war, 231–2; general-specialist care balance, 233; Haldane's reforms, 215, 216; horse-drawn ambulances, 216–17; Keogh's reforms, 215–16; medical officer fatalities, 222; Medium First Aid Kit, 221; *Military Hygiene and Sanitation for Soldiers* (1908 manual), 216; mortality levels from wounds, 215; motorised ambulances, 216, 217; network of clearing hospitals behind lines, 215, 216, 221, 222; plastic surgery, 233–9; preventative, 216; private charities and, 217–18; private hospitals attached to army, 242–3; RAMC formed (1898), 215; Regimental Aid Posts (advanced dressing stations), 222, 228, 232; role of bacteria, 214, 219–20, 221; shrapnel wounds, 219–20, 221; Sir Almroth Wright's criticisms, 226–7; specialisms within CCSs, 223, 230–1; speed of treatment of wounded men, 218–19, 221, 222; stretcher-bearers, 221, 222, 226, 228; surgery, 213–14, 218, 220–1, 224, 226–7, 228, 229, 230, 233–9, 339; survival rates, 214–15, 233, 240; Territorial Force and, 215, 216, 218, 221; treatment of German prisoners, 232; volunteers at the front, 216, 217, 218, 234; X-rays, 231

Mills, William, 159–60, 378

Milne, E.A., 352

Milner, Lord, 285, 317, 320

Minerva cars, 44

mines under enemy lines, 166, 167, 302, 305–6

Ministry of Munitions, 91, 170, 172, 188, 200, 208, 315, 337; Anti-Aircraft Experimental Station, 352; establishment of, 156; Inventions Department, 156

Model T Ford, 30

Moir, Ernest, 156

Molyneux, Edward, 127

Mond, Sir Alfred, 331, 332

Montgomery, Rev. William, 127, 139–40, 142

Moore, Arthur, 272–3

Moore-Brabazon, John, 14, 41–2, 43–5, 48, 49–50, 59, 63, 73–4, 84, 336, 379; photography and, 84–5, 87, 89, 92–3

Morestin, Hippolyte, 234
The Morning Post, 161, 277
Mors cars, 44
Morse, Samuel, 17, 18
motor boats, 28
motor cars: American, 30; assembly
    plants, 29; British, 27–8, 29, 30,
    43, 45–6; coach manufacturers
    and, 29; electric starter systems,
    29; French, 27, 43, 45; fuel supply
    industry, 29; German, 27; growth
    of motor industry, 30; Northcliffe
    and, 49; racing of, 28–9, 43–4;
    A.V. Roe and, 46; speed limits in
    Britain, 27; spread of in
    Edwardian Britain, 28, 29–30, 35,
    356; standardising of
    components, 30
motor-boat racing, 49
motorcycle engines, 63
Mott, Frederick W., 249
Moulton, Lord, 168
munitions: anti-aircraft weapons,
    156, 352, 353; armour plating, 38,
    115, 120, 196, 197, 205; 'declinist'
    view of inter-war Britain, 353–4;
    explosions and health risks at
    factories, 171; Germany
    outproduces Allies (in 1914), 154;
    grenades, 158–60; Gun
    Ammunition Filling
    Department, 170; munitions
    factories in Britain, 170–2, 296,
    311, 357; muskets, 158; naval
    heavy guns, 169, 296; production
    of, 208; rifles, 37, 78, 81; tanks
    and, 198–9, 201, 202, 205;
    torpedoes, 72, 296; Trench
    Warfare Department, 155–6, 162,
    200; *see also* artillery; machine
    guns; Ministry of Munitions
Murray, Gilbert, 280
mustard gas, 186–8, 189, 190, 219,
    354
Myers, Charles Samuel, 23, 242–3,
    244, 246, 248–50, 252, 254–5, 256,
    379-80

Namier, Lewis, 283, 323, 380-1
Namur, Belgium, 161
Nash, Paul, 313
National Broadcasting Company
    (NBC), 338
National Health Insurance
    Commission, 219
National Hospital for Nervous
    Diseases, London, 23, 252, 256,
    259–60
National Insurance, 22
National Insurance Commission, 279,
    282
National Memorial Arboretum,
    Staffordshire, 266
National Physical Laboratory,
    Teddington, 6, 59, 158, 350, 354
National Security Agency, US, 339–40
National War Aims Committee
    (NWAC), 319–20, 330
Nauen transmitting station, 104–5,
    140
naval engagements, 112–16, 118–23,
    138, 238, 352
Naval War Staff, 68
navigation, aerial, 79–80, 345, 355
Nelson & Sons, Thomas, 285, 286–7,
    289
Nernst, Walther, 341–2, 346
Neuve-Chapelle, Artois region, 87,
    89, 117, 156, 287
Nevinson, Christopher, 313
New College, Oxford, 10, 127
New Guinea, 242, 256
New York, 16, 298
New Zealand, 233–4
Nicholson, General Sir William, 66–7
Nightingale, Florence, 215
nitric acid, 20, 167
nitroglycerine, 166
Nobel, Alfred, 166
Nobel Prizes for science, 6, 15, 20,
    150, 152, 166, 174, 342, 352
Nonne, Max, 262
Northcliffe, Lord: aviation and, 32,
    49–50, 58, 62, 381-2; early papers
    and magazines, 24–5; influence

Northcliffe, Lord: – *continued*
of, 25, 287; innovations in
newspaper industry, 24–5;
Ministry of Information and,
322–3, 324, 326–8, 329–30; war
coverage and, 272–3, 275
North West Frontier, 273
nuclear physics, 14, 149–50, 155, 353,
357
Nulli Secundus (airship), 56
Nuttall, J.M., 153
nylon, 338

The Observer, 25
O'Gorman, Mervyn, 63, 64, 71, 342–3
oil industry, 98, 123, 349
Oliver, Henry Francis, 101, 109, 115,
118, 119, 120, 121–2
Omdurman, Battle of (1898), 218
omnibuses, motor, 29–30, 356
operational research, 352, 353
Ordnance Board, 155, 162
Orford Ness, Suffolk, 344, 347, 351
Oriel College, Oxford, 346
Osborne Naval College, 102
Ostend, 197
Ostwald, Friedrich, 20
Otto, Nikolaus, 27
Ottoman Turks, 97, 98, 117, 138
Owen, Wilfred, 258–9
Oxford University, 35, 127, 225, 247,
283, 285, 288; science at, 10, 33,
186, 344–5, 346, 348–9

pacifism, 278, 280, 319
Page, Dr Walter Hines, 143–4, 145,
146
Palestine, 97–8, 311, 324–5
Pan American Airways, 338
Panhard et Lavassor, 27, 45
Pankhurst, Christabel, 274
Para Mantois optics company, 87
Paris, 23, 207–8, 234
Parkes, Alexander, 19
Parsons, Sir Charles, 154
Pathé, 291–2
Pear, Tom, 244, 255

Pearse, Padraig, 132
Peddie, William, 349
Pemberton Billing, Noel, 90, 399
penicillin, discovery of, 338–9
pensions, state, 13
Perkin, William, 19
Pershing, General, 146
Persia (Iran), 123
Petavel, Joseph Ernest, 33, 59
petroleum, 20, 27, 28, 31, 201
Peugeot, Armand, 27
Peugeot cars, 45
pharmaceutical industry, 20, 21, 213,
219, 338
phosgene, 185, 189, 354
photographers, 11, 288, 292, 299, 323,
333
photography, 20, 52, 83–5, 86; aerial
reconnaissance, 83–5, 86, 87–9,
91–3, 95–7, 98, 347; box brownie
cameras, 84, 300; cameras used
for aerial reconnaissance, 86, 87,
89, 91, 92–3, 98; at the front, 298;
half-tone printing process, 25;
Imperial War Museum and, 332,
333; optical lenses, 84, 86–7, 157,
299; RFC School of Photography,
91–2; X-ray films and plates, 21
Physical Societies, 15
physics, 5–6, 7, 8, 168, 344–5; nuclear,
14, 149–50, 155, 353, 357;
quantum theory, 14, 353
Pickford, Mary, 312
pigeons, 163
Planck, Max, 14, 341
pneumatic tyres, 26, 28
poison gas, 9, 173–4, 175–7, 179–90;
British development and use of,
179–80, 181–3, 184, 185, 188, 203,
219, 288, 337; Chemical Warfare
Committee, 188; chlorine, 173,
175–8, 180, 181, 184, 189, 354;
first use of (April 1915), 173,
176–7, 288; gas masks, 177, 181,
184, 185–6; Geneva Protocol
(1925), 403; Hague Convention
(1899/1907), 174, 175;

improvised protection against, 177–8; Livens Projector, 187–8; mustard gas, 186–8, 189, 190, 219, 354; phosgene, 185, 189, 354; tear gas experiments, 174–5, 188; total casualties, 189–90; World War Two, 403

Pommern (German warship), 121

Ponsonby, Arthur, *Falsehood in Wartime*, 333

Poole, Lieutenant E.S., 265

population growth, 22, 35, 38

Porton Down, Salisbury Plain, 184, 354

Portsmouth, 130, 347, 354

Post Office, 1–2, 102, 106, 135, 336–7

press and mass communications: advertising, 35; allowed to report from the front, 288–9; anti-German hysteria, 274–5, 276, 278, 282; atrocity and barbarism stories, 274–5, 278; censorship and, 11, 271–2; debate on 'shell shock', 244, 247; Defence of the Realm Act (DORA) and, 271–2; developments in printing, 24, 25; explosion of tabloid press, 25, 35; in France, 288; Harmsworth's innovations, 24–5; illustrated magazines, 272; Lloyd George and, 322–4; news agencies, 283–4, 316, 318; patriotic fervour and, 75; photographers, 288; Press Bureau and, 271–3, 280, 287–8, 316; quasi-correspondents ('Eye-Witnesses'), 280–2; reporting of Gillies' work, 237; speed of distribution, 24; *see also* entries for individual newspapers

propaganda: Admiralty and, 294, 295–7, 316; atrocity and barbarism stories, 274–5, 278, 330, 333; *The Battle of the Somme*, 301–10, 314; *Britain Prepared*, 294–8, 309; British campaigns in neutral countries, 278–9, 280,

282–5, 286–7, 289, 316, 320–1, 324; British use of cinema, 294–8, 301–12, 314, 318; Department of Information, 317–19, 320, 321; directed at enemy countries, 318; directed at Germany, 327–30, 333–4; discredited in post-war Britain, 333, 334; Robert Donald and, 316–17, 320; Enemy Propaganda Department, Crewe House, 326–30; German campaigns in neutral countries, 278; German use of cinema, 292–3, 310; Lawrence of Arabia and, 324–5; Ministry of Information and, 322–4, 325–32; National War Aims Committee (NWAC), 319–20, 330; Nazi Germany and, 334; recruitment posters, 269–71; in Second World War, 334; war artists and, 313, 318, 332, 333; War Cabinet and, 316, 317, 320, 324, 327; War Propaganda Bureau (Wellington House), 278–9, 280, 282–5, 286–7, 289, 294, 297–8, 309, 312–13, 316, 318, 320–1

Prothero, George, 127

Prussian military victories (1866-71), 38

psychoanalysis, 14, 255

psychology, 10, 23, 242–3, 244; interpretation of dreams, 255, 256; the libido and, 339; Maghull Hospital and, 255–6; William Rivers and, 242, 255–9; the unconscious mind, 23–4, 339; *see also* shell shock

psychotherapy, 23, 237, 255, 261

public health: improved sanitation, 21–2; industrial production and, 10, 171; mass immunisation projects, 21–2; Medical Research Committee established (1913), 22; purification of water supplies, 21, 22; the 'TB Penny', 22

public opinion, 12, 273–4, 284, 288, 314; attitudes to shell shock, 244, 265–6; NWAC and, 330; war-weariness and, 319, 322, 330
Punch, 49

HMS *Queen Elizabeth*, 295
HMS *Queen Mary*, 109–10, 120, 130
Queen Mary's Hospital, Roehampton, 232–3
Queen's Hospital, Sidcup, 233, 237–9

radar, development of, 350, 351–2, 353, 354–5, 357
radio (wireless telegraphy): Admiralty and, 104, 105–6, 155, 163; advances in during War, 10, 336–7; broadcast services, 336–7, 338, 357; closing of amateur radio stations, 105; distress signals, 19; electric telegraph and, 17; German Navy and, 104–6, 108–9, 110–17; improved signalling and reception, 105; interception of diplomatic cables, 110, 111, 125, 126–7, 131–2, 138, 139–44, 145–6; interception of signals, 2, 8, 102, 103–6, 109–14, 116–17, 118–19, 121–3, 138, 163; international agreements, 19; Marconi and, 18; radio companies, 18–19, 336–7, 338; Royal Navy on-board radios, 104, 155, 163; Royal Society War Committee and, 7; Russian, 103–4; thermionic valves, 18, 336; use of at sea, 6, 18, 19; Watson-Watt and, 349–50
radiology, 21
railways, 37, 78, 224, 226, 228, 230, 231, 232, 253, 356
Ramsay, Sir William, 174
Ransome & Rapier, 161–2
Rawlinson, General, 229–30
Rayleigh, Lord, 5-7, 33, 35, 59, 150, 382
reconnaissance, military: *The Art of*

Reconnaissance (army manual), 70; cavalry and, 5, 81; 'sound ranging', 151–3, 206; *see also* aerial reconnaissance
Red Cross, 234, 324
Rees, Sir Milsom, 234
Regensburg (German warship), 121
Reith, John, 337
Repington, Colonel Charles à Court, 66, 287
Reuters, 283–4, 316, 318
Reynolds, Osborne, 33
Rheims Cathedral, 274
Richardson, Captain, 106–7
Richthofen, Manfred von (the 'Red Baron'), 94
Rivers, William, 242, 255–9, 382-3
Roberts, David, 196
Robertson, Miss, 128
Robey & Co., Lincoln, 196
Robinson, Harold, 151, 153
Rochester, Kent, 72
rockets and flares, 163
Roddam, Miss Olive, 128
Roe, Alliott Verdon, 46–8, 55
Roger, Alexander, 155–6
Rolland, Romain, 278
Rolls, Charles, 43, 44, 45–6, 62; death of (1910), 62–3, 84, 281
Rolls-Royce, 45–6, 281
Romania, 74
Röntgen, Wilhelm, 21, 149
Roosevelt, Franklin D., 298
Rosyth, 114, 119
Rothschild, Lord, 62
Rotter, Charles, 108–9
Rotterdam, 136–8
Rows, Ronald, 255
Royal Air Force (RAF), 332, 334, 335, 355; aerial reconnaissance and, 96–7; Allied counter-offensive (July 1918) and, 97, 208; creation of (1 April 1918), 95; 'declinist' view of inter-war Britain, 353–4; openness to scientific ideas, 355–6
Royal Aircraft Establishment, 355

Royal Aircraft Factory (formerly
Balloon Factory), 71, 73–4, 90–1,
342–3, 350
Royal Army Medical Corps (RAMC),
217, 221; College at Millbank,
186, 215, 216; dramatic
improvements during war, 239;
formation of (1898), 215; Medical
Research Committee and, 226–7;
numerical strength of, 239; opens
Queen's Hospital, Sidcup, 237–8;
shell shock and, 244, 246–7, 252
Royal Artillery, Inventions Branch,
159
Royal Automobile Club (RAC), 45
Royal Commission on Awards to
Inventors, 160, 209–10
Royal Engineers, 52, 180, 187, 194,
347
Royal Flying Corps (RFC): aerial
photographic unit, 86; 'Bloody
April' 1917, 95; Brabazon joins,
84; casualties during Somme, 93;
crossing to Amiens (13 August
1914), 2–5, 78–9; established
(April 1912), 69–70; film
cameramen with, 296; Flying
School, 69, 347; hospital at
Hampstead, 257–8; inner tubes
as lifebelts, 3, 4; life expectancy
of pilots, 95; Lindemann and,
343–4; mobilised (August 1914),
78; photography and, 84–5, 87–9,
91–3; reconnaissance flights, 72,
80–3, 87–8; roundel identification
mark, 82; School of Photography,
91–2; structure of Military Wing,
70; support vehicles, 79; Tizard
and, 347–8; Trenchard becomes
commander, 89; wartime
expansion of, 89, 93
Royal Horse Artillery, 151
Royal Institution, 346
Royal Marines, 197
Royal Naval Air Service (RNAS),
72–3, 95, 197, 354
Royal Navy: Churchill and, 67–9; as

conservative-minded, 35;
'declinist' view of inter-war
Britain, 353–4; dreadnought
battleships, 38, 119; film
cameramen with, 294, 295–7;
Fisher reforms, 102; Hall's
reforms, 109–10; move from coal
to oil, 68, 123; North Sea patrols,
112; officers and 'other ranks',
36–7; on-board radios, 104, 155,
163; patriotic pride in, 75;
Portland Naval Review (1912),
68; Reserve, 197; size of, 38;
Voluntary Service, 127; see also
Admiralty
Royal Society, 5–7, 59, 102, 156, 157,
168–9, 184, 186, 188, 216, 242,
342, 352; War Committee of, 7–8,
86, 156, 179, 180, 181, 219, 349
Royce, Frederick Henry, 45
Russell, Bertrand, 14, 150, 278
Russia, 15, 74, 75, 77, 175, 309; The
Battle of the Somme shown in,
309–10; capture of the
Magdeburg, 107–8, 116; cinema
industry, 291; departure from
war (1918), 95, 207, 321–2;
February 1917 revolution,
309–10; machine guns and, 192;
October 1917 Revolution, 95,
207, 319, 321–2; poison gas
casualties, 189; population
growth, 38; radio
communications, 103–4; Russo-
Japanese War (1904-5), 38, 116,
161, 194, 295; Soviet state, 324
Rutherford, Ernest, 14, 149, 151, 155,
157, 352–3, 383–4

Saltpetrière hospital, Paris, 23
salvarsan, 21
Samson, Lieutenant Charles, 67, 68,
69
Sandhurst Royal Academy, 36
sanitation, 21–2, 214, 215; Military
Hygiene and Sanitation for Soldiers
(1908 manual), 216

Santos-Dumont, Alberto, 42, 49, 50
Sargant, Mr Justice, 209, 210
Sargent, John Singer, 313
Sassoon, Siegfried, 160, 257–8, 259
Sayonara steam yacht, 130–1
Scapa Flow, 107, 111–12, 114, 119, 288, 295
Scarborough, 112, 113
Scheer, Admiral Reinhard, 118, 119, 120–1
Schlieffen Plan, 77
School of Military Engineering, Chatham, 53
science: achievements and breakthroughs during War, 9–10, 335–40; biophysics, 352; common view of role in War, 9; 'declinist' view of inter-war Britain, 353–4; education system and, 32–3, 34–5, 36; fluid mechanics, 33; genetics, 14; neurology, 22–3, 252–3 *see also* shell shock; physiology, 10, 168, 181, 261, 352; public sector in inter-war period, 354, 355; pure versus applied, 10–11; revolution in early twentieth-century, 14–15; ruling class as biased against, 32–3, 34–5, 36; vocational training in, 33–4; *see also* physics; psychology; radio (wireless telegraphy)
scientists, 5–9; Board of Invention and Research, 154–5; 'boffin' term, 11, 340; 'Census' of in Britain (1916), 168; French, 178; German, 341–2; 'H Department' at Farnborough ('Chudleigh Mess'), 342–3, 350; links with government and industry, 10, 11, 154–7; Munitions Inventions Department, 156; Second World War boffins, 11, 340–53, 356; Trench Warfare Department, 155–6, 162, 200; *see also* chemists
Scotland Yard's Special Branch, 130
Seacole, Mary, 215

Seale Hayne Hospital, Devon, 254–5
seaplanes, 69, 72; design and building of, 72–3
Second World War, 123, 153, 186, 209, 334, 339, 357; area bombing during, 346; Battle of Britain, 351, 353; Chain Home system, 351–2, 353; Great War science and, 340–53, 356
Secret Intelligence Service (SIS, MI6), 125
Secret Service Bureau, 125
Selfridges, Oxford Street, 51
Sennett, Mack, 291
Seton-Watson, Robert, 326, 384–5
Seydlitz (German warship), 115
Shackleton, Sir Ernest, 311, 318
Shaw, George Bernard, 133, 278, 280
Shaw, William Napier, 59
Sheffield, Gary, 264–5
shell shock: as affecting anyone, 245–6; anxiety and stress of trench warfare, 243–4, 246–7, 248, 249, 258, 261; army categorisation of, 247–8, 253; army's attitude to, 244–6, 250–2, 253, 261, 264–6; 'Aviators' Neurasthenia', 257–8; Battle of Passchendaele (1917) and, 253; Battle of the Somme (1916) and, 250–2, 253, 264–5; coining of term, 243; debate over causes of, 244, 246–8, 249–50, 253–5, 261–2; early cases, 241–2; electric shock treatment, 259–60, 262, 263; evacuations to England, 241–2, 248, 252, 254–5, 258, 259; executions for cowardice or desertion, 264–6; as ill-chosen term, 249–50; Samuel Myers and, 243, 244, 246, 248–50, 252, 254–5; 'nervous and mental shock' diagnoses, 241–2; neurasthenia category for officers, 245, 248, 257–8; paralysis and 'the shakes', 241, 243, 245, 254, 260; posthumous pardons for men

executed, 266; principle of
 'proximity' and, 248–9, 252, 261;
 public attitudes to, 244, 265–6;
 William Rivers and, 256–8;
 suspicion of malingering and,
 246; treatments for, 244, 248–9,
 252, 254–9, 262–3
Shephard, Ben, 250
Shepherd, Ernest, 177
shipbuilding, 55
Short Brothers, 44–5, 46, 55, 62, 67, 70,
 72, 336
Signal Corps, US Army, 312
Silvertown factory explosion, 171
Simon, Lieutenant, 130
Sinn Fein, 129–30, 131–2
Skoda, 161
Slade School of Fine Arts, 235
Sligo, Lord, 131
Sloggett, Sir Arthur, 217–18, 227, 228,
 229–30, 252, 253
Smith, George Albert, 295
Smith, Grafton Elliott, 255, 385
Snagge, Sir Harold, 324
social class: Max Aitken and, 281;
 cavalry and, 151; cinema and,
 292; convalescence and, 225;
 early aviators and, 43, 44, 45,
 46–7; mental disorders and, 245,
 248; nervous conditions and,
 22–3; officers and 'other ranks',
 36–7; ruling class as biased
 against science, 32–3, 34–5, 36;
 shortage of servants, 171
socialism, 319, 327
Somervell, Howard, 229
Somerville College, Oxford, 225
Sopwith, Thomas, 46, 55
Sopwith Aviation Company, 46, 55,
 70, 72–3, 91, 95, 348
'sound ranging', 151–3, 206
HMS *Southampton*, 120
Spain, 77, 133–4
Spandau machine guns, 95
spectroscopy, X-ray, 150
Spencer, Stanley, 235, 313
spies, 134–5; German, 110, 116

St Andrew's University, 219, 349
St Bartholomew's Hospital, London,
 218, 234, 242, 255–6
St John Ambulance Brigade, 216, 217
St Mary's Hospital, Paddington,
 226
The *Star*, 305, 307
Starling, Professor, 184
steel helmets, 157–8
steel industry, 157
Stein, General von, 333
Stern, Albert Gerald, 197, 198, 200,
 201
Stevens Institute of Technology, USA,
 34
Stevenson, Frances, 306
Stokes, Wilfred, 161–3, 385–6
Stonehenge, 347
Strachan, Hew, 164, 395–6
Strand Magazine, 195
Strange, Louis, 173
street lighting, 16, 356
Strutt, John William *see* Rayleigh,
 Lord
submarines, 37; Germany's
 unrestricted warfare, 122,
 138–40, 142, 146; sound location
 of, 155; weaponry against, 353;
 see also U-boats, German
Sudan, 218, 269, 273
suffragettes, 14, 274
sugar, 20
sulphuric acid, 19
Sunday Express, 322
supersonic airliners, 10
Sweden, 140, 141
Swettenham, Sir Frank, 287–8
Swinton, Colonel Ernest Dunlop,
 194–5, 200, 202, 203, 204, 210,
 280–1, 386; *The Defence of Duffer's
 Drift* (novel), 194
Switzerland, 297
Sykes, Major Frederick, 69, 70
syphilis, treatments for, 21
Syria, 98

Tank Corps, 205, 210

tanks: agricultural machinery and, 196, 197–8; Allied counter-offensive (July 1918), 208; armour plating, 196, 197, 201, 205; British army's obstructiveness, 198, 199, 204, 210; at Cambrai, 205–7, 208, 209, 210; caterpillar tracks, 196, 197, 198; Churchill and, 196–7, 198–200, 210; development of, 193–4, 195–201, 209; first use of (the Somme, 1916), 203–4; German anti-tank tactics, 206; guns on, 198–9, 201, 202, 205; Holt tractors, 196, 197–8; during last months of war, 208–9; manufacture and supply of, 201–2; 'Mother' prototype, 200–1; origin of name 'tank', 201; at Passchendaele, 205; second generation of, 205; in Second World War, 209; training of commanders and crew, 202, 205; H.G. Wells forsees, 195–6, 199

Tawney, R.H., 214

Telefunken, 18

telegraph wires, 1–2, 8, 17, 101, 104, 138

telephony, 7, 17–18, 163, 357

television, 357

tetanus, 220

HMS *Theseus*, 107–8

Thetford, Norfolk, 202

Thomas, Lowell, 324–5

Thompson, Sylvanus, 34

Thomson, Sir Basil, 130, 133, 134–5

Thomson, Sir J.J., 154, 155, 156

Thorpe, Professor, 179–80

The Times, 24, 25, 66, 117, 154, 156, 179, 217, 272–3, 287, 288, 306, 307, 326, 351

Tisdall, Claire, 253

Titanic, sinking of (1912), 19

Tizard, Sir Henry, 346–9, 350, 351, 353, 354, 355, 356, 386-7

TNT (trinitrotoluene), 166–7, 170, 171

Tong, Edward 'Teddy', 293–4, 300

Tonks, Henry, 235, 387-8

Topical Company, 291–2, 311

torpedoes, 72, 296, 354

Tower of London, 331

Toynbee, Arnold, 283, 323, 388

trade unions, 14, 274, 319–20

trams, electric, 56

Trenchard, Colonel Hugh, 85–6, 89

Trevelyan, G.M., 280

Trinity College, Cambridge, 33, 45, 149, 150, 352

Trinity College, Dublin, 252

Trippe, Juan, 338

Tritton, William Ashbee, 200, 210

tuberculosis, 22

Tullibardine, Marquess of, 62

tungsten, 157

typhoid, 21–2, 214, 216

U-boats, German, 118, 132, 133, 143, 322, 328; Battle of Dogger Bank and, 114, 115, 116; unrestricted submarine warfare and, 122, 138–40, 142, 146

Ulster Volunteer Force, 129, 320

HMS *Undaunted*, 108

underground railway, electric, London, 35, 269

United States of America (USA): Max Aitken and, 282; American Expeditionary Force, 146; 'armed neutrality' policy, 142; army actions in war, 208; *The Battle of the Somme* shown in, 309; *Britain Prepared* shown in, 297–8; broadcasting industry, 338; cinema industry, 291, 312, 357; civil aviation industry, 338; Civil War, 37, 191, 214, 215; Department of Information and, 318; development of poison gases, 189; development of the tank in, 196, 197; electricity industry in, 16; entry into war (6 April 1917), 97, 145, 207, 322, 328; exports of explosives, 172; German propaganda in, 278; Germans' diplomatic mission in,

125, 129, 131–2, 138, 139–42; Irish Americans, 129, 141, 298; motor industry, 30; post-War private enterprise, 338; scientists in, 15; shell shock and, 261; university-industry links, 34; use of cinema, 312; Wright brothers, 32, 42, 45, 51–2, 53, 62, 64; Zimmermann telegram and, 139–42, 143–6

universities: exclusive nature of, 33; inter-war work with military, 355; links with government and industry, 10, 11, 33, 219; medical research, 22; Officer Training Corps at, 216; provision of drugs to army, 219; pure versus applied science debate, 10–11; scientific studies in, 33; University Grants Committee, 157; in USA, 34

University College Hospital, London, 229

University College, London, 10

Urban, Charles, 294–8, 309, 312, 388–9

urban growth, 35

Vergemere (luxery yacht), 133–4

HMS *Vernon*, 354

Vickers-Maxim, 55–6, 64, 70, 90, 95, 154, 192, 196, 202, 205, 296, 335–6

Victor bomber, 48

Vienna, 14, 23, 76, 191

Vincent, Dr Clovis, 262

vocational training, 33–4

Voisin, Gabriel, 42–3

Vorticist movement, 313

Wales, Prince of, 239

Walpole, Hugh, 324

Walton, Sergeant, 264

War Office, 6, 7, 11, 35, 64, 86, 117, 155, 168, 180, 315, 316; allows reporting from the front, 288; cinema and, 292, 293–4, 295, 298, 300–1, 303, 304, 307, 310, 332; code breakers at, 126; Commercial and Scientific Advisory Committee, 156;

development of the tank and, 197–8, 199, 200; MI7, 318, 320; Military Intelligence Division, 103; press regulation and, 271, 280–1; shell shock and, 241–2, 250, 255, 265; treatment of the wounded and, 216, 218, 221; Trench Warfare Department, 155–6, 162, 200

War Trade Intelligence Department, 135–6

Warner, Pelham, 318

Wassmuss, Wilhelm, 124–5

Watson-Watt, Robert, 349–52, 353, 389–90

weather conditions, 80, 96, 113

Weizmann, Chaim, 169–70, 390–1

Wells, H.G., 36, 51, 154, 195–6, 199, 280, 327–8

Welsh, T., 294

Westminster, Duke of, 62

White, George, 56

Whitehead, A.N., 14

Whittle, Frank, 355

Wickham-Legg, L.G., 127

Wickham-Steed, Henry, 326

Wilhelm II, Kaiser, 25, 139, 292–3, 341

Wilhelmshaven, 112, 116, 118, 119, 121, 123

Wilkinson, Joseph Brooke, 293, 294, 300

Willoughby, L.A., 127

Wilson, Admiral Sir Arthur, 112, 113–14, 115, 116, 121–2

Wilson, President Woodrow, 139, 141, 142, 144–5, 146, 319

Wilson, Sir Henry, 216–17

Wilson, Walter, 200, 210

Wiltshire, Harold, 249

Wimperis, Henry, 354–5

wind tunnels, 61

Wolseley motor car company, 55, 63–4

women, 12, 27, 76, 171–2, 217, 245, 270, 332; Admiralty recruitment of, 127–8; chemists at Woolwich Arsenal, 168; in munitions factories, 170–1, 296, 311, 357; the

women – *continued*
    'new' woman, 14; newspapers
    and journals for, 24, 25; as nurses
    at base camp Hospitals, 224–5;
    nurses at CCSs, 223; suffrage
    debate, 13, 14, 274; war work,
    170–2, 296, 311, 332, 357
Woods optics company, Derby, 86
Woolwich, 155, 168, 354; Royal
    Laboratory, 159; Royal Military
    Academy, 53
Wormwood Scrubs, 197
wounded, treatment of *see* military
    medicine
Wren, Christopher, 6
Wright, Sir Almroth, 226–7
Wright, Wilbur and Orville, 32, 42,
    45, 51–2, 53, 62, 64

X-rays, 21, 149–50

Yealland, Lewis, 259–60
Young, George, 126–7
Ypres, Flanders, 175, 176–9, 186–7,
    193
Yugoslav National Council, 327

Zeiss, 86
Zeppelins, 56; bases in occupied
    Belgium, 197; bombing raids
    along the English coast, 156,
    180–1
Zimmermann, Arthur: telegram and,
    139–42, 143–6
Zionism, 170
the Zoetrope, 26
Zyklon B, 403

About the Author

Taylor Downing is a television producer and writer. He was educated at Cambridge University and went on to become managing director and head of history at Flashback Television, an independent production company. His most recent books include *Spies in the Sky*, *Churchill's War Lab*, *Cold War* (with Sir Jeremy Isaacs) and *Night Raid*.

ML   4-15